AF873689

Clustering Challenges

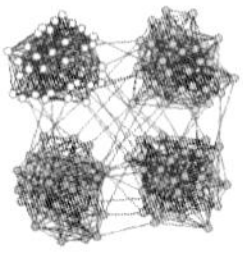

in

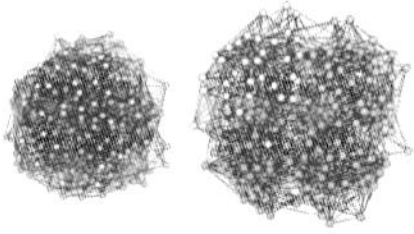

Biological Networks

CLUSTERING CHALLENGES

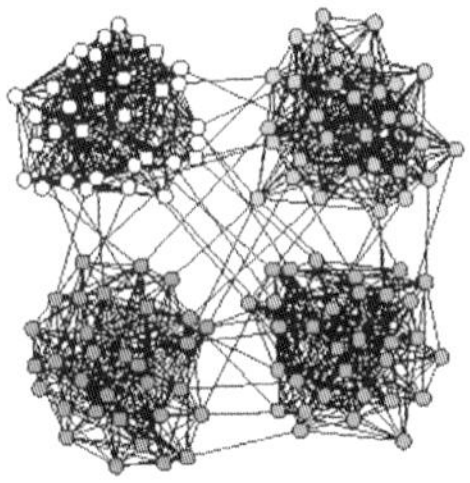

IN

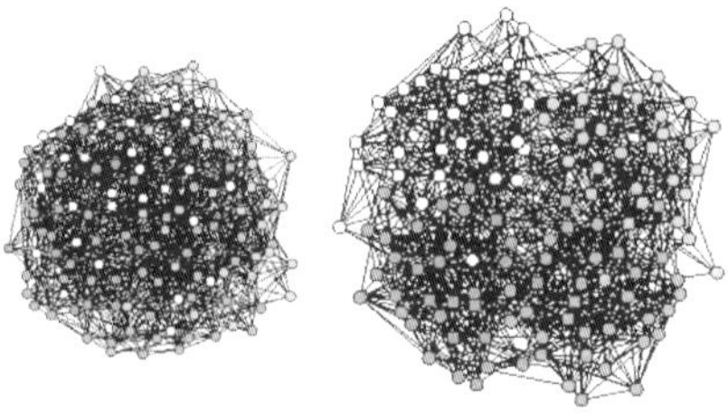

BIOLOGICAL NETWORKS

Editors

Sergiy Butenko
Texas A&M University, USA

W Art Chaovalitwongse
Rutgers University, USA

Panos M Pardalos
University of Florida, USA

World Scientific

NEW JERSEY • LONDON • SINGAPORE • BEIJING • SHANGHAI • HONG KONG • TAIPEI • CHENNAI

Published by

World Scientific Publishing Co. Pte. Ltd.
5 Toh Tuck Link, Singapore 596224
USA office: 27 Warren Street, Suite 401-402, Hackensack, NJ 07601
UK office: 57 Shelton Street, Covent Garden, London WC2H 9HE

British Library Cataloguing-in-Publication Data
A catalogue record for this book is available from the British Library.

CLUSTERING CHALLENGES IN BIOLOGICAL NETWORKS

Copyright © 2009 by World Scientific Publishing Co. Pte. Ltd.
All rights reserved. This book, or parts thereof, may not be reproduced in any form or by any means, electronic or mechanical, including photocopying, recording or any information storage and retrieval system now known or to be invented, without written permission from the Publisher.

For photocopying of material in this volume, please pay a copying fee through the Copyright Clearance Center, Inc., 222 Rosewood Drive, Danvers, MA 01923, USA. In this case permission to photocopy is not required from the publisher.

ISBN-13 978-981-277-165-0
ISBN-10 981-277-165-4

Printed in Singapore by Mainland Press Pte Ltd

"The whole is more than the sum of its parts"

Aristotle (Greek philosopher, 384 BC - 322 BC)

Preface

Clustering can be defined as the partitioning of a data set into subsets (called *clusters*), so that each subset consists of elements that are similar with respect to some similarity criterion. The measure of similarity can be the distance between the data points (used in *distance-based clustering*) or some descriptive concept (as in *conceptual clustering*), and can be chosen differently depending on the type of the data set of interest and the purpose of clustering. The typical objectives include the data classification and reduction, detection of natural modules based on their properties, and the determination of outliers. Clustering algorithms have been successfully used to analyze the data sets arising in many important applications of diverse origin, including biology. In fact, applications of clustering in biology can be traced back to Aristotle's *History of Animals*, in which he classified plants and animals according to complexity of their structure and function.

Many biological systems can be conveniently modeled using graphs or networks, with vertices representing the data points and edges connecting pairs of vertices corresponding to data points that are related in a certain way. Network clustering and cluster detection algorithms represent an important tool in structural analysis of such networks. For example, in gene networks, the vertices correspond to genes and the edges represent functional relations between these genes that are identified using the comparative genomics methods. Solving clustering problems in gene networks allows to identify groups of genes with similar expression patterns. This information is crucial for understanding the nature of genetic diseases. Other examples of biological networks include the protein interaction networks, metabolic networks, and signaling networks.

Network clustering problems present a number of formidable research challenges, many of which are still to be addressed. On the one hand, developing a proper mathematical model describing the clusters that are interesting from biological perspective may be very tricky. On the other hand, most known optimization problems on graphs used as the basis for network clustering appear to be NP-hard, making it extremely difficult to solve large-scale instances of such problems to optimality. Based on the objective that clustering aims to achieve for a particular application, one has to choose an appropriate graph-theoretic definition of a cluster, formulate the corresponding optimization problem, and develop a network clustering algorithm for the developed model. From the practical perspective, the effectiveness of any clustering algorithm has to be confirmed through empirical evidence, and this process is complicated by possible errors in the data used to construct the network.

This volume presents a collection of papers, several of which have been presented at DIMACS Workshop on Clustering Problems in Biological Networks that took place at Rutgers University on May 9 - 11, 2006. It consists of two parts, with the first part containing surveys of selected topics and the second part presenting original research contributions. While clustering in biological networks represents the central theme of this volume, some of the chapters deal with other related problems in computational biology that may not necessarily fall within the vaguely defined network clustering domain.

This book will be a valuable source of material to faculty, students, and researchers in mathematical programming, data analysis and data mining, as well as people working in computer science, engineering and applied mathematics. In addition, the book can be used as a supplement to any course in data mining or computational/systems biology.

We would like to thank the authors of the chapters, the anonymous referees and the staff of World Scientific for their cooperation, without which the publication of this volume would not have been possible. We also acknowledge the support of the National Science Foundation in organizing the DIMACS Workshop mentioned above.

Sergiy Butenko, W. Art Chaovalitwongse, and Panos M. Pardalos
September 2008

Contents

PART 1

Surveys of Selected Topics

Chapter 1

Fixed-Parameter Algorithms for Graph-Modeled Data Clustering

Falk Hüffner*

Institut für Informatik, Friedrich-Schiller-Universität Jena
Ernst-Abbe-Platz 2, D-07743 Jena, Germany
hueffner@minet.uni-jena.de

Rolf Niedermeier

Institut für Informatik, Friedrich-Schiller-Universität Jena
Ernst-Abbe-Platz 2, D-07743 Jena, Germany
niedermr@minet.uni-jena.de
Web: http://theinf1.informatik.uni-jena.de/

Sebastian Wernicke†

Institut für Informatik, Friedrich-Schiller-Universität Jena
Ernst-Abbe-Platz 2, D-07743 Jena, Germany
wernicke@minet.uni-jena.de

Fixed-parameter algorithms can efficiently find optimal solutions to some NP-hard problems, including several problems that arise in graph-modeled data clustering. This survey provides a primer about practical techniques to develop such algorithms; in particular, we discuss the design of *kernelizations* (data reductions with provable performance guarantees) and *depth-bounded search trees*. Our investigations are circumstantiated by three concrete problems from the realm of graph-modeled data clustering for which fixed-parameter algorithms have been implemented and experimentally evaluated, namely CLIQUE, CLUSTER EDITING, and CLIQUE COVER.

*Supported by the Deutsche Forschungsgemeinschaft, Emmy Noether research group PIAF (fixed-parameter algorithms), NI 369/4.
†Supported by the Deutsche Telekom Stiftung.

1.1. Introduction

The central idea behind graph-modeled data clustering is to depict the similarity between a set of entities as a graph: Each vertex represents an entity—such as a gene or protein—and two vertices are connected by an edge if the entities that they represent have some (context-specific) similarity; for instance, two genes have a similar expression profile or two proteins have a high sequence similarity. Groups of highly connected vertices in the resulting graph represent clusters of mutually similar entities. Hence, detecting these dense groups can identify clusters in the graph-encoded data.

Graph-modeled data clustering has been shown to have useful applications in many areas of bioinformatics, including the analysis of gene expression [8, 16, 65, 66], proteins [45, 46], gene networks [68], allergic reactions [9], and marine ecosystems [54]. There is a catch, however: Most problems that are concerned with the detection of cluster structures in a graph are known to be NP-hard, that is, there is probably no algorithm that can solve all instances efficiently [32]. Thus, whenever such a problem is encountered and large instances need to be solved, it is common to employ heuristic algorithms [58], approximation algorithms [5, 67], or similar techniques. These usually come with some disadvantages: The solutions are not guaranteed to be optimal or there are no useful guarantees concerning the running time of the algorithm. Further, approximation algorithms and—to some extent—heuristic algorithms are not suited to cope with enumerative tasks. There are many scenarios where these disadvantages seem too severe, that is, where we need to solve a combinatorially hard problem both optimally and yet at the same time somewhat efficiently. For some combinatorial problems, this can be achieved by means of *fixed-parameter algorithms* [25, 30, 60]. These are based on the observation that not all instances of an NP-hard problem are equally hard to solve; rather, this hardness depends on the particular *structure* of a given instance. Opposed to "classical" computational complexity theory—which sees problem instances only in terms of their size—fixed-parameter algorithms and the underlying theory of *fixed-parameter tractability (FPT)* reflect such differences in structural hardness by expressing them through a so-called *parameter*, which is usually a nonnegative integer variable denoted k.

Whenever the parameter k turns out to be small, fixed-parameter algorithms may solve an NP-hard problem quite fast (sometimes even in linear time)—with provable bounds on the running time and guaranteeing the optimality of the solution that is obtained. More precisely, a size-n instance of a fixed-parameter tractable problem can be solved in $f(k) \cdot p(n)$ time, where f is a function solely depending on k, and $p(n)$ is a polynomial in n.

The purpose of this survey is twofold: First, we provide a primer about some important and practically relevant techniques for the design of fixed-parameter algorithms in the realm of graph-modeled data clustering (Sec. 1.2); in particular, Sec. 1.2.1 exhibits *kernelizations* (data reductions with provable performance guarantees) and Sec. 1.2.2 discusses *depth-bounded search trees*. Second, we present three concrete case studies from the realm of graph-modeled data clustering where fixed-parameter algorithms have been devised, implemented, and successfully tested:

- CLIQUE (Sec. 1.3.1). Using techniques that were originally developed for the fixed-parameter tractable VERTEX COVER problem, it is possible to detect a size-$(n-k)$ clique in an n-vertex and m-edge graph in $O(1.3^k + kn + m)$ time [12, 15]. So-called k-*isolated cliques*, that is, i-vertex cliques that have less than $k \cdot i$ edges to vertices that lie outside of them, can be exhaustively enumerated in $O(4^k \cdot k^2 m)$ time [44, 49, 50].
- CLUSTER EDITING (Sec. 1.3.2). In this problem, the assumption is that the input graph has an underlying cluster structure that is a disjoint union of cliques, which has been distorted by adding and removing at most k edges. For an n-vertex input graph, this underlying cluster structure can be found in $O(1.92^k + n + m)$ time [33, 34, 63].
- CLIQUE COVER (Sec. 1.3.3). The assumed underlying cluster structure in this problem is an overlapping union of cliques, that is, the task is to cover the edges of a given graph with a minimum number of cliques. Fixed-parameter algorithms allow for optimal problem solutions within a running time that is competitive with common heuristics [35, 36].

Practical experiments that we discuss in the case studies suggest that the presented fixed-parameter algorithms are capable of solving many real-world instances in reasonable time. In particular, they perform much better on real-world data than the provable worst-case bounds suggest. Thus, for some NP-hard clustering problems, fixed-parameter tractability theory offers algorithms which are both efficient and capable of delivering optimal solutions. It should hence be part of the algorithmic toolkit for coping with graph-based clustering problems.

We conclude our survey with advice on employing fixed-parameter algorithms in practice (Sec. 1.4.1) and with a list of specific challenges for future research (Sec. 1.4.2).

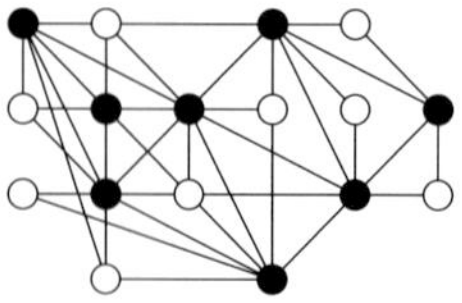

Fig. 1.1. A graph with a size-8 vertex cover (cover vertices are marked black, the solution size is optimal).

1.2. Fixed-Parameter Tractability Basics and Techniques

In this section, we introduce the basics of fixed-parameter tractability, in particular exhibiting two techniques that are of major practical importance[a] and have by now facilitated many success stories in bioinformatics, namely

- kernelizations, that is, data reductions with provable performance guarantees (Sec. 1.2.1) and
- depth-bounded search trees (Sec. 1.2.2).

Both techniques are introduced by means of a single natural and easy to grasp problem, namely the NP-hard VERTEX COVER problem.

> VERTEX COVER
> INPUT: An undirected graph $G = (V, E)$ and a nonnegative integer k.
> TASK: Find a subset of vertices $C \subseteq V$ with k or fewer vertices such that each edge in E has at least one of its endpoints in C.

This problem is illustrated in Fig. 1.1 and is—among many other applications—of central importance to practically solving the CLIQUE problem that we discuss in Sec. 1.3.1.[b]

Throughout this work, we assume basic knowledge from algorithmics [19, 48] and graph theory [24, 71]. For a given undirected graph $G = (V, E)$, we always use n to denote the number of its vertices and m to denote the number of its edges. For $v \in V$, we use $N_G(v)$ to denote the neighbor set $\{v \in V \mid \{u, v\} \in E\}$ and $N_G[v]$ to denote the closed neighborhood $N_G(v) \cup \{v\}$, omitting the indices whenever they are clear from the context.

The core approach of fixed-parameter tractability [25, 30, 60] is to consider *parameterized problems*—that is, problems that consist of the instance I and a pa-

[a]A broader view on fixed-parameter algorithm design techniques can be found in Ref. 60.
[b]VERTEX COVER is the *Drosophila* of fixed-parameter research in that many initial discoveries that influenced the whole field originated from studies of this single problem (e.g., see Guo et al. [40]).

rameter k—and ask whether there is an algorithm that confines the combinatorial explosion that is involved in solving the problem to the parameter.

Definition 1.1. An instance of a parameterized problem consists of a problem instance I and a parameter k. A parameterized problem is *fixed-parameter tractable* if it can be solved in $f(k) \cdot |I|^{O(1)}$ time, where f is a computable function solely depending on the parameter k, and not on the input size $|I|$.

For NP-hard problems, $f(k)$ will of course not be polynomial, since otherwise we would have an overall polynomial-time algorithm.

As parameterized complexity theory points out, there are problems that are likely not to be fixed-parameter tractable [25, 30, 60]. It is important to note in this respect that a problem can have *various* parameterizations such as the size of the solution that is sought after or some structural parameter that characterizes the input. A problem that is not fixed-parameter tractable with respect to some parameter may still be so with respect to others. Also, the choice of the parameter can greatly affect the efficiency of the algorithm that is obtained.

Besides the classic reference [25], two new monographs are available on parameterized complexity, one focusing on theoretical foundations [30] and one focusing on techniques and algorithms [60].

1.2.1. *Kernelizations*

Before firing up a computationally expensive algorithm to solve a combinatorially hard problem, one should always try to perform a *reduction* on the input data, the idea being to quickly presolve those parts of the input data that are relatively easy to cope with and thus to shrink the input to those parts that form the "really hard" core of the problem. Costly algorithms need then only be applied to the reduced instance. In some practical scenarios, data reduction may even reduce a seemingly hard problem to triviality [35, 56, 70].

Clearly, practitioners are likely to already be aware of data reduction rules. The reason why they should also consider fixed-parameter tractability in this context is that fixed-parameter theory provides a way to use data reduction rules not only in a heuristic way, but to *prove their power* by so-called *kernelizations*. These run in polynomial time and give an upper bound on the size of a reduced instance that solely depends on the parameter value, that is, they come with a performance guarantee both concerning their running time as well as their effectiveness. Having a quantitative measure for the performance of a data reduction can moreover help to guide the search for further improved data reductions in a constructive way [39].

1.2.1.1. *An Introductory Example*

Consider our running example VERTEX COVER. To reduce the input size for a given instance of this problem, it is clearly permissible to remove isolated vertices, that is, vertices with no adjacent edges. This leads to a first simple data reduction rule.

> REDUCTION RULE VC1. Remove all isolated vertices.

In order to cover an edge in the graph, one of its two endpoints *must* be in the vertex cover. If one of these is a degree-1 vertex, then the other endpoint has the potential to cover more edges than the degree-1 vertex, leading to a second reduction rule.

> REDUCTION RULE VC2. For degree-1 vertices, put their neighboring vertex into the cover.[c]

Note that this reduction rule assumes that we are only looking for *one* optimal solution to the VERTEX COVER instance we are trying to solve; there may exist other minimum vertex covers that do include the reduced degree-1 vertex.

After having applied the easy rules VC1 and VC2, we can further do the following in the fixed-parameter setting where we ask for a vertex cover of size at most k.

> REDUCTION RULE VC3. If there is a vertex v of degree at least $k + 1$, put v into the cover.

The reason this rule is correct is that if we did not take v into the cover, then we would have to take every single one of its $k + 1$ neighbors into the cover in order to cover all edges adjacent to v. This is not possible because the maximum allowed size of the cover is k.

After exhaustively performing the rules VC1–VC3, no vertex in the remaining graph has a degree higher than k, meaning that choosing a vertex into the cover can cause at most k edges to become covered. Since the solution set may be no larger than k, the remaining graph can have at most k^2 edges if it is to have a solution. By rules VC1 and VC2, every vertex has degree at least two, which implies that the remaining graph can contain at most k^2 vertices.

---

[c]"Put into the cover" means adding the vertex to the solution set and removing it and its incident edges from the instance.

1.2.1.2. *The Kernelization Concept*

Abstractly speaking, what have we done in the previous section? After applying a number of rules *in polynomial time* to an instance of VERTEX COVER, we arrived at a reduced instance whose size can *solely be expressed in terms of the parameter* k. Since this can be easily done in $O(n)$ time, we have found a data reduction for VERTEX COVER with guarantees concerning its running time as well as its effectiveness. These properties are formalized in the concepts of a *problem kernel* and the corresponding *kernelization* [25].

Definition 1.2. Let $\mathcal{L}$ be a parameterized problem, that is, $\mathcal{L}$ consists of input pairs (I, k), where I is the problem instance and k is the parameter. A *reduction to a problem kernel* (or *kernelization*) means to replace an instance (I, k) by a *reduced* instance (I', k') called *problem kernel* in polynomial time such that

(1) $k' \leq k$,
(2) I' is smaller than $g(k)$ for some function g only depending on k, and
(3) (I, k) has a solution if and only if (I', k') has one.

While this definition does not formally require that it is possible to reconstruct a solution for the original instance from a solution for the problem kernel, all kernelizations we are aware of easily allow for this.

The methodological approach of kernelization, including various techniques of data reduction, is best learned by the concrete examples that we discuss in Sec. 1.3; there, we will also discuss kernelizations for VERTEX COVER that even yield a kernel with a *linear* number of vertices in k.

To conclude this section, we state some useful general observations and remarks concerning Definition 1.2 and its connections to fixed-parameter tractability. Most notably, there is a close connection between fixed-parameter tractable problems and those problems that have a problem kernel—they are exactly the same.

Theorem 1.1 (Cai et al. [11]). *Every fixed-parameter tractable problem is kernelizable and vice-versa.*

Unfortunately, the practical use of this theorem is limited: the running times of a fixed-parameter algorithm directly obtained from a kernelization is usually not practical; and, in the other direction, the theorem does not constructively provide us with a data reduction scheme for a fixed-parameter tractable problem. Hence, the main use of Theorem 1.1 is to establish the fixed-parameter tractability or amenability to kernelization of a problem—or show that we need not search any further (e.g., if a problem is known to be fixed-parameter intractable, we do not need to look for a kernelization).

Rule VC3 explicitly needed the value of the parameter k. We call this a *parameter-dependent* rule as opposed to the parameter-independent rules VC1 and VC2, which are oblivious to k. Of course, one typically does not know the actual value of k in advance and then has to get around this by iteratively trying different values of k.[d] While, in practice, one would naturally prefer to avoid this extra outer loop, assuming explicit knowledge of the parameter clearly adds some leverage to finding data reduction rules and is hence frequently encountered in kernelizations.

1.2.2. *Depth-Bounded Search Trees*

After preprocessing the given input data of a problem by a kernelization and cutting away its "easy parts," we are left with the "really hard" problem kernel to be solved. A standard way to explore the huge search space of a computationally hard problem is to perform a systematic exhaustive search. This can be organized in a tree-like fashion, which is the main subject of this section.

Certainly, search trees are no new idea and have been extensively used in the design of exact algorithms (e.g., see Ref. 22, 26, 31, 57, 72). The main contribution of fixed-parameter theory to search tree approaches is the consideration of search trees whose depth is bounded by the parameter, usually leading to search trees that are much smaller than those of naïve brute-force searches. Additionally, the speed of search tree exploration can (provably) be improved by exploiting kernelizations [59].

An extremely simple search tree approach for solving VERTEX COVER is to just take one vertex and branch into two cases: either this vertex is in the vertex cover or not. This leads to a search tree of size $O(2^n)$. As we outline in this section, we can do much better than that and obtain a search tree whose depth is upper-bounded by k, giving a size bound of $O(2^k)$. Since usually $k \ll n$, this can draw the problem into the zone of feasibility even for large graphs (as long as k is small).

The basic idea is to find a small subset of the input instance in polynomial time such that at least one element of this subset *must* be part of an optimal solution to the problem. In the case of VERTEX COVER, the most simple such subset is any two vertices that are connected by an edge. By definition of the problem,

[d] In general, the constraint $k < n$ is easily established. As Dehne et al. [23] point out in their studies of CLUSTER EDITING, it depends on the concrete problem which search strategy for the "optimum" value of k is most efficient to employ in practice.

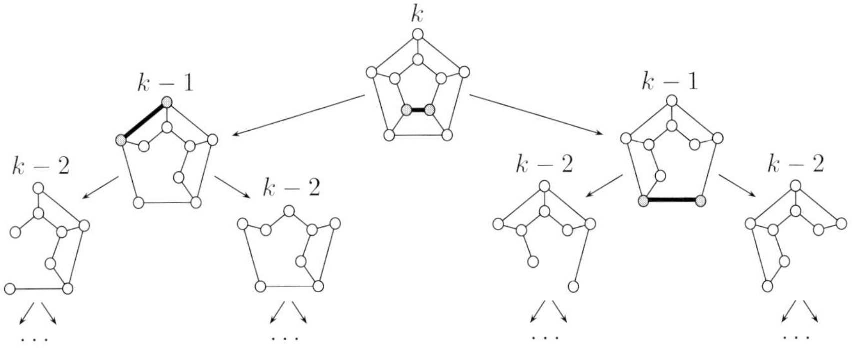

Fig. 1.2. Simple search tree for finding a vertex cover of size at most k in a given graph. The size of the tree is upper-bounded by $O(2^k)$.

one of these two vertices *must* be part of a solution. Thus, a simple search-tree algorithm to solve VERTEX COVER on a graph G proceeds by picking an arbitrary edge $e = \{v, w\}$ and recursively searching for a vertex cover of size $k - 1$ both in $G - v$ and $G - w$.[e] That is, the algorithm *branches* into two subcases knowing one of them must lead to a solution of size at most k—if one such solution exists.

As shown in Fig. 1.2, these recursive calls of the simple VERTEX COVER algorithm can be visualized as a tree structure. Because the depth of the recursion is upper-bounded by the parameter value and we always branch into two subcases, the size of this tree is upper-bounded by $O(2^k)$. This means that the size of the tree is *independent of the size of the initial input instance* and only depends on the value of the parameter k.

The main idea behind fixed-parameter algorithmics is to get the combinatorial explosion as small as possible. For our VERTEX COVER example, one can easily achieve a size-$o(2^k)$ search tree by distinguishing more detailed branching cases rather than just picking single endpoints of edges to be in the cover.[f] An example for such an "improved" search-tree is given in our case study of CLUSTER EDITING in Sec. 1.3.2. The currently "best" search trees for VERTEX COVER are of size $O(1.28^k)$ [15] and mainly achieved by extensive case distinguishing. However, it should be noted for practical applications that it is always concrete implementation and testing that has to decide whether the administrative overhead

[e]For a vertex $v \in V$, we define $G - v$ to be the graph G with v and the edges incident to v removed.
[f]Note that analogously to the case of data reduction, most of these branchings assume that only *one* minimum solution is sought after. Since some graphs can have 2^k minimum vertex covers, a size-$o(2^k)$ search tree for enumerating *all* minimum vertex covers requires the use of *compact solution representations* as outlined by Damaschke [21] and is beyond the scope of this work.

caused by distinguishing more and more cases pays off. A simpler algorithm with slightly worse bounds on the search tree size often turns out to be preferable in practice. Here, recent progress with the analysis of search tree algorithms using multivariate recurrences [27] might help: with this method, it was shown that some simple algorithms perform in fact much better than previously proved [31]. Also, new algorithms were developed guided by the new analysis methods [31]; however, there is no practical experience yet with these approaches.

In combination with data reduction (see Sec. 1.3.1), the use of depth-bounded search trees has proven itself quite useful in practice, allowing to find vertex covers of more than ten thousand vertices in some dense graphs of biological origin [3]. Search trees also trivially allow for a parallel implementation: when branching into subcases, each processor in a parallel setting can further explore one of these branches with no additional communication required. Cheetham et al. [14] expose this in their parallel VERTEX COVER solver to achieve a near-optimum (i.e., linear with the number of processors employed) speedup on multiprocessor systems. Finally, it is generally beneficial to augment search tree algorithms with admissible heuristic evaluation functions in order to further increase their performance and memory efficiency by cutting away search tree parts that cannot lead to good solutions [29, 51].

1.3. Case Studies from Graph-Modeled Data Clustering

This section surveys fixed-parameter algorithms and experimental results for three important NP-complete problems from the realm of graph-modeled data clustering, namely CLIQUE, CLUSTER EDITING, and CLIQUE COVER. The purpose of these case studies is twofold: First, they serve to teach in more detail the methodological approaches of designing kernelizations and depth-bounded search trees. Second, the encouraging experimental results that are known for these problems underpin the general usefulness of fixed-parameter algorithms for optimally solving NP-hard problems in graph-modeled data clustering.

1.3.1. *Clique*

A "classical" combinatorial problem that is closely connected to graph-modeled data clustering is to find a clique in a graph, that is, a subset of vertices that are fully connected.

> CLIQUE
> INPUT: An undirected graph $G = (V, E)$ and a nonnegative

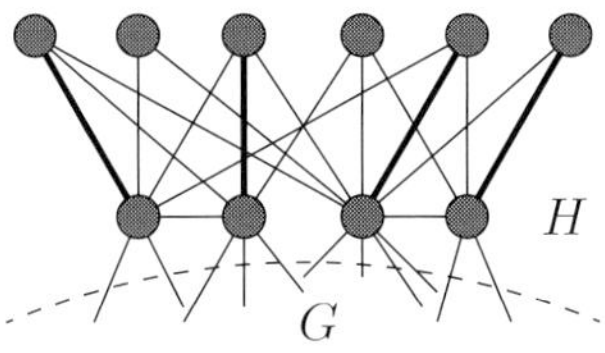

Fig. 1.3. A graph G with a crown $I \cup H$. The thick edges constitute a maximum matching of size $|H|$ in the bipartite graph that is induced by the edges between I and H.

integer k.
TASK: Find a k-vertex clique in G.

It is also a common task to *enumerate* maximal cliques in a graph, that is, all cliques that are not a proper subset of any other clique.

CLIQUE is NP-hard [32] and hard to approximate in polynomial time [42]. In a similar sense as it is generally assumed that P $\neq$ NP, it is strongly believed that CLIQUE is not fixed-parameter tractable when parameterized by the size of the cliques that are sought after [25]. Nevertheless, CLIQUE has a close connection to the fixed-parameter tractable VERTEX COVER problem that we used as our running example to introduce fixed-parameter techniques: If an n-vertex graph contains a size-k clique, then its *complement graph*[g] contains a size-$(n-k)$ vertex cover and vice versa. This can be made use of when seeking after or enumerating cliques.

1.3.1.1. *Finding Maximum Cardinality Cliques*

The catch when solving CLIQUE for a graph G by means of finding a minimum-cardinality vertex cover for the complement graph G' is that if the maximum size k of a clique in G is rather small compared to its total number of vertices n, then G' will have a rather large minimum-size vertex cover. Therefore, one has to rely on effective data reduction rules that preprocess the complement graph G' so that depth-bounded search tree algorithms become practically applicable for the reduced graph that remains. One kernelization for VERTEX COVER that has proven itself to be of particular practical importance in this respect is the so-called *crown reduction* [1], which generalizes the VERTEX COVER data reduction rule VC2 (the elimination of degree-1 vertices by taking their neighbors into the cover) and thus leads to a data reduction that requires no explicit knowledge of the parameter k and yields a kernel with a number of vertices *linear* in k.

[g]That is, the graph that contains exactly those edges that are not contained in the original graph.

A *crown* in a graph consists of an independent set I (that is, no two vertices in I are connected by an edge) and a set H containing all vertices adjacent to I. In order for $I \cup H$ to be a crown, there has to be a size-$|H|$ matching in the bipartite graph induced by the edges between I and H (i.e., one in which every vertex of H is matched). An example for a crown structure is given in Fig. 1.3. If there is a crown $I \cup H$ in the input graph G, then we need *at least* $|H|$ vertices to cover all edges in the crown. But since all edges in the crown can be covered by taking *at most* $|H|$ vertices into the cover (as I is an independent set), there is a minimum-size vertex cover for G that contains all the vertices in H and none of the vertices in I. We may thus delete any given crown $I \cup H$ from G, reducing k by $|H|$.

It turns out that finding crowns can be achieved in polynomial time by computing maximum matchings [18]. The size of the thus reduced instance is upper-bounded via the following theorem.

Theorem 1.2 (Abu-Khzam et al. [1]). *A graph that is crown-free and has a vertex cover of size at most k can contain at most $3k$ vertices.*

There are several kernelizations for VERTEX COVER that achieve a kernel of $O(k)$ vertices; some of these even yield an at-most-$2k$-vertex kernel, e.g., see Ref. 1, 60. However, it has been found that crown reductions often offer a good balance between the polynomial time that is required to compute the kernel and the size that the reduced graphs usually turn out to have in practice [1–3].

Some quite successful implementations for solving CLIQUE rely on kernelization techniques (especially crown reductions) for VERTEX COVER that are combined with depth-bounded search trees [1, 2, 74]. The exploration of the search trees is usually highly optimized, for instance, by using efficient data structures and ensuring proper load balancing in parallel scenarios; details of these techniques are described, e.g., by Abu-Khzam et al. [1] and Zhang et al. [74] Even attempts to implement parts of the algorithms in hardware have been reported [3].

With the combination of kernelizations and depth-bounded search trees, it is currently possible to find cliques that consist of over 400 vertices in some dense graphs of biological origin within hours [3].

1.3.1.2. *Enumerating Maximal Cliques*

Instead of finding a single maximum-size clique in a graph, one would often like to enumerate all cliques of maximal size. In graph-modeled data clustering, this can have mainly two reasons: First, an enumeration obviously identifies *all* clusters that are present in the data. Second, with an enumerative solution one can include expert knowledge as to what cliques that are present in the input graph are (biologically) "meaningful."

Analogously to the task of finding a maximum-size clique, the task of enumerating maximal cliques in a graph is equivalent to enumerating minimal vertex covers for its complement graph. To enumerate all maximal cliques in a graph, one can therefore rely on kernelizations for VERTEX COVER that are suited for finding *all* vertex covers up to a certain size [17] and on enumerative depth-bounded search trees such as discussed by Damaschke [21]. However, so far there is no empirical evidence of the practical viability of this approach.

An interesting result concerning the enumeration of maximal cliques that makes a more "indirect" use of VERTEX COVER was recently shown by Ito et al. [44]. It is based on an *alternative parameterization* other than the clique size. This parameterization is based on the observation that the hardness of finding a large clique in a graph is determined by the *isolation* of that clique, meaning that if we restrict ourselves to finding cliques that have only a few edges to "external" vertices outside of the clique, then this is a much easier task compared to finding cliques that have many edges to external vertices. The intuitive reason for this is that isolated cliques are better distinguishable from the remaining graph. To quantify the isolation of a clique, Ito et al. [44] introduced the notion of an *isolation factor* k. An i-vertex clique is said to be k-isolated if it has less than $k \cdot i$ edges to external vertices. It turns out that enumerating all k-isolated cliques is fixed-parameter tractable with respect to k. Komusiewicz et al. [49, 50] pointed out an error in the algorithm and provided a corrected version.

Theorem 1.3. (Ito et al. [44], Komusiewicz et al. [49, 50]). *All k-isolated cliques in an m-edge graph can be enumerated in $O(4^k \cdot k^2 m)$ time.*

The underlying algorithm of this result is based on a search tree for VERTEX COVER; the main achievement lies in showing that the isolation factor k can also serve as a bound for the depth of this search tree, which is achieved by a parameter-dependent data reduction. This nicely demonstrate the benefit of alternative parameterizations for a problem. Ito et al. [44] mentioned that some preliminary experiments suggest that the detection of isolated cliques is quite efficient in practice.

1.3.2. *Cluster Editing*

The CLUSTER EDITING problem is based on the assumption that the input graph is a disjoint union of cliques—a so-called *cluster graph*—that has been perturbed

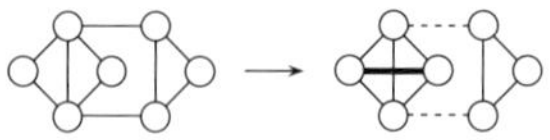

Fig. 1.4. Illustration for the CLUSTER EDITING problem: By removing two edges from and adding one edge to the graph on the left (that is, $k = 3$), we can obtain a graph that consists of two disjoint cliques.

by adding or removing edges. CLUSTER EDITING appears as an important problem in the analysis of data from synthetic genetic arrays [8, 23].

> CLUSTER EDITING
> **Input:** An undirected graph $G = (V, E)$ and a nonnegative integer k.
> **Task:** Modify G to consist of disjoint cliques by adding or deleting at most k edges.

Figure 1.4 illustrates this problem. CLUSTER EDITING is NP-hard [53, 66], and its minimization version can be approximated in polynomial time within a factor of 4 [13]. A randomized expected factor-3 approximation algorithm was given by Ailon et al. [4]. CLUSTER EDITING is also a special case of the CORRELATION CLUSTERING problem occurring in machine learning [6].

In what follows, we concentrate on a search tree-based fixed-parameter approach towards exactly solving CLUSTER EDITING. Here, the overall strategy is based on an easy-to-see observation, namely that a cluster graph has a very special structure: If two vertices are connected by an edge, then their neighborhoods must be the same. Hence, whenever we encounter two connected vertices u and v in the input graph G that are connected by an edge and where one vertex, say u, has a neighbor w that is not connected to v, we call $\{u, v, w\}$ a *conflict triple* of vertices because it compels us to do one of three things: Either remove the edge $\{u, v\}$, or connect v with w, or remove the edge $\{u, w\}$. Each of these three modifications counts with respect to the parameter k and, therefore, exhaustively branching into these cases for at most k forbidden substructures, we obtain a search tree of size $O(3^k)$ to solve CLUSTER EDITING.

The search tree size can be significantly reduced. More specifically, a more sophisticated branching strategy gives a search tree size of $O(2.27^k)$ [34], which—using computer-generated branching rules—has been further improved to a size of $O(1.92^k)$ [33]. The computer-generated result, however, is based on a quite complicated branching with lots of case distinctions that might not be of practical value. In the following, we describe the key observations and fundamental ideas behind the improved search trees for CLUSTER EDITING. As a remark, there also exist size-$o(3^k)$ search trees for *enumerating* all solutions to a given instance of CLUSTER EDITING [20].

The basic approach to obtain the improved search trees for CLUSTER EDITING is to do a case distinction, where we provide for every possible situation additional branching steps. The analysis of successive branching steps, then, yields a better worst-case bound on the search tree size. To this end, we make use of two annotations for unordered vertex pairs:

"*permanent*": In this case, $\{u, v\} \in E$ and it is not allowed to delete $\{u, v\}$;
"*forbidden*": In this case, $\{u, v\} \notin E$ and it is not allowed to add $\{u, v\}$.

Clearly, if an edge $\{u, v\}$ is deleted, then the vertex pair is made *forbidden*. If an edge $\{u, v\}$ is added, then the vertex pair is made *permanent*.

We distinguish three main situations that may apply when considering the conflict triple $\{u, v, w\}$:

(C1) Vertices v and w do not share a common neighbor, that is, $\forall x \in V, x \neq u$: $\{v, x\} \notin E$ or $\{w, x\} \notin E$.
(C2) Vertices v and w have a common neighbor $x \neq u$ and $\{u, x\} \in E$.
(C3) Vertices v and w have a common neighbor $x \neq u$ and $\{u, x\} \notin E$.

Regarding case (C1), the following lemma shows that a branching into two subcases suffices.

Lemma 1.1. *Given a graph $G = (V, E)$, a nonnegative integer k, and a conflict triple $u, v, w \in V$ of vertices that satisfy case (C1) from above, adding the edge $\{v, w\}$ cannot yield a better solution than deleting one of the edges $\{u, v\}$ or $\{u, w\}$.*

Proof. Consider a clustering solution G' for G where we did add $\{v, w\}$ (see Fig. 1.5 for an example). We use $N_{G \cap G'}(v)$ to denote the set of vertices that are neighbors of v in G and in G'. Without loss of generality, assume that $|N_{G \cap G'}(w)| \leq |N_{G \cap G'}(v)|$. We then construct a new graph G'' from G' by deleting all edges adjacent to w. It is clear that G'' is also a clustering solution for G. We compare the cost of the transformation $G \rightarrow G''$ to that of the transformation $G \rightarrow G'$:

- -1 for not adding $\{v, w\}$,
- $+1$ for deleting $\{u, w\}$,
- $-|N_{G \cap G'}(v)|$ for not adding all edges $\{w, x\}$, $x \in N_{G \cap G'}(v)$,
- $+|N_{G \cap G'}(w)|$ for deleting all edges $\{w, x\}$, $x \in N_{G \cap G'}(w)$.

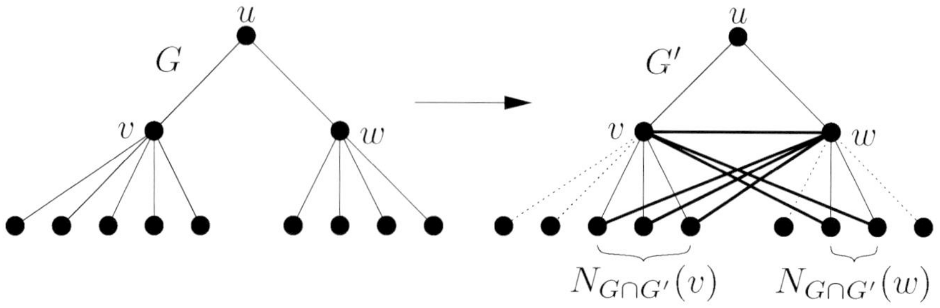

Fig. 1.5. In case (C1), adding the edge $\{v, w\}$ does not need to be considered. Here, G is the given graph and G' is a clustering solution of G that adds the edge $\{v, w\}$. The dashed lines denote edges being deleted to transform G into G', and the bold lines denote edges being added. Observe that the drawing only shows that parts of the graphs (in particular, edges) which are relevant for our argument.

Here, we omitted possible vertices which are neighbors of w in G' but not in G: they would only increase the cost of transformation $G \to G'$.

In summary, the cost of $G \to G''$ is not higher than the cost of $G \to G'$, that is, we do not need more edge additions and deletions to obtain G'' from G than to obtain G' from G. □

As a consequence of Lemma 1.1, the search tree only has to branch into two instead of three subcases in case (C1). Making use of the markers "permanent" and "forbidden," the standard branching into three subcases can also be avoided in cases (C2) and (C3). Each of these cases, however, requires specific considerations [34] which have to be omitted here.

Besides a search tree strategy for CLUSTER EDITING, also data reductions yielding problem kernels are known for this problem. The first result was a problem kernel with $O(k^2)$ vertices [34] and this has recently been improved to a problem kernel with only $O(k)$ vertices [28, 38]. For a practical solving algorithm, the search tree has to be combined with the kernelization [23, 34, 59]. By splitting up cases (C2) and (C3) further using a computer, we arrive at the following theorem.

Theorem 1.4 (Gramm et al. [33], Protti et al. [63]). CLUSTER EDITING *can be solved in* $O(1.92^k + n + m)$ *time.*

For a recent implementation of the above strategy, experiments indicated that the fixed-parameter approach outperforms a solution based on linear programming and appears to be of practical use [23]. The implementation can solve certain synthetic instances with $n = 100$ and 40 edit operations within an hour.

There are several problems closely related to CLUSTER EDITING that deserve similar studies. Among these are the more general CORRELATION CLUSTERING problem [6, 13] and the BICLUSTER EDITING problem [63, 69], the latter also known to be fixed-parameter tractable [63].

1.3.3. *Clique Cover*

The model assumption for CLIQUE COVER is that the data forms *overlapping* cliques, as opposed to the disjoint cliques that were the underlying model for CLUSTER EDITING. In general, this clustering model is applicable whenever various items carry an unknown subset of *features*, and two items will be measured as similar whenever they have at least one feature in common. Then, the set of items that carry one particular feature forms a clique. Therefore, finding a set of cliques that covers all edges gives the most parsimonious explanation of the data under this model: each clique corresponds to one feature. Guillaume and Latapy [37] argue that this model is very widely applicable to discover underlying structure in complex real-world networks.

Formally, the CLIQUE COVER problem is defined as follows:

> CLIQUE COVER
> **Input:** An undirected graph $G = (V, E)$ and a nonnegative integer k.
> **Task:** Find a set of at most k cliques in G such that each edge in E has both its endpoints in at least one of the selected cliques.

CLIQUE COVER is NP-hard [61], and there is strong evidence that it cannot be approximated to a constant factor [55]. Therefore, the standard approach to solving CLIQUE COVER in practice so far is to employ heuristics [47, 52, 62, 64]. Behrisch and Taraz [7] give simple greedy algorithms for CLIQUE COVER that provide asymptotically optimal solutions for certain random intersection graphs. However, because of the fundamental inapproximability of the problem, the results of these algorithms can become nearly arbitrarily bad for particular inputs. This makes fixed-parameter algorithms, which provide optimal solutions, attractive. A natural parameter is the size of the feature set k, since it seems reasonable to assume that there are much fewer features than items.

Gramm et al. [35] presented a parameterized approach to CLIQUE COVER, which is based on data reduction. In fact, the fixed-parameter tractability of CLIQUE COVER with respect to k is based on a *problem kernel*. The first two rules that lead to this kernel are easy to see:

> REDUCTION RULE CC1. Remove isolated vertices and vertices that are only adjacent to covered edges.

REDUCTION RULE CC2. If there is an edge $\{u, v\}$ whose endpoints have exactly the same closed neighborhood, that is, $N[u] = N[v]$, then delete u. To reconstruct a solution for the unreduced instance, add u to every clique containing v.

Rules CC1 and CC2 together suffice to show the problem kernel (the basic underlying observation was already made by Gyárfás [41]).

Theorem 1.5 (Gyárfás [41] and Gramm et al. [35]). *A* CLIQUE COVER *instance reduced with respect to Rules CC1 and CC2 contains at most* 2^k *vertices or, otherwise, has no solution.*

Proof. Consider a graph $G = (V, E)$ that is reduced with respect to Rules CC1 and CC2 and has a clique cover $C_1, \ldots, C_k$ of size k. We assign to each vertex $v \in V$ a binary vector b_v of length k where bit i, $1 \leq i \leq k$, is set iff v is contained in clique C_i. If we assume that G has more than 2^k vertices, then there must be $u \neq v \in V$ with $b_u = b_v$. Since Rule CC1 does not apply, every vertex is contained in at least one clique, and since $b_u = b_v$, u and v are contained in the same cliques. Therefore, u and v are connected. As they also share the same neighborhood, Rule CC2 applies, in contradiction to our assumption that G is reduced with respect to Rule CC2. Consequently, G cannot have more than 2^k vertices. □

By Theorem 1.1, this implies that CLIQUE COVER is fixed-parameter tractable with respect to parameter k. Unfortunately, the worst-case size of a reduced instance is still exponential, as opposed to the polynomially-sized kernels that are known for VERTEX COVER and CLUSTER EDITING.

As an example of an advanced data reduction rule, we now formulate a generalization of Rule CC2. While one can show that this rule finds a strict superset of the reduction opportunities of Rule CC2, it does not seem possible to use it to improve the worst-case problem kernel size bound in Theorem 1.5. Nevertheless, Rule CC2 improves the running time of solving CLIQUE COVER in practice, as described below.

To discuss the advanced data reduction rule, we need some additional terminology: For a vertex v, we partition the set of vertices that are connected by an edge to v into *prisoners* p with $N(p) \subseteq N(v)$ and *exits* x with $N(x) \setminus N(v) \neq \emptyset$. We say that the prisoners *dominate* the exits if every exit x has an adjacent prisoner. An illustration of the concept of prisoners and exits is given in Fig. 1.6.

REDUCTION RULE CC3. Consider a vertex v that has at least

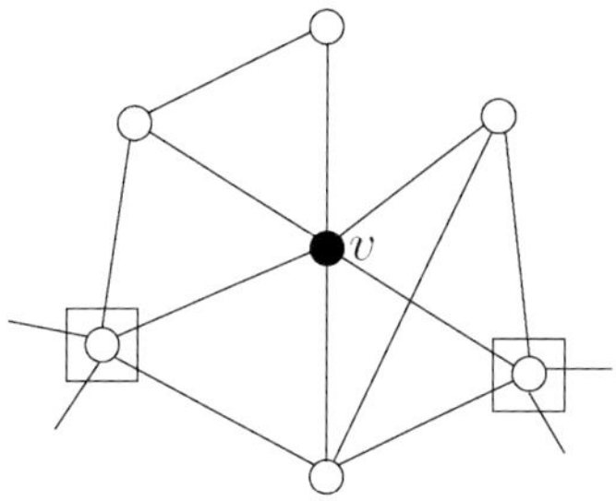

Fig. 1.6. An illustration of the partition of the neighborhood of a vertex v. The two vertices with rectangles are exits, the other white ones are prisoners.

one prisoner. If each prisoner is connected to at least one vertex other than v via an edge, and the prisoners dominate the exits, then delete v. To reconstruct a solution for the unreduced instance, add v to every clique containing a prisoner of v.

Lemma 1.2. *Reduction Rule CC3 is correct.*

Proof. By definition, every neighbor of v's prisoners is also a neighbor of v itself. If a prisoner of v participates in a clique C, then $C \cup \{v\}$ is also a clique in the graph. Therefore, it is correct to add v to every clique containing a prisoner in the reduced graph. Next, we show that all edges adjacent to v are covered by the cliques resulting by adding v to the cliques containing v's prisoners. We consider separately the edges connecting v to prisoners and edges connecting v to exits. Regarding an edge $\{v, w\}$ to a prisoner w, vertex w has to be part of a clique C of the solution for the instance after application of the rule. Therefore, the edge $\{v, w\}$ is covered by $C \cup \{v\}$ in the solution for the unreduced instance. Regarding an edge $\{v, x\}$ to an exit x, the exit x is dominated by a prisoner w and therefore x has to be part of a clique C with w. Hence, the edge $\{v, x\}$ is covered by $C \cup \{v\}$ in the solution for the unreduced instance. □

Concerning experimental results, Gramm et al. [35] implemented an algorithm to optimally solve CLIQUE COVER that is based on four data reduction rules (CC1–CC3 and one additional rule) and a simple branching strategy. The implementation was able to solve within a few seconds 14 real-world instances from an application in graphical statistics, with up to 124 vertices and more than 2 700 edges [36]. Further experiments on random graphs showed that in particular sparse instances could be solved quickly. The algorithm relied mainly on the data reduction rules: many instances were reduced to an empty instance before the branching even began. By way of contrast, when the data reduction was not successful, running times increased sharply. Additional experiments for the feature model mentioned at the beginning of this section showed that random instances with 100 items and up to 30 features could be solved within a few minutes.

Two reduction rules of Gramm et al. (Rule CC3 and another rule that we do not discuss here) have so far not been proved to improve the size of the problem kernel and are also quite slow to compute. For some instances, however, both rules were beneficial, meaning that the obtained speedups for them instances were much larger than observed slowdowns for other instances. This suggests the general application of these rules, possibly with an additional heuristic to disable them based on instance properties.

As a final note, unlike the CLUSTER EDITING model, CLIQUE COVER does not take perturbed data into account. In practice, probably edges within feature clusters are missing, and spurious edges exist. However, this can be handled with a post-processing: as long as there are not too many errors, spurious edges will be covered by size-2 cliques, which can be easily filtered; and a clique missing an edge will be optimally covered by two cliques, and so in a post-processing one can check for all pairs of cliques whether they could be merged by adding a small number of edges.

1.4. Conclusion

We conclude our survey with a list of guidelines for the practical design of fixed-parameter algorithms and some open challenges in the field.

1.4.1. *Practical Guidelines*

The following list sums up the experiences of our and other research groups who have implemented fixed-parameter algorithms:

Fixed-Parameter Tractability in General

(1) Do not despair of bad upper bounds for fixed-parameter algorithms that are given in theoretically oriented papers—the analysis is worst-case and often much too pessimistic.
(2) Fixed-parameter algorithms are the better the smaller the parameter value is. Hence, parameterizations with small parameter values should be sought after.
(3) Most existing fixed-parameter algorithms in the literature are concerned with optimization problems on unweighted graphs. Solving enumerative

problems and problems on weighted graphs therefore usually requires some additional work to be done.

(4) Exponential memory consumption usually turns out to be more harmful in practice than exponential time, especially in enumerative tasks. Using memory-saving techniques generally pays off, even if this means some decrease in time efficiency.

Data Reductions and Kernelizations

(1) One should always start by designing data reduction rules because these are helpful in combination with basically any algorithmic technique that is subsequently applied.
(2) The *order* of applying data reduction rules can significantly influence the practical effectiveness and efficiency of the overall data reduction. Experimental work is important to find good orderings.
(3) Even if a data reduction has no provable performance guarantee, it can still turn out to be very effective in practice.

Depth-Bounded Search Trees

(1) The branching in a search tree can produce new opportunities for data reduction. Therefore, data reductions should be applied repeatedly during the course of the whole algorithm and not only as a preprocessing step [59]. To achieve maximum efficiency, the exact frequency of applying data reductions may require some tuning.
(2) Search tree algorithms can be parallelized rather easily.
(3) Complicated case distinctions for the branching should be avoided when a simpler search strategy is available that yields almost the same worst-case running time bounds. The simpler strategy usually turns out to be faster.

1.4.2. *Challenges*

While there has been substantial work on fixed-parameter algorithms for clustering problems and several examples show the potential of this approach, the field is still quite young, and there remain a couple of challenges for future research:

- The CLIQUE model is often too restricted in applications; one would rather prefer a notion of "dense subgraph" (e.g., Ref. 10, 43, 73). Except for Ref. 44, 49, 50, we are not aware of fixed-parameter approaches for such scenarios.

- For simplicity, many fixed-parameter approaches drop the requirement to be able to handle weighted problems or to handle enumeration. Extensions of known results in this direction are desirable.
- Of our three case studies, CLIQUE COVER seems to be the least explored problem. While kernels with a linear number of vertices are known for VERTEX COVER and CLUSTER EDITING, the only known kernel for CLIQUE COVER is of exponential size. Also, no search tree with a fixed-parameter bound on its size is known for CLIQUE COVER except for a trivial brute-force exploration of the problem kernel.
- Some works consider the variant of CLUSTER EDITING where there are *don't care*-edges that have zero editing cost [6]. It is not yet known whether a fixed-parameter algorithm exists for this problem.

A particularly important challenge for future work is to bring progress from fixed-parameter algorithmics to a broader audience by providing easily accessible software tools that are finely tuned by algorithm engineering and additional tools such as heuristics and parallelization.

References

[1] F. N. Abu-Khzam, R. L. Collins, M. R. Fellows, M. A. Langston, W. H. Suters, and C. T. Symons. Kernelization algorithms for the Vertex Cover problem: theory and experiments. In *Proc. 6th ALENEX*, pages 62–69. SIAM, 2004.

[2] F. N. Abu-Khzam, M. A. Langston, and W. H. Suters. Fast, effective vertex cover kernelization: A tale of two algorithms. In *Proc. 3rd AICCSA*. ACS/IEEE, 2005. 16 pages.

[3] F. N. Abu-Khzam, M. A. Langston, P. Shanbhag, and C. T. Symons. Scalable parallel algorithms for FPT problems. *Algorithmica*, 45(3):269–284, 2006.

[4] N. Ailon, M. Charikar, and A. Newman. Aggregating inconsistent information: ranking and clustering. In *Proc. 37th STOC*, pages 684–693. ACM, 2005.

[5] G. Ausiello, P. Crescenzi, G. Gambosi, V. Kann, A. Marchetti-Spaccamela, and M. Protasi. *Complexity and Approximation: Combinatorial Optimization Problems and Their Approximability Properties*. Springer, 1999.

[6] N. Bansal, A. Blum, and S. Chawla. Correlation clustering. *Machine Learning*, 56(1):89–113, 2004.

[7] M. Behrisch and A. Taraz. Efficiently covering complex networks with cliques of similar vertices. *Theoretical Computer Science*, 355(1):37–47, 2006.

[8] A. Ben-Dor, R. Shamir, and Z. Yakhini. Clustering gene expression patterns. *Journal of Computational Biology*, 6(3–4):281–297, 1999.

[9] M. Benson, M. A. Langston, M. Adner, B. Andersson, Å. Torinsson Naluai, and L. O.

Cardell. A network-based analysis of the late-phase reaction of the skin. *The Journal of Allergy and Clinical Immunology*, 118(1):220–225, 2006.

[10] S. Butenko and W. E. Wilhelm. Clique-detection models in computational biochemistry and genomics. *European Journal of Operational Research*, 173(1):1–17, 2006.

[11] L. Cai, J. Chen, R. Downey, and M. R. Fellows. Advice classes of parameterized tractability. *Annals of Pure and Applied Logic*, 84:119–138, 1997.

[12] L. Sunil Chandran and Fabrizio Grandoni. Refined memorisation for Vertex Cover. *Information Processing Letters*, 93(3):125–131, 2005.

[13] M. Charikar, V. Guruswami, and A. Wirth. Clustering with qualitative information. *Journal of Computer and System Sciences*, 71(3):360–383, 2005.

[14] J. Cheetham, F. K. H. A. Dehne, A. Rau-Chaplin, U. Stege, and P. J. Taillon. Solving large FPT problems on coarse-grained parallel machines. *Journal of Computer and System Sciences*, 67(4):691–706, 2003.

[15] J. Chen, I. A. Kanj, and G. Xia. Improved parameterized upper bounds for Vertex Cover. In *Proc. 31st MFCS*, volume 4162 of *LNCS*, pages 238–249. Springer, 2006.

[16] E. J. Chesler, L. Lu, S. Shou, Y. Qu, J. Gu, J. Wang, H. C. Hsu, J. D. Mountz, N. E. Baldwin, M. A. Langston, D. W. Threadgill, K. F. Manly, and R. W. Williams. Complex trait analysis of gene expression uncovers polygenic and pleiotropic networks that modulate nervous system function. *Nature Genetics*, 37:233–242, 2005.

[17] M. Chlebík and J. Chlebíková. Improvement of Nemhauser–Trotter theorem and its applications in parameterized complexity. In *Proc. 9th SWAT*, volume 3111 of *LNCS*, pages 174–186. Springer, 2004.

[18] B. Chor, M. R. Fellows, and D. W. Juedes. Linear kernels in linear time, or how to save k colors in $O(n^2)$ steps. In *Proc. 30th WG*, volume 3353 of *LNCS*, pages 257–269. Springer, 2004.

[19] T. H. Cormen, C. E. Leiserson, R. L. Rivest, and C. Stein. *Introduction to Algorithms*. MIT Press, 2nd edition, 2001.

[20] P. Damaschke. On the fixed-parameter enumerability of Cluster Editing. In *Proc. 31st WG*, volume 3787 of *LNCS*, pages 283–294. Springer, 2005.

[21] P. Damaschke. Parameterized enumeration, transversals, and imperfect phylogeny reconstruction. *Theoretical Computer Science*, 351(3):337–350, 2006.

[22] M. Davis, G. Logemann, and D. W. Loveland. A machine program for theorem-proving. *Communications of the ACM*, 5(7):394–397, 1962.

[23] F. K. H. A. Dehne, M. A. Langston, X. Luo, S. Pitre, P. Shaw, and Y. Zhang. The Cluster Editing problem: Implementations and experiments. In *Proc. 2nd IWPEC*, volume 4169 of *LNCS*, pages 13–24. Springer, 2006.

[24] R. Diestel. *Graph Theory*. Springer, 3rd edition, 2005.

[25] R. G. Downey and M. R. Fellows. *Parameterized Complexity*. Springer, 1999.

[26] L. Drori and D. Peleg. Faster exact solutions for some NP-hard problems. *Theoretical Computer Science*, 287(2):473–499, 2002.

[27] D. Eppstein. Quasiconvex analysis of backtracking algorithms. In *Proc. 15th SODA*, pages 788–797. ACM/SIAM, 2004.

[28] M. R. Fellows, M. A. Langston, F. Rosamond, and P. Shaw. Polynomial-time linear kernelization for Cluster Editing. Manuscript, 2006.

[29] A. Felner, R. E. Korf, and S. Hanan. Additive pattern database heuristics. *Journal of Artificial Intelligence Research*, 21:1–39, 2004.

[30] J. Flum and M. Grohe. *Parameterized Complexity Theory*. Springer, 2006.

[31] F. V. Fomin, F. Grandoni, and D. Kratsch. Some new techniques in design and analysis of exact (exponential) algorithms. *Bulletin of the EATCS*, 87:47–77, 2005.

[32] M. R. Garey and D. S. Johnson. *Computers and Intractability: A Guide to the Theory of NP-Completeness*. W. H. Freeman, 1979.

[33] J. Gramm, J. Guo, F. Hüffner, and R. Niedermeier. Automated generation of search tree algorithms for hard graph modification problems. *Algorithmica*, 39(4):321–347, 2004.

[34] J. Gramm, J. Guo, F. Hüffner, and R. Niedermeier. Graph-modeled data clustering: Exact algorithms for clique generation. *Theory of Computing Systems*, 38(4):373–392, 2005.

[35] J. Gramm, J. Guo, F. Hüffner, and R. Niedermeier. Data reduction, exact, and heuristic algorithms for clique cover. In *Proc. 8th ALENEX*, pages 86–94. SIAM, 2006. Long version to appear under the title "Data reduction and exact algorithms for clique cover" in *ACM Journal of Experimental Algorithmics*.

[36] J. Gramm, J. Guo, F. Hüffner, R. Niedermeier, H.-P. Piepho, and R. Schmid. Algorithms for compact letter displays: Comparison and evaluation. *Computational Statistics & Data Analysis*, 2006. To appear.

[37] J.-L. Guillaume and M. Latapy. Bipartite structure of all complex networks. *Information Processing Letters*, 90(5):215–221, 2004.

[38] J. Guo. A more effective linear kernelization for cluster editing. In *Proc. ESCAPE*, LNCS. Springer, 2007. To appear.

[39] J. Guo and R. Niedermeier. Invitation to data reduction and problem kernelization. *ACM SIGACT News*, 38(1):31–45, 2007.

[40] J. Guo, R. Niedermeier, and S. Wernicke. Parameterized complexity of generalized Vertex Cover problems. In *Proc. 9th WADS*, volume 3608 of *LNCS*, pages 36–48. Springer, 2005. Long version to appear under the title "Parameterized complexity of Vertex Cover variants" in *Theory of Computing Systems*.

[41] A. Gyárfás. A simple lower bound on edge coverings by cliques. *Discrete Mathematics*, 85(1):103–104, 1990.

[42] J. Håstad. Clique is hard to approximate within $n^{1-\epsilon}$. *Acta Mathematica*, 182(1):105–142, 1999.

[43] K. Holzapfel, S. Kosub, M. G. Maaß, and H. Täubig. The complexity of detecting fixed-density clusters. *Discrete Applied Mathematics*, 154(11):1547–1562, 2006.

[44] H. Ito, K. Iwama, and T. Osumi. Linear-time enumeration of isolated cliques. In *Proc. 13th ESA*, volume 3669 of *LNCS*, pages 119–130. Springer, 2005.

[45] H. Kawaji, Y. Takenaka, and H. Matsuda. Graph-based clustering for finding distant relationships in a large set of protein sequences. *Bioinformatics*, 20(2):243–252, 2004.

[46] H. Kawaji, Y. Yamaguchi, H. Matsuda, and A. Hashimoto. A graph-based clustering method for a large set of sequences using a graph partitioning algorithm. *Genome Informatics*, 12:93–102, 2001.

[47] E. Kellerman. Determination of keyword conflict. *IBM Technical Disclosure Bulletin*, 16(2):544–546, 1973.

[48] J. Kleinberg and E. Tardos. *Algorithm Design*. Addison Wesley, 2005.

[49] C. Komusiewicz. Various isolation concepts for the enumeration of dense subgraphs.

Diplomarbeit, Institut für Informatik, Friedrich-Schiller-Universität Jena, 2007.
[50] C. Komusiewicz, F. Hüffner, H. Moser, and R. Niedermeier. Isolation concepts for enumerating dense subgraphs. In *Proc. 13th COCOON*, volume 4598 of *LNCS*, pages 140-150. Springer, 2007.
[51] R. E. Korf, M. Reid, and S. Edelkamp. Time complexity of iterative-deepening-A*. *Artificial Intelligence*, 129:199–218, 2001.
[52] L. T. Kou, L. J. Stockmeyer, and C.-K. Wong. Covering edges by cliques with regard to keyword conflicts and intersection graphs. *Communications of the ACM*, 21(2):135–139, 1978.
[53] M. Křivánek and J. Morávek. NP-hard problems in hierarchical-tree clustering. *Acta Informatica*, 23(3):311–323, 1986.
[54] M. A. Langston, A. D. Perkins, D. J. Beare, R. W. Gauldie, P. J. Kershaw, J. B. Reid, K. Winpenny, and A. J. Kenny. Combinatorial algorithms and high performance implementations for elucidating complex ecosystem relationships from North Sea historical data. In *Proc. International Council for the Exploration of the Sea Annual Science Conference*, 2006.
[55] C. Lund and M. Yannakakis. On the hardness of approximating minimization problems. *Journal of the ACM*, 41:960–981, 1994.
[56] S. Mecke and D. Wagner. Solving geometric covering problems by data reduction. In *Proc. 12th ESA*, volume 3221 of *LNCS*, pages 760–771. Springer, 2004.
[57] K. Mehlhorn. *Data Structures and Algorithms, Volume 2: NP-Completeness and Graph Algorithms*. EATCS Monographs on Theoretical Computer Science. Springer, 1984.
[58] Z. Michalewicz and B. F. Fogel. *How to Solve it: Modern Heuristics*. Springer, 2nd edition, 2004.
[59] R. Niedermeier and P. Rossmanith. A general method to speed up fixed-parameter-tractable algorithms. *Information Processing Letters*, 73:125–129, 2000.
[60] R. Niedermeier. *Invitation to Fixed-Parameter Algorithms*. Oxford University Press, 2006.
[61] J. B. Orlin. Contentment in graph theory: Covering graphs with cliques. *Indagationes Mathematicae (Proceedings)*, 80(5):406–424, 1977.
[62] H.-P. Piepho. An algorithm for a letter-based representation of all-pairwise comparisons. *Journal of Computational and Graphical Statistics*, 13(2):456–466, 2004.
[63] F. Protti, M. D. da Silva, and J. L. Szwarcfiter. Applying modular decomposition to parameterized Bicluster Editing. In *Proc. 2nd IWPEC*, volume 4169 of *LNCS*, pages 1–12. Springer, 2006.
[64] S. Rajagopalan, M. Vachharajani, and S. Malik. Handling irregular ILP within conventional VLIW schedulers using artificial resource constraints. In *Proc. CASES*, pages 157–164. ACM, 2000.
[65] S. Seno, R. Teramoto, Y. Takenaka, and H. Matsuda. A method for clustering expression data based on graph structure. *Genome Informatics*, 15(2):151–160, 2004.
[66] R. Shamir, R. Sharan, and D. Tsur. Cluster graph modification problems. *Discrete Applied Mathematics*, 144(1–2):173–182, 2004.
[67] V. V. Vazirani. *Approximation Algorithms*. Springer, 2001.
[68] B. H. Voy, J. A. Scharff, A. D. Perkins, A. M. Saxton, B. Borate, E. J. Chesler, L. K. Branstetter, and M. A. Langston. Extracting gene networks for low-dose radiation

using graph theoretical algorithms. *PLoS Computational Biology*, 2(7):e89, 2006.

[69] L. Wang, Y. Lin, and X. Liu. Approximation algorithms for bi-clustering problems. In *Proc. 6th WABI*, volume 4175 of *LNBI*, pages 310–320. Springer, 2006.

[70] K. Weihe. Covering trains by stations or the power of data reduction. In *Proc. 1st ALEX*, pages 1–8, 1998.

[71] D. B. West. *Introduction to Graph Theory*. Prentice Hall, 2nd edition, 2001.

[72] G. J. Woeginger. Exact algorithms for NP-hard problems: A survey. In *Proc. 5th IWCO*, volume 2570 of *LNCS*, pages 185–208. Springer, 2003.

[73] H. Yu, A. Paccanaro, V. Trifonov, and M. Gerstein. Predicting interactions in protein networks by completing defective cliques. *Bioinformatics*, 22(7):823–829, 2006.

[74] Y. Zhang, F. N. Abu-Khzam, N. E. Baldwin, E. J. Chesler, Michael A. Langston, and Nagiza F. Samatova. Genome-scale computational approaches to memory-intensive applications in systems biology. In *Proc. 18th SC*, page e12. IEEE Computer Society, 2005.

Chapter 2

Probabilistic Distance Clustering: Algorithm and Applications

C. Iyigun

RUTCOR–Rutgers Center for Operations Research, Rutgers University
640 Bartholomew Rd., Piscataway, NJ 08854-8003, USA
Email: iyigun@rutcor.rutgers.edu

A. Ben-Israel

RUTCOR–Rutgers Center for Operations Research, Rutgers University
640 Bartholomew Rd., Piscataway, NJ 08854-8003, USA
Email: adi.benisrael@gmail.com

The probabilistic distance clustering method of the authors [2, 8], assumes the cluster membership probabilities given in terms of the distances of the data points from the cluster centers, and the cluster sizes. A resulting extremal principle is then used to update the cluster centers (as convex combinations of the data points), and the cluster sizes (if not given.) Progress is monitored by the joint distance function (JDF), a weighted harmonic mean of the above distances, that approximates the data by capturing the data points in its lowest contours. The method is described, and applied to clustering, location problems, and mixtures of distributions, where it is a viable alternative to the Expectation–Maximization (EM) method. The JDF also helps to determine the "right" number of clusters for a given data set.

2.1. Introduction

We take data points to be vectors $\mathbf{x} = (x_1, \ldots, x_n) \in \mathbb{R}^n$, and a **dataset** $\mathcal{D}$ consisting of N data points $\{\mathbf{x}_1, \mathbf{x}_2, \ldots \mathbf{x}_N\}$. A **cluster** is a set of data points that are similar, in some sense, and **clustering** is a process of partitioning a data set into disjoint clusters.

In **distance clustering** (or **d–clustering**), "similarity" is interpreted in terms of a **distance** function $d(\mathbf{x}, \mathbf{y})$ in $\mathbb{R}^n$, such as

$$d(\mathbf{x}, \mathbf{y}) = \|\mathbf{x} - \mathbf{y}\|, \ \forall\, \mathbf{x}, \mathbf{y} \in \mathbb{R}^n, \tag{2.1}$$

where $\|\cdot\|$ is a norm. In particular, the **Mahalanobis distance** with the norm,

$$\|\mathbf{u}\| = \langle \mathbf{u}, \Sigma^{-1}\mathbf{u}\rangle^{1/2}, \tag{2.2}$$

where Σ is the covariance matrix of the data involved.

Example 2.1. A data set in $\mathbb{R}^2$ with $N = 1100$ data points is shown in Fig. 2.1. The data on the left was simulated from a normal distribution $N(\boldsymbol{\mu}, \Sigma)$, with:

$$\boldsymbol{\mu}_1 = (2, 0),\ \Sigma_1 = \begin{pmatrix} 0.0005 & 0 \\ 0 & 0.05 \end{pmatrix},\ (100 \text{ points}),$$

and the data on the right consist of 1000 points, simulated in a circle of diameter 1 centered at $\boldsymbol{\mu}_2 = (3, 0)$ according to a radially symmetric distribution with $\text{Prob}\{\|\mathbf{x} - \boldsymbol{\mu}_2\| \leq r\} = 2r,\ 0 \leq r \leq \frac{1}{2}$.

This data will serve as illustration in Examples 2.2–2.3 below.

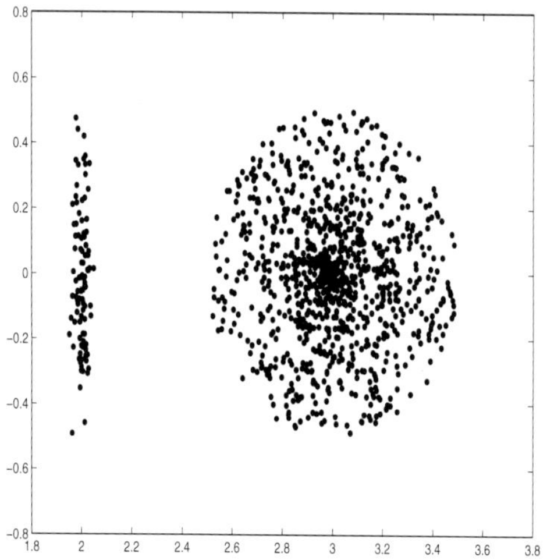

Fig. 2.1. A data set in $\mathbb{R}^2$

In d–clustering each point is typically assigned to the cluster with the nearest center. After each assignment, the cluster centers may change, requiring a re–classification of the data points. A d–clustering algorithm will therefore iterate between centers and re–assignments. The best known such method is the **k–means clustering algorithm**, see Hartigan [5].

In **probabilistic clustering** the cluster membership is expressed by probabilities $p_k(\mathbf{x}) = \text{Prob}\,\{\mathbf{x} \in \mathcal{C}_k\}$, that a data point $\mathbf{x}$ belongs to the cluster $\mathcal{C}_k$. In **probabilistic d–clustering** these probabilities depend on the relevant distances.

Probabilistic d–clustering adjusted for the cluster size q, is called **probabilistic {d,q}–clustering**.

An algorithm for probabilistic {d,q}–clustering, called the **PDQ Algorithm**, is presented in Sec. 2.2. It updates the cluster centers, and the cluster sizes (if not given) using the **extremal principle** of Sec. 2.2.3.

A byproduct of this approach is the **joint distance function**, see Sec. 2.2.2, a function that captures the data in its low level sets, and provides a continuous approximation of a discrete data set, using few parameters.

In Sec. 2.4 we apply the PDQ algorithm to estimation of the parameters of Gaussian mixtures, and report the results in Sec. 2.5.

The PDQ algorithm is also effective for solving capacitated multi–facility location problems, see Sec. 2.6.

Another application of the PDQ algorithm is for determining the "right" number of clusters, see Sec. 2.7.

For other approaches to probabilistic clustering see the surveys in Höppner et al. [7], Tan et al. [14]. A unified optimization framework for clustering was given by Teboulle [15].

2.2. Probabilistic {d,q}–Clustering

Let a data set $\mathcal{D} \subset \mathbb{R}^n$ be partitioned into K clusters $\{\mathcal{C}_k : k = 1, \cdots, K\}$,

$$\mathcal{D} = \bigcup_{k=1}^{K} \mathcal{C}_k \tag{2.3}$$

let $\mathbf{c}_k$ be the **center** (in some sense) of the cluster $\mathcal{C}_k$, and let q_k be the **cluster size**, which is known in some applications, and an unknown to be estimated in others. In what follows the cluster sizes, or their estimates, are assumed given wherever appearing in the right hand side of a formula.

With each data point $\mathbf{x} \in \mathcal{D}$, and a cluster $\mathcal{C}_k$ with center $\mathbf{c}_k$, we associate:

- a **distance** $d_k(\mathbf{x}, \mathbf{c}_k)$, also denoted $d_k(\mathbf{x})$, or just d_k if $\mathbf{x}$ is understood,
- a **probability** of membership in $\mathcal{C}_k$, denoted $p_k(\mathbf{x})$, or just p_k.

In general, the distance functions $d_k(\cdot)$ are different for different clusters. In particular, different Mahalanobis distances

$$d_k(\mathbf{x}) = \langle \mathbf{x} - \mathbf{c}_k, \Sigma_k^{-1}(\mathbf{x} - \mathbf{c}_k)\rangle^{1/2}, \tag{2.4}$$

using estimates $\{\mathbf{c}_k, \Sigma_k\}$ for the cluster in question.

There are several ways to model the relationship between distances and probabilities [2, 8]. The following assumption is our basic principle.

Principle. For each $\mathbf{x} \in \mathcal{D}$ and cluster $\mathcal{C}_k$, the probability $p_k(\mathbf{x})$ satisfies

$$\frac{p_k(\mathbf{x})\, d_k(\mathbf{x})}{q_k} = \text{constant, say } D(\mathbf{x}), \text{ depending on } \mathbf{x}\,. \tag{2.5}$$

Cluster membership is thus more probable the closer the data point is to the cluster center and the bigger the cluster.

2.2.1. *Probabilities*

From the above principle, and the fact that probabilities add to 1 we get

Theorem 2.1. *Let the cluster centers $\{\mathbf{c}_1, \mathbf{c}_2, \ldots, \mathbf{c}_K\}$ be given, let $\mathbf{x}$ be a data point, and let $\{d_k(\mathbf{x}) : k = 1, \ldots, K\}$ be its distances from the given centers. Then the membership probabilities of $\mathbf{x}$ are*

$$p_k(\mathbf{x}) = \frac{\prod\limits_{j \neq k} \dfrac{d_j(\mathbf{x})}{q_j}}{\sum\limits_{i=1}^{K} \prod\limits_{j \neq i} \dfrac{d_j(\mathbf{x})}{q_j}}\,, \quad k = 1, \ldots, K. \tag{2.6}$$

Proof. Using (2.5) we write for i, k

$$p_i(\mathbf{x}) = \left(\frac{p_k(\mathbf{x}) d_k(\mathbf{x})}{q_k}\right) \Big/ \left(\frac{d_i(\mathbf{x})}{q_i}\right).$$

Since $\sum\limits_{i=1}^{K} p_i(\mathbf{x}) = 1$,
$$p_k(\mathbf{x}) \sum_{i=1}^{K} \left(\frac{d_k(\mathbf{x})/q_k}{d_i(\mathbf{x})/q_i}\right) = 1.$$

$$\therefore\ p_k(\mathbf{x}) = \frac{1}{\sum\limits_{i=1}^{K} \left(\dfrac{d_k(\mathbf{x})/q_k}{d_i(\mathbf{x})/q_i}\right)} = \frac{\prod\limits_{j \neq k} d_j(\mathbf{x})/q_j}{\sum\limits_{i=1}^{K} \prod\limits_{j \neq i} d_j(\mathbf{x})/q_j}\,. \qquad \square$$

In particular, for K=2,

$$p_1(\mathbf{x}) = \frac{d_2(\mathbf{x})/q_2}{d_1(\mathbf{x})/q_1 + d_2(\mathbf{x})/q_2}\,, \quad p_2(\mathbf{x}) = \frac{d_1(\mathbf{x})/q_1}{d_1(\mathbf{x})/q_1 + d_2(\mathbf{x})/q_2}\,, \tag{2.7}$$

and for $K = 3$,

$$p_1(\mathbf{x}) = \frac{d_2(\mathbf{x})d_3(\mathbf{x})/q_2q_3}{d_1(\mathbf{x})d_2(\mathbf{x})/q_1q_2 + d_1(\mathbf{x})d_3(\mathbf{x})/q_1q_3 + d_2(\mathbf{x})d_3(\mathbf{x})/q_2q_3}\,, \tag{2.8}$$

etc.

2.2.2. *The Joint Distance Function*

We denote the constant in (2.5) by $D(\mathbf{x})$, a function of $\mathbf{x}$. Since the probabilities

$$p_k(\mathbf{x}) = D(\mathbf{x}) / \left(\frac{d_k(\mathbf{x})}{q_k}\right)\,, \; k = 1, \ldots, K,$$

add to 1 we get,

$$D(\mathbf{x}) = \frac{\prod_{j=1}^{K} \frac{d_j(\mathbf{x})}{q_j}}{\sum_{i=1}^{K} \prod_{j \neq i} \frac{d_j(\mathbf{x})}{q_j}}\,. \tag{2.9}$$

$D(\mathbf{x})$ is called the **joint distance function** (abbreviated **JDF**) of $\mathbf{x}$, and is, up to a constant, the harmonic mean of the K weighted distances $\{d_k(\mathbf{x})/q_k\}$, see [1].

In particular, for $K = 2$,

$$D(\mathbf{x}) = \frac{d_1(\mathbf{x})\, d_2(\mathbf{x})/q_1q_2}{d_1(\mathbf{x})/q_1 + d_2(\mathbf{x})/q_2}\,, \tag{2.10}$$

and $K = 3$,

$$D(\mathbf{x}) = \frac{d_1(\mathbf{x})\, d_2(\mathbf{x})\, d_3(\mathbf{x})/q_1q_2q_3}{d_1(\mathbf{x})\, d_2(\mathbf{x})/q_1q_2 + d_1(\mathbf{x})\, d_3(\mathbf{x})/q_1q_3 + d_2(\mathbf{x})\, d_3(\mathbf{x})/q_2q_3}\,. \tag{2.11}$$

Example 2.2. Figure 2.2(a) shows level sets of the JDF (2.10) for the data of Example 2.1.

2.2.3. *An Extremal Principle*

The principle (2.5) may be derived from an extremal principle. For notational simplicity we consider here the case of 2 clusters, the results in the general case are analogous.

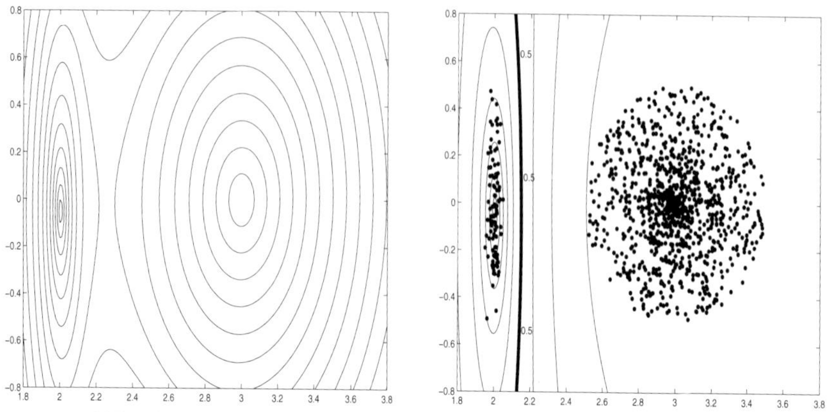

(a) Level sets of the JDF

(b) Level sets of cluster probabilities

Fig. 2.2. Results returned by the PDQ algorithm (Algorithm 1 below) for the data of Example 2.1

Let $\mathbf{x}$ be a given data point with distances $d_1(\mathbf{x})$, $d_2(\mathbf{x})$ to the cluster centers, and assume the cluster sizes q_1, q_2 known. Then the probabilities in (2.7) are the optimal solutions of the extremal problem

$$\begin{aligned} \min \quad & \frac{d_1(\mathbf{x})\, p_1^2}{q_1} + \frac{d_2(\mathbf{x})\, p_2^2}{q_2} \\ \text{s.t.} \quad & p_1 + p_2 = 1 \\ & p_1,\, p_2 \geq 0 \end{aligned} \tag{2.12}$$

Indeed, the Lagrangian of this problem is

$$L(p_1, p_2, \lambda) = \frac{d_1(\mathbf{x})\, p_1^2}{q_1} + \frac{d_2(\mathbf{x})\, p_2^2}{q_2} + \lambda(1 - p_1 + p_2) \tag{2.13}$$

and zeroing the partials $\partial L / \partial p_i$ gives the principle (2.5).

Substituting the probabilities (2.7) in (2.13) we get the optimal value of (2.12),

$$L^*(p_1(\mathbf{x}), p_2(\mathbf{x})) = \frac{d_1(\mathbf{x})\, d_2(\mathbf{x}) / q_1 q_2}{d_1(\mathbf{x})/q_1 + d_2(\mathbf{x})/q_2} \tag{2.14}$$

which is again the JDF (2.10).

The corresponding extremal problem for the data set $\mathcal{D} = \{\mathbf{x}_1,\, \mathbf{x}_2,\, \ldots,\, \mathbf{x}_N\}$

is

$$\begin{aligned} \min \quad & \sum_{i=1}^{N} \left(\frac{d_1(i)\, p_1(i)^2}{q_1} + \frac{d_2(i)\, p_2(i)^2}{q_2} \right) \\ \text{s.t.} \quad & p_1(i) + p_2(i) = 1 \ , \\ & p_1(i),\, p_2(i) \geq 0 \ , \ i = 1, \cdots, N \ , \end{aligned} \tag{2.15}$$

where $p_1(i), p_2(i)$ are the cluster probabilities at $\mathbf{x}_i$ and $d_1(i), d_2(i)$ are the corresponding distances. The problem separates into N problems like (2.12), and its optimal value is

$$\sum_{i=1}^{N} \frac{d_1(i)\, d_2(i)/q_1 q_2}{d_1(i)/q_1 + d_2(i)/q_2} \tag{2.16}$$

the sum of the joint distance functions of all points.

2.2.4. *An Extremal Principle for the Cluster Sizes*

Taking the cluster sizes as variables in the extremal principle (2.15),

$$\begin{aligned} \min \quad & \sum_{i=1}^{N} \left(\frac{d_1(i)\, p_1(i)^2}{q_1} + \frac{d_2(i)\, p_2(i)^2}{q_2} \right) \\ \text{s.t.} \quad & q_1 + q_2 = N \\ & q_1,\, q_2 \geq 0 \end{aligned} \tag{2.17}$$

with $p_1(i), p_2(i)$ assumed known, we have the Lagrangian

$$L(q_1, q_2, \lambda) = \sum_{i=1}^{N} \left(\frac{d_1(i)\, p_1(i)^2}{q_1} + \frac{d_2(i)\, p_2(i)^2}{q_2} \right) + \lambda(q_1 + q_2 - N) \tag{2.18}$$

Zeroing the partials $\partial L / \partial q_k$ gives,

$$q_k^2 = \frac{1}{\lambda} \left(\sum_{i=1}^{N} d_k(i)\, p_k(i)^2 \right) , \ k = 1, 2 \ , \tag{2.19}$$

and since $q_1 + q_2 = N$,

$$\frac{q_k}{N} = \frac{\left(\sum\limits_{i=1}^{N} d_k(i)\, p_k(i)^2 \right)^{1/2}}{\left(\sum\limits_{i=1}^{N} d_1(i)\, p_1(i)^2 \right)^{1/2} + \left(\sum\limits_{i=1}^{N} d_2(i)\, p_2(i)^2 \right)^{1/2}} \ , \ k = 1, 2 \ . \tag{2.20}$$

2.2.5. *Centers*

Dealing first with the case of 2 clusters, we rewrite (2.15) as a function of the cluster centers,

$$f(\mathbf{c}_1, \mathbf{c}_2) = \sum_{i=1}^{N} \left(\frac{d_1(\mathbf{x}_i, \mathbf{c}_1)\, p_1(\mathbf{x}_i)^2}{q_1} + \frac{d_2(\mathbf{x}_i, \mathbf{c}_2)\, p_2(\mathbf{x}_i)^2}{q_2} \right) \tag{2.21}$$

and look for centers $\mathbf{c}_1, \mathbf{c}_2$ minimizing f.

Theorem 2.2. *Let the distance functions d_1, d_2 in (2.21) be elliptic,*

$$d(\mathbf{x}, \mathbf{c}_k) = \langle (\mathbf{x} - \mathbf{c}_k), Q_k(\mathbf{x} - \mathbf{c}_k) \rangle^{1/2}, \; k = 1, 2, \tag{2.22}$$

where Q_1, Q_2 are positive definite, so that

$$\begin{aligned} f(\mathbf{c}_1, \mathbf{c}_2) = \sum_{i=1}^{N} \Bigg(& \sqrt{\langle (\mathbf{x}_i - \mathbf{c}_1), Q_1(\mathbf{x}_i - \mathbf{c}_1) \rangle}\, \frac{p_1(\mathbf{x}_i)^2}{q_1} \\ & + \sqrt{\langle (\mathbf{x}_i - \mathbf{c}_2), Q_2(\mathbf{x}_i - \mathbf{c}_2) \rangle}\, \frac{p_2(\mathbf{x}_i)^2}{q_2} \Bigg) , \end{aligned} \tag{2.23}$$

and let the probabilities $p_k(\mathbf{x}_i)$ and cluster sizes q_k be given. If the minimizers $\mathbf{c}_1, \mathbf{c}_2$ of (2.23) do not coincide with any of the data points $\mathbf{x}_i$, they are given by

$$\mathbf{c}_1 = \sum_{i=1}^{N} \left(\frac{u_1(\mathbf{x}_i)}{\sum\limits_{t=1}^{N} u_1(\mathbf{x}_t)} \right) \mathbf{x}_i \,, \quad \mathbf{c}_2 = \sum_{i=1}^{N} \left(\frac{u_2(\mathbf{x}_i)}{\sum\limits_{t=1}^{N} u_2(\mathbf{x}_t)} \right) \mathbf{x}_i \,, \tag{2.24}$$

where

$$\begin{aligned} u_1(\mathbf{x}_i) &= \frac{\left(\dfrac{d_2(\mathbf{x}_i, \mathbf{c}_2)}{q_2} \right)^2 \dfrac{1}{d_1(\mathbf{x}_i, \mathbf{c}_1)}}{\left(\dfrac{d_1(\mathbf{x}_i, \mathbf{c}_1)}{q_1} + \dfrac{d_2(\mathbf{x}_i, \mathbf{c}_2)}{q_2} \right)^2} \,, \\ u_2(\mathbf{x}_i) &= \frac{\left(\dfrac{d_1(\mathbf{x}_i, \mathbf{c}_1)}{q_1} \right)^2 \dfrac{1}{d_2(\mathbf{x}_i, \mathbf{c}_2)}}{\left(\dfrac{d_1(\mathbf{x}_i, \mathbf{c}_1)}{q_1} + \dfrac{d_2(\mathbf{x}_i, \mathbf{c}_2)}{q_2} \right)^2} \,, \end{aligned} \tag{2.25}$$

or equivalently, in terms of the probabilities (2.7),

$$u_1(\mathbf{x}_i) = \frac{p_1(\mathbf{x}_i)^2}{d_1(\mathbf{x}_i, \mathbf{c}_1)} \,, \quad u_2(\mathbf{x}_i) = \frac{p_2(\mathbf{x}_i)^2}{d_2(\mathbf{x}_i, \mathbf{c}_2)} \,. \tag{2.26}$$

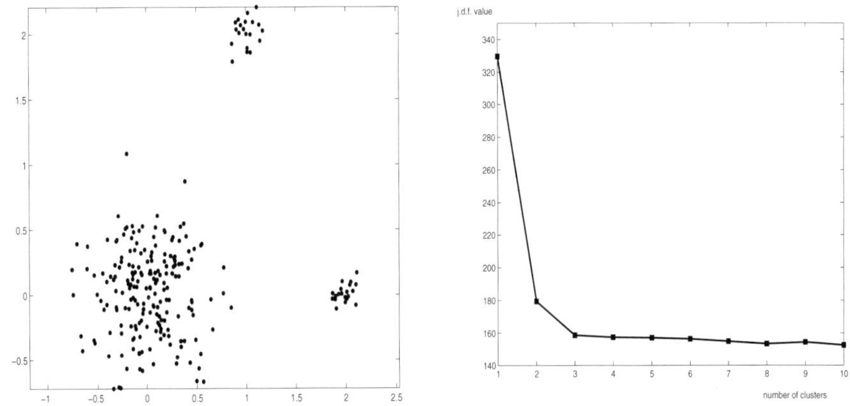

(a) A data set with $K = 3$ clusters (b) The JDF as a function of K

Fig. 2.10. Note the change of slope of the JDF at $K = 3$

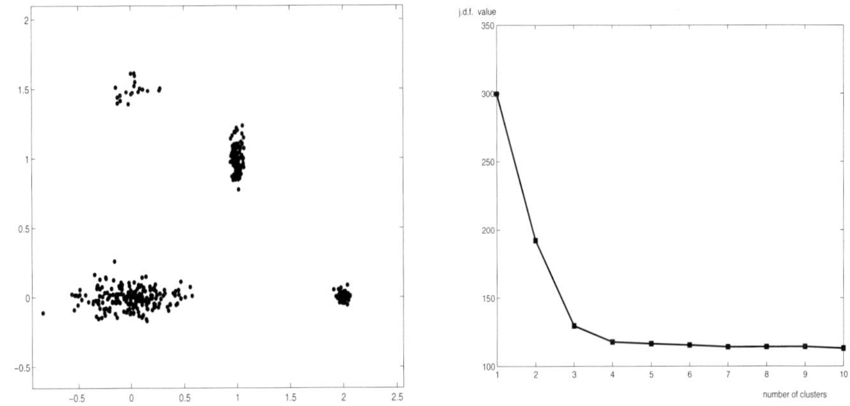

(a) A data set with $K = 4$ clusters (b) The JDF as a function of K

Fig. 2.11. Note the change of slope of the JDF at $K = 4$

References

[1] M. Arav. Contour approximation of data and the harmonic mean. *Mathematical Inequalities & Applications*, to appear.

[2] A. Ben–Israel, and C. Iyigun. Probabilistic distance clustering, *Journal of Classification*, to appear.

[3] L. Cooper. Heuristic methods for location–allocation problems. *SIAM Review*, 6, 37-53, 1964.

[4] M. Halkidi, Y. Batistakis, and M. Vazirgiannis, Cluster validity methods, *ACM SIG-*

MOD Record, 31(2): 40–45, 2002.

[5] J. Hartigan, *Clustering Algorithms*. John Wiley, 1975.

[6] T. Hastie, R. Tibshirani, and J. H. Friedman, *The Elements of Statistical Learning*. Springer, 2003.

[7] F. Höppner, F. Klawonn, F. Kruse and T. Runkler, *Fuzzy Cluster Analysis*. John Wiley, 1999.

[8] C. Iyigun and A. Ben–Israel. Probabilistic distance clustering adjusted for cluster size. *Probability in the Engineering and Informational Sciences*, to appear.

[9] C. Iyigun and A. Ben–Israel. A Probabilistic distance clustering method for multi–facility location problems, in preparation.

[10] C. Iyigun and A. Ben–Israel. A probabilistic distance clustering for mixtures of distributions, in preparation.

[11] C. Iyigun and A. Ben–Israel. A New clustering validity criterion based on contour approximation of data, in preparation.

[12] A. K. Jain, and R. C. Dubes. *Algorithms for Clustering Data*. Prentice Hall, 1988.

[13] R. Love, J. Morris and G. Wesolowsky. *Facilities Location: Models and Methods*. North-Holland, 1988.

[14] P. Tan, M. Steinbach, and V. Kumar. *Introduction to Data Mining*. Addison Wesley, 2006.

[15] M. Teboulle. A unified continuous optimization framework to center-based clustering methods. *Journal of Machine Learning*, 8: 65-102, 2007.

[16] P. Tsui. *EM–GM Algorithm MATLAB Code*. PAMI Research Group, University of Waterloo, 2006.

[17] E. Weiszfeld, Sur le point par lequel la somme des distances de n points donnes est minimum. *Tohoku Math. J.*, 43: 355-386, 1937.

Chapter 3

Analysis of Regulatory and Interaction Networks from Clusters of Co-expressed Genes

E. Yang

Biomedical Engineering Department
Rutgers University, Piscataway, NJ 08854, USA

A. Misra

Department of Environmental and Occupational Medicine
Rutgers University, Piscataway, NJ 08854, USA

T. J. Maguire

Biomedical Engineering Department
Rutgers University, Piscataway, NJ 08854, USA

I. P. Androulakis*

Biomedical Engineering Department
Rutgers University, Piscataway, NJ 08854, USA

Extracting biological insight from high-throughput genomic studies of human diseases remains a major challenge, primarily due to our inability to recognize, evaluate and rationalize the relevant biological processes recorded from vast amounts of data.

We will discuss an integrated framework combining fine-grained clustering of temporal gene expression data, selection of maximally informative clusters, based of their ability to capture the underlying dynamic transcriptional response, and the subsequent analysis of the resulting network of interactions among genes in individual clusters. The latter are developed based on the identification of common regulators among the genes in each cluster through mining literature data. We characterize the structure of the networks in terms of fundamental graph properties, and explore biologically the implications of the scale-free character of the resulting graphs. We demonstrate the biological importance of the highly connected hubs of the networks and show how these can be further exploited as targets for potential therapies during the early onset of inflammation and for

*Corresponding Author

characterizing the mechanism of action of anti-inflammatory drugs. We conclude by identifying two possible challenges in network biology, namely, the nature of the interactions and the potentially limited information content of the temporal gene expression experiments, and discuss expected implications.

3.1. Identification of Intervention Targets: Regulatory and Interaction Networks

At any given time, a cell will only express a small fraction of the thousands of genes in the organism's genome. Expressed genes reflect the structure and functional capacities of the cell as well as the ability of the cell to respond to external stimuli. In a complex organism, external stimuli to a great extent take the form of chemical messages whose purpose is to coordinate the function of the complex society of cells [1]. The transcription of genes is tightly regulated by DNA-binding proteins (transcription factors, TF) that attach to the promoter region and regulate the expression of the corresponding genes. If the state of the cell is defined by the genes that are expressed within it, and if the expression is the result of the coordinated action of a group of transcription factors, then losing even a single transcription factor can, and will have, a profound effect on the state of the cell [2]. Targeting expression by controlling the regulatory process through the corresponding transcription factors is emerging as a viable alternative for the identification of drug targets [3, 4] and the control of disease conditions [5]. Although it is realized that gene expression is regulated at multiple levels (transcriptionally, translationally and post-translationally) an effective and common means of regulating it occurs at the transcription level [2].

In recent years, significant efforts have been made experimentally and computationally to identify transcription factors, their target genes and the interaction mechanisms that regulate gene expression [6, 7]. An important technique for elucidating binding interactions is chromatin immunoprecipitation (ChiP) experiments [8]. Computational methods are currently emerging to provide TF predictions where experimental data is not available [9]. However, physical binding of a TF is a necessary but not sufficient condition for transcription initiation or regulation. Due to various complex post-translational modifications as well as interactions among multiple TFs, the measured expression level of regulatory genes does not reflect the actual activity of the TFs themselves. Therefore, regulator transcription levels are generally not appropriate measures of transcription factor activity (TFA). Recently, methods combining TF-gene connectivity data and gene expression measurements have emerged as a method to quantify these regulatory interactions. Prominent examples are the decomposition-based methods which combine ChiP and microarray data through the inversion of regression techniques to estimate TFAs [10–12, 14]. Statistical, regression and decomposition techniques have been proposed and successfully applied to reverse engineer regulatory networks [15–22]. The main goal of reverse engineering is to identify the activation program of transcription modules under particular conditions [23] so as to hypothesize how activation/deactivation of expression can be induced/suppressed [24]. Aside from the development of descriptive models that correlate TFA and expression of target genes, a critical question becomes how to identify those TFs that significantly contribute to regulation and should be modulated. Along these lines Gao et al. [12] speculated that the mRNA profile of the target gene should be similar to the reconstructed TFA for the regulating proteins, whereas Sun et al. [22] claimed that accurate binding information should lead to robust TFA reconstructions.

In addition to regulatory networks, it is becoming increasingly clear that interaction networks (pathway and signaling) need to be combined with gene expression data in order to establish an integrated, systems-wide, view of biological processes and response [25–32]. Furthermore, it is becoming increasingly clear that the structure of the interactions emanating from recorded responses captures critical properties indicative of the state of the cellular system [33, 33–42]. The computational analysis of regulatory, signaling and interaction pathways promises the identification of critical targets (aka "hubs") with the ability to maximally impact the cellular response [35]. Therefore, the systematic reconstruction of the different types of networks (regulatory, signaling, interaction) is a critical prerequisite for the deciphering the mechanisms that drive cells to carry out appropriate functions. The existence of "hubs", that is the existence of a small sub-set of highly interconnected nodes, promises profound implications due to the speculated nature of these nodes. The implications of the existence of such nodes has been speculated primarily in terms of their essential function [37–43]. It is of particular interest to analyze the robustness characteristics of this type of scale-free network structures in terms of their robustness [44].

The purpose of this paper is to present a concurrent analysis of gene expression data in an attempt to construct interaction networks that could be used as the template for identifying putative points of intervention, Fig. 3.1. In order to do so, we need to understand and characterize the structure these networks and identify the potential implications of partial, or complete, inhibition of specific nodes in the network.

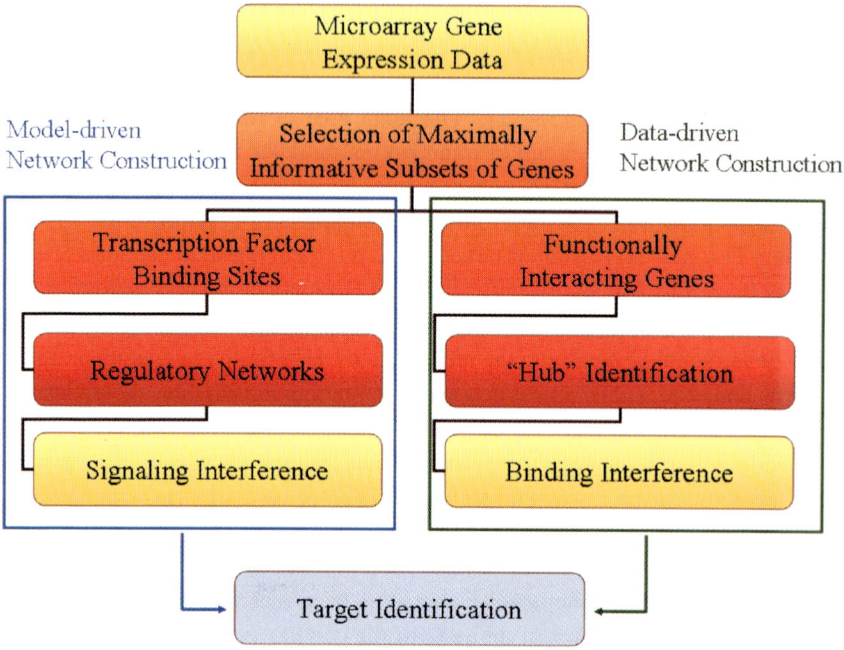

Fig. 3.1. Functional and regulatory networks.

3.1.1. *Identification of Informative Temporal Expression Patterns*

In order to develop context-specific networks, that is networks characteristic of cellular responses to specific external stimuli, we need to first identify a subset of relevant co-expressed genes whose transcription profiles are maximally affected by the specific experiment performed. We recently proposed an integrated methodology is composed of a number of steps outlined in Fig. 3.2, [45] for the analysis of temporal gene expression measurements. Specifically, our analysis consists of the following steps:

(a) Fine-grained clustering. Each gene's expression is characterized as a waveform over time and a characteristic attribute is defined for each time course. This identifier is then used to compare waveforms to each other and group similar waveforms, based on the similarity of the identifiers, to characteristic expression motifs. We have adopted the basic formalism of Symbolic Aggregate approXimation of the time series discussed in [46]. SAX is based on the premise of transforming a time series into a corresponding sequence of symbols. Each series is first normalized via the z-score given as $\bar{Y}_{j,t} = \frac{Y_{j,t}-\mu}{\sigma}$ with j=genes, t=time. An

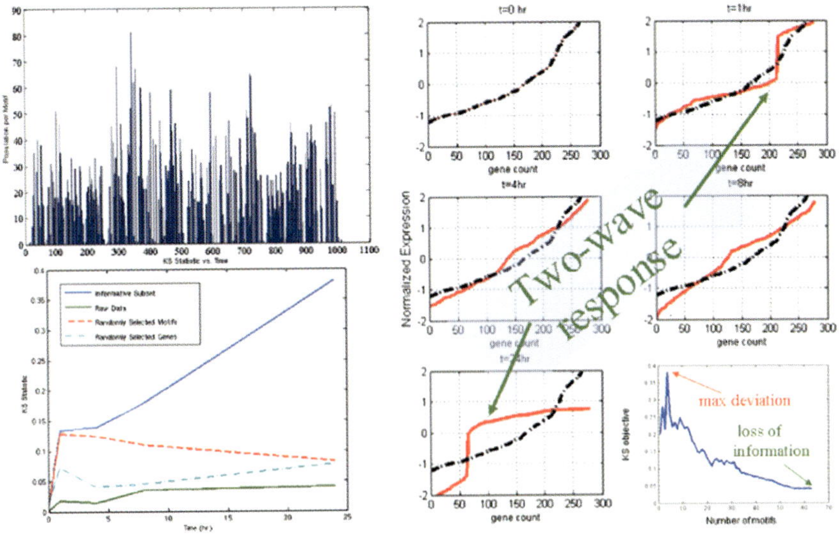

Fig. 3.2. Basic elements of the process for selecting informative expression profiles.

equiprobable discretization is applied where the breakpoints are defined such that the area defined by the boundaries of the breakpoint are equal. This method of discretization was selected because empirical evidence suggests that the z-score normalized sub-patterns should have a highly Gaussian distribution [46], thereby equally distributing a set of randomly generated signals throughout the hash space. Coefficients below the smallest breakpoint are "mapped" to the first symbol of a chosen alphabet. Other points are "mapped" accordingly within their respective intervals. A more extensive discussion and visualization of this process and can be found in [46].

(b) Identification of informative expression motifs. This novel symbolic representation makes it possible to further simplify the time series in order to uniquely characterize the overall dynamic response of each transcriptional profile with a single identifier [47]. After the alphabet has been generated, it is condensed into a single value using the function proposed by [48]: $hash(c, w, a) = 1 + \Sigma_{j=1}^{w}[ord(c_j) - 1] * a^{w-j}$, where a is the size of the alphabet, w is length of the word, and c is the "letter" sequence to which the expression profile is assigned. The parameter a is selected such that the population distribution of the motifs exhibits significant non-exponential distribution signaling the presence of significant differences in the population of expression profiles. Genes with similar normalized expression profiles "hash" to similar motif values to generate a distribution of motif values. This allows for the identification of (i) overpopulated motifs, and (ii) genes sharing similar motif values. Hence we have achieved a fine-grained "clustering" of the data where the number of potential clusters is dependent upon the definition of the hashing function.

(c) Quantification of transcription state. We define the transcription state of the system as the CDF of expression values of a select subset of motifs (based on the corresponding genes) and we will track this quantity as it evolves over time relative to the control state (distribution at t=0hr). We characterize each motif for its ability to represent the overall transcription dynamics of the system. In order to do so we define a new term, transcriptional state that quantifies the deviation of the aggregate distribution of expression values from a control state. An optimization framework is defined which characterizes expression motifs for their strength in replicating the entire system. Thus, we are able to rank the expression motif for their contribution to the overall state change of the system. The minimum number of expression motifs required to accurately represent the dynamic response of the system defines the set of informative genes, i.e., genes maximally affected by the specific experimental perturbation. To quantify the hypothesis that informative subsets of genes should give rise to a distribution of expression values maximally affected by the experiment, the Kolmogorov-Smirnov (KS) [49] test for evaluating whether or not two arbitrary distributions are different, is employed. Informative subsets are the ones with the ability to capture significant deviations from the base distribution. The KS statistic is defined as: $D = \max_{1 \leq i \leq n} |F(Y_{gi}) - F(Y_{gi}(0))|$, where $F(Y_{gi}(0))$ is the cumulative distribution of the expression values at time t=0 This statistic allows a metric that defines the magnitude of the difference between two distributions to be computed. Since the data is presented as a time series, at each time point a value for the KS statistic is obtained. Therefore, the overall metric becomes .With the definition of the transcriptional state and the ability to quantify the deviations from the control (sham) state we are now in the position to define a rigorous methodology for selecting maximally informative expression motifs. The application of the KS test over time allows us to quantify just how much the CDF of a particular sub-set of genes deviates from the corresponding CDF at time t=0 (control/sham). We currently implement a greedy algorithm that adds peaks in the order of their population and select the subset with the greatest deviation. The greedy heuristic was selected to minimize the combinatorial complexity of the problem, and we feel that is an adequate approximation due to fact that the greater over-representation of a motif, the more important this motif is. A detailed discussion of the methodology is presented in [50]. In order to fully explore the methods we focus on two distinct experimental protocols to assess the potential for identifying experiment-specific properties of the resulting regulatory and functional networks.

3.2. Analysis of Regulatory Networks

3.2.1. *Expression Data*

The microarray data was obtained from an experiment that was conducted to examine the behavior of a bolus injection of corticosteroids upon temporal gene expression profile of liver in vivo. This dataset was specifically chosen due to the *a priori* knowledge that corticosteroids have powerful transcriptionally mediated effects upon the rat experimental model. The data collection and preliminary analysis were previously presented in [51]. Male adrenalectomized (ADX) Wistar rats (Rattus rattus) weighing 225-250 g were obtained from Harlan Sprague-Dawley. Rats (3) were sacrificed by exsanguination under anesthesia at 0.25, 0.5, 0.75, 1, 2, 4, 5, 5.5, 6, 7, 8, 12, 18, 30, 48, and 72 hr after dosing. The sampling time points were selected based on previous studies describing Glucocorticoid Receptor (GR) dynamics and enzyme induction in liver and skeletal muscle. Four cannulated vehicle treated rats were sacrificed as controls (data represented as time zero). Total RNA was isolated from liver tissue and hybridized to a U34A GeneChip array.

3.2.2. *Regulatory Network Construction and Analysis*

As a result of the gene selection step on the aforementioned data set describing the administration of corticosteroid 529 probes were isolated in 12 clusters. These 529 probes corresponded to 454 individual genes of which 339 genes had reliable sequence information about the promoter region so that transcription factor prediction could be performed.

Transcriptional networks are comprised of the links between pairs of genes: those which code for transcription factors and the genes that they regulate. These links are abstractions of the regulatory activity of transcription factors binding on a given promoter region and either up or down regulate the levels of gene expression. The construction of this network can be used in a variety of ways such as the quantification of the aggregate behavior of the system after the perturbation of a single node, or the identification of possible key intervention points with which to mediate the response of an organism to an experimental perturbation. Transcriptional networks can be represented by bi-partite networks in which sets of transcription factors are shown to directly regulate a set of genes Fig. 3.3. While transcriptional networks have elements such as feed forward loops, feedback loops, and input cascades which are not explicitly visible in a bi-partite representation, it can be shown that this representation can be converted into an equivalent DAG (Directed Acyclic Graph) with no loss in generality. In our current case, we have treated the transcription factors as separate from the set of genes, and have conducted the graph analyses based on the outgoing connectivity of the transcription factors and the incoming connectivity of the selected genes. The attraction with representing transcriptional networks as a bi-partite networks is the existence of numerous algorithms such as NCA and PLS [14, 17], that allow for the efficient quantification of the links between the factors and the genes they regulate. Analysis of the network properties of bi-partite networks is also simplified due to the separation of nodes with outgoing connections and nodes with incoming connection.

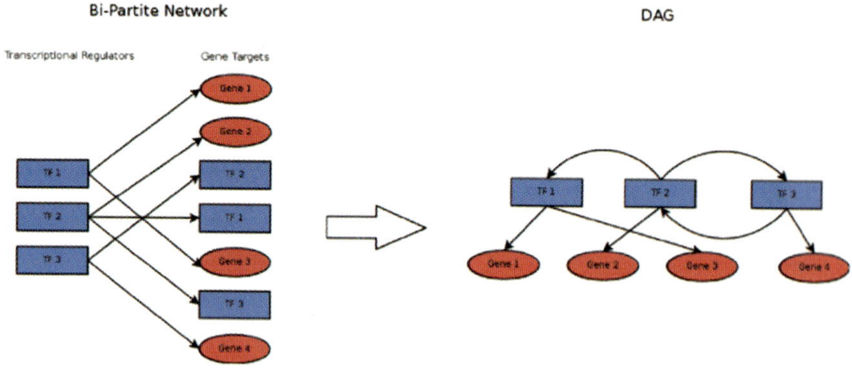

Fig. 3.3. A bi-partite representation of a transcriptional network and its associated DAG. There is no loss in generality in terms of the possible networks that can be represented. The representation as a bi-partite network however allows for efficient quantification of the network through various algorithms such as nca and pls.

The construction of transcriptional network falls under two primary paradigms, the first of which is to use the results of multiple experiments to construct Boolean or Bayesian networks [52–54] which infer the regulation of genes based on their related expression levels by making connections between genes that show either correlated or anti-correlated expression profiles. The second category of methods is the use of regulatory interactions inferred from other data such as transcriptional binding interactions inferred via Chip-Chip experiments or transcription factor predictions [8]. Both methods have been successfully implemented in the study of yeast. Constructing the gene regulatory networks in mammalian systems is a far more difficult process than in yeast. This is due to the fact that most of the regulatory interactions have not been previously mapped out, the relative cost of running live animal experiments. Therefore, an insufficient number of perturbed states have been obtained for one to infer comprehensive binary or Boolean networks from the available data. Duc to the fact that the construction of mammalian transcriptional networks is data-constrained as opposed to studies in yeast, most of the interactions must comprise of predicted links obtained through various transcription factor binding site prediction algorithms such as CONSITE, FOOTER, AlignACE, and CORG [55–58]. The problem with utilizing these algorithms for the prediction of transcription factor binding sites lies primarily in the large number of false positives and false negatives. Work has been previously done on ways to improve the prediction based on the concept of phylogenetic footprinting in which false positives are eliminated by looking at evolutionarily conserved segments in the promoter region. However, despite these attempts, the algorithms are still relatively inaccurate [59]. Due to the relative inaccuracies of the transcription factor prediction methods, gene expression data and clustering were used to identify the subnet that was active under the experimental conditions and to perform gross organization of genes into clusters before network construction. What one hopes to obtain from the construction of the network is the identification of possible biological systems affected via corticosteroid administration as well as the identification of possible points of intervention which allow for the control of specific biological processes independently. Therefore, one may be able to ascertain whether or not it is possible to affect the inflammatory response of the liver without triggering currently concurrent responses in metabolism or the immune response.

The transcriptional network was constructed using transcription factor binding site prediction via CORG. CORG was selected over other tools such as CONSITE because of a built in facility for extracting promoter regions of a specific gene as well as automatic phylogenetic footprinting which allows for the analysis of the network characteristics as the false positive links are pruned. The network generated via CORG is a standard bi-partite network because it finds the feed-forward interactions of transcription factors and their respective genes. However, as mentioned before, it can be transformed into a standard DAG if the output genes can be associated with a transcription factor. Given the large number of uncertainties in the network construction step due to the use of transcription factor prediction, the analysis will focus upon the global properties of the links rather than the biological significance of each link. The network is given in Fig. 3.4, and is a representation of a bi-partite network where the nodes in blue are the associated transcription factors and the nodes in green represent the final genes. We have treated the transcription factors as separate from the set of genes, and have conducted the graph analyses based on the outgoing connectivity of the transcription factors and the incoming connectivity of the selected genes. we have focused primarily upon the outgoing connectivity of the transcription factors because these represent the underlying transcriptional network.

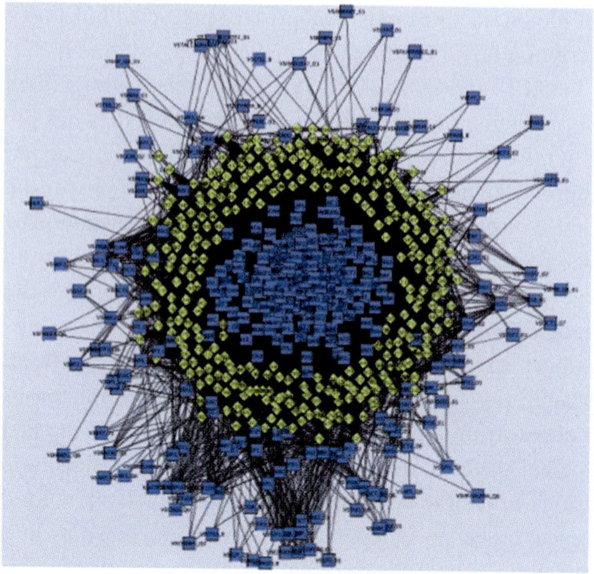

Fig. 3.4. Computationally constructed regulatory network (genes in green, transcription factors in blue).

The predicted transcription factor binding sites are essentially the links within our bi-partite network. The aggregate properties of the linkages are given in Fig. 3.5 and Fig. 3.6. The differences between the results obtained as depicted in Fig. 3.5 and Fig. 3.6 is the extent in which phylogenetic footprinting was used. In Fig. 3.5, phylogenetic footprinting was done on rat and mouse, two species which are evolutionarily very similar, while in Fig. 3.6, the phylogenetic footprinting was done on species which were much further evolutionarily specifically between rat and human. This analysis was done on each individual cluster separately to determine whether or not the process of clustering had an effect upon the generated network.

The aggregate connectivity of the input nodes i.e. the transcription factors shows a strong exponential decay characteristic. This is in agreement with previous evaluations upon network topology which have suggested that most biological networks are indeed scale free. What is more surprising is that even with the removal of a large number of connections via phylogenetic footprinting, the

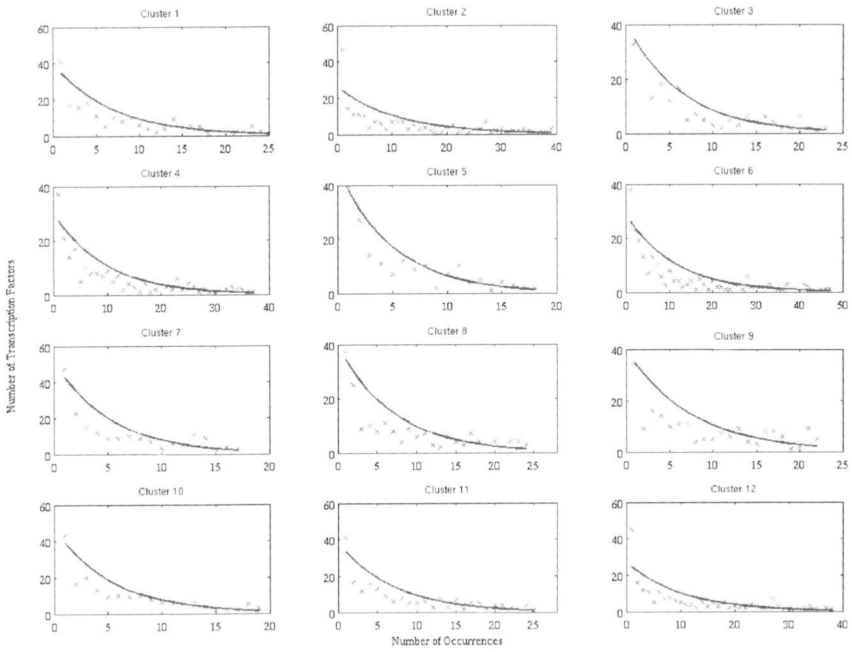

Fig. 3.5. The distribution of transcription factor connectivity from mouse vs. rat. What is evident is the exponential distribution of transcription factor connectivity. This strongly suggests that the power law distribution is driven by the distribution of highly selective vs. highly promiscuous transcription factors.

overall network character does not change. Looking at the remaining connections between the human and rat case it is apparent that even though many fewer transcription factor binding sites are found at much lower frequencies, it still exhibits a gross exponential curve.

What this suggests is that the scale free nature of transcriptional networks is not driven primarily by the sequence of the promoter region, but rather by some overall quality of the transcription factors. This is verified in Fig. 3.7, where a set of randomly selected genes was subjected to the same analysis. The strong exponential character that is present despite an essentially random sampling of promoter regions. What is found is that many of the transcription factors are highly promiscuous such as STAT5 and STAT6, having a consensus sequence which is 8 bases long but only specifically recognizes 4 to 5 base pairs in the binding sequence, which would correlate to a hit every 256 to 512 bases evaluated respectively. Given that the promoter regions being analyzed are generally several thousand bases, a predicted link between a gene and transcription factors such as

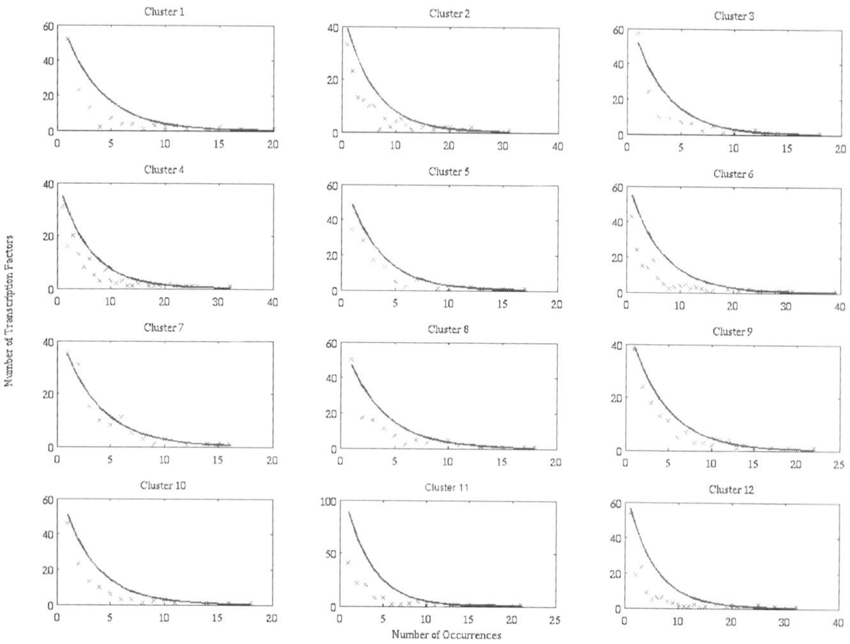

Fig. 3.6. The distribution of transcription factor connectivity from human vs. rat. The overall power law distribution remains suggesting that while phylogenetic footprinting has eliminated a number of false positives, the conserved promoter region does not differ in character to the overall promoter region.

STAT5 and STAT6 is generally not surprising. What is questionable is whether or not the aggregate properties of a predicted link provides an estimate for the prevalence of a transcription factor. In the case for STAT5, it has been shown experimentally that STAT5 indeed binds widely to genomic sequences in ChIP (Chromatic Immuno-Precipitation) experiments [60].

In their experiment, Nelson et al. had observed that around 17% of the STAT5 transcription factors lies within 1k of the TSS (Transcription Start Site) [60]. Nelson et al., took around 7k of bases for analysis with 1k upstream for the promoter region and 6k of exonic/intronic sequences. With this in mind the result is not surprising. Since 1k upstream of the start site represents 1/6th of the exonic/intronic sequence, its appearance 17% of the time is not surprising. This result suggests that each transcription factor has an overall binding selectivity with an expected appearance every n-bases. Therefore, if the position weight matrices are accurate in terms of their relative promiscuity, we believe that the aggregate behavior of the predicted transcription factor binding is a good substitute for empirical exper-

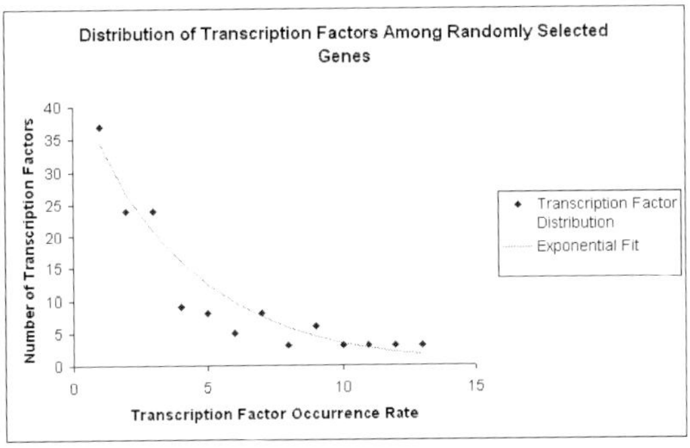

Fig. 3.7. A power law distribution of transcription factors amongst a randomly selected group of genes.

imental data. However, the question of which of these binding sites are active under a given experimental procedure in complex mammalian systems has not been adequately answered by either computational or experimental means.

From the analysis, our contention is that the primary biological aspect that drives transcriptional networks to be scale free is the distribution of transcription factors by their promiscuity rather than their respective promoter regions. If it was driven by properties of their respective promoter regions, we would expect a different character in the overall all connectivity property when comparing random genes, and two different iterations of phylogenetic footprinting. We believe this to be a significant outcome because it suggests that the creation and evolutionary conservation of transcriptional links is not driven by mutations in the promoter region, but rather by mutations in the coding regions of the transcription factors themselves. While this does not preclude mutations in the promoter region playing a significant role in genetic diseases or differences in phenotype within a species by changing the binding affinities of transcription factors, we feel that the evidence points to the hypothesis that the specificity of a transcription factor mediated by its amino acid sequence is the primary determinant of the transcriptional network structure.

It has been shown that the distribution of promiscuous to selective transcription factors follows an exponential distribution thereby allowing one to account for the scale free nature of biological networks. However, there are slight discrepancies in the sub-networks as evidenced by the presence of transcription factors that are highly conserved amongst the genes within a cluster despite their selective binding sites. Other transcription factor binding matrices such as PAX4 and USF1, are much more selective in their binding with 8-10 exact matches in their recognition sequence but appear as often as the more promiscuous transcription

Table 3.1. Transcription factors conserved more than 95% of the time between mouse and rat. In contrast to the human/rat case, all of the clusters show transcription factors conserved more than 95% of the time as well transcription factors which are highly conserved and not found in a random sampling of genes.

Cluster	Transcription Factors					
1	STAT 5	STAT 6	TEF-1			
2	STAT 5	STAT 6	TEF-1	CDX		
3	STAT 6	TEF-1	AP2 ALPHA			
4	STAT 5					
5	STAT 5	STAT 6				
6	STAT 5	STAT 6	TEF-1			
7	STAT 5	STAT 6	USF1			
8	STAT 5	STAT 6	TEF-1	GE II	CDX	GATA6
9	STAT 5	STAT 6	TEF-1	GE II	CDXA	AP2 ALPHA PAX4
	GATA6	CIZ	SRY	USF1		
10	STAT 5	STAT 6	GATA6			
11	STAT 5	TEF-1	STAT 6	CDX	AP2 ALPHA	
12	STAT 5	STAT 6	TEF-1			

factor like STAT5 or STAT6. There is the question of whether the deviations from the power-law distribution due to these transcription factors is a significant characteristic of transcriptional network, or whether these deviations are within acceptable deviations from the parameterized exponential curve. However, given that these links are not present during a random sampling of promoter regions, we believe that these links represent additional information provided by the gene selection and classification step. Given that even after clustering, the transcription factors in Table 3.1 are still relatively few, they could have been missed in a global assessment of network properties, but still may have an important effect upon the robustness of the network. These highly connected nodes which show up only in a set of highly correlated genes may serve to function as a means for preserving the connectivity structure of a network given the removal of a highly connected hub. Recent experimental evidence [61] has suggested that in fact the removal of highly connected hubs is not as lethal in biological organisms as it is to purely scale free networks such as the Internet. Our computational assessment seems to agree with this assessment, and it suggests that the process of clustering has allowed for the possible identification of more hubs than would be suggested under a purely scale free topography.

Another interesting observation is that while the whole transcriptional network appears to be scale-free in nature, each of the small co-expressed subsets also appears to be part of individual scale-free sub-networks. It has been shown that if in network construction, newer pathways had starting points in pre-existing pathways much like the way new hyperlinks on the Internet tend to link to older more popular pages, then the overall structure of the network is scale free [62, 63]. The presence of scale-free subnets suggests that the creation of individual pathways is therefore an iterative process by which pre-existing pathways are augmented and modified to provide an evolutionarily beneficial response. It would suggest that biological mechanisms are not "irreducibly" complex and that the network structure is consistent with an iterative evolutionary process.

We hypothesize that transcription factor binding sites which represent points of intervention have the following qualities:

(1) They are localized to a high degree to a specific cluster or process
(2) They are prevalent to a high degree in spite of their selectivity

In Fig. 3.8, we show the relative localizations of the biological process ontologies into each individual clusters. What is evident is that given our clustering and selection, we have isolated a set of genes which show a co-functional relationship along with their co-expression. What this signifies is that each set of co-expressed genes has a consistent scale-free network, and each of these scale free networks appear to be regulating a specific but loosely independent set of biological processes. The term loosely independent set is used to denote the fact that while not all biological processes are localized to only one cluster, there is a predilection of ontologies to be located in a single cluster as evidenced by the diagonally dominant plot shown in Fig. 3.8.

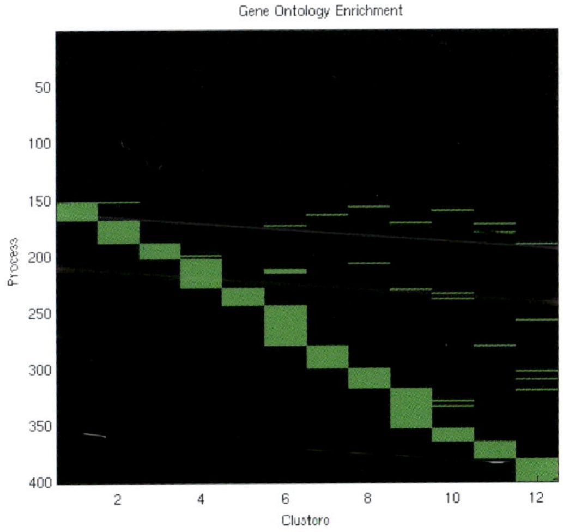

Fig. 3.8. Biological process enrichment of the extracted genes. The processes have been thresholded ($p < .05$).

After establishing the fact that each individual cluster appears to have a set of co-functional genes, the transcription factors that could possibly co-regulate these genes can be identified. These transcription factors therefore may provide points of possible transcriptional intervention to alter significant biological processes. Therefore it may be possible to alter the inflammatory response without changing the immune response, or the metabolic responses. It is our hypothesis that if a transcription factor can be associated with a primary ontology and if these transcription factors are associated with a large majority of genes in the given clusters, then it could provide a possible point of intervention. A list of these possible transcription factors are given in Table 1.

The transcription factors that comprised the highly connected nodes, i.e., their around found to be regulating all the clusters, can be broken down into two sets, the set of transcription factors that are highly connected throughout the genome and the set of transcription factors that are highly connected only to the genes that were selected as relevant. For instance, transcription factors such as STAT 5 and STAT 6 were selected as highly connected. However, these transcription factors are also highly represented in a set of randomly selected genes belying their important biological role, and while these transcription factors undoubtedly play a role in the corticosteroid response, they do not represent valid points at which the corticosteroid response can be mediated due to their widespread effects on many other systems, as evidenced by their presence in every functional and co-expressed cluster. The other transcription factors such as CDX (Caudal-type homeodomain protein), AP2-Alpha (activating enhancer binding protein 2), USF(Upstream Stimulating Factor), and PAX4(Paired Box Gene 4) represent hubs that are highly connected to the selected genes and may present themselves as possible targets to mediate the response. Coupling the biological information obtained via iHOP [64], it becomes possible to ascertain the role that each hub may play. These four transcription factors were specifically chosen for further analysis from the set of hub genes because additional information as to the up or down-regulation of the proteins that code for these transcription factors is available in the form of mRNA expression data. Therefore, it becomes possible to rationalize their biological activity with their observed activity and the consequence upon corticosteroid response. For instance, the down-regulation of USF may point to the decrease in lipid and glucose metabolism by the liver [65], leading to the observed increase in level of circulating free fatty acid and glucose in the bloodstream after an infusion of corticosteroid in a rat. The up-regulation of AP2-Alpha hints at a possible mechanism by which corticosteroids may suppress cellular proliferation given its role in suppressing the growth of malignant tumors [66].

Two of the transcription factors that are highly connected in the network pose an interesting question. CDX and PAX-4 are not known to be expressed in the liver, a fact which is supported by the expression levels found in our microarray. However, biological processes associated with these two transcription factors seem to correlate well with the clinical observations of corticosteroid administration. PAX-4 is normally associated with the differentiation of beta islet cells in the pancreas [67] and consequently whose expression levels in the pancreas ought to change in response to possible steroid induced diabetes due to the lowered sensitivity of the overall organism to insulin. CDX on the other hand has also been characterized as a regulator of cancer cell proliferation and is often up-regulated in rapidly proliferating cells [68]. Given the presence of these transcription factors the question arises as to whether a consistent set of genes is relevant over multiple tissues and that the different expression profiles of these relevant genes are controlled via the presence or absence of different transcription factors, or that these transcription factor binding sites represent the presence of a currently unknown transcription factor that binds to a very similar recognition sequence as CDX and PAX-4.

3.3. Analysis of Interaction Networks

3.3.1. *Expression Data*

Male Sprague-Dawley rats were subjected to a cutaneous 3rd degree burn injury consisting of a full skin thickness scald burn of the dorsum, calculated to be 20% of the rat's total body surface area [69]. Liver samples were obtained at 5 time points (0, 1, 4, 8, and 24h post burn). RNA extracted from the extracted livers was isolated and subsequently hybridized to a U34A GeneChip that had 8,799 probes represented on each chip. The control for this experiment was obtained at time 0, which was prior to the injury. It has been previously shown that time had no significant effect upon the response of rats to the sham treatment [70].

3.3.2. *Interaction Network Construction and Analysis*

In vivo estimation of protein interactions is undoubtedly the best way of determining biological interactions within the cell [71]. However, such an approach is rather impractical in exploratory analyses. Therefore, in silico methods have been proposed that, effectively, mine the available literature for established relationships between proteins exemplified by software tools such as Ingenuity Pathways Analysis (http://www.ingenuity.com/) , GeneGo Systems Reconstruction (http://www.genego.com/) and Ariadne Genomics PathwayAssist (http://www.ariadnegenomics.com/) are already available and able to construct such maps based on information extracted from the scientific literature and/or experimental data [72]. In this work, we adopted the framework implemented by PathwayAssist [73, 75] for the construction of protein interaction maps. PathwayAssist effectively mines over 100,000 events of regulation, interaction and modification between protein, cell processes and small molecules, thus allowing the construction of complex interaction networks. It must be emphasized that although the results might be sensitive to the specific software used, the essence of the network properties is not. In order to computationally construct the interaction network, the genes comprising the 9 most populated motifs (per the selection procedure earlier outlined) when analyzed using Ariadne Genomics PathwayAssist. The nodes that were selected were such that only direct connections between them could be drawn, therefore establishing either direct links between nodes, or indirect links using only nodes from the original set of informative genes.

Despite the apparent complexity of the resulting network, Fig. 3.9, it is interesting to realize the validation of the emergence of non-random connections as previously speculated for protein interaction networks. The scale-free nature of the resulting protein interaction network emanating from the analysis of informative genes identified by SLINGSHOTS a gene selection algorithm developed by our group is further established by observing the distribution of arcs per node of the network depicted in Fig. 3.10. Clearly, only a small fraction of nodes has a characteristically large number of direct neighbors, which is an indication of the emergence of a small number of highly connected nodes.

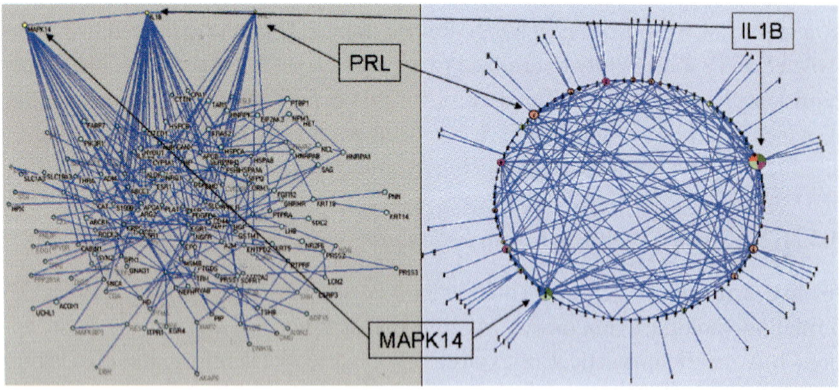

Fig. 3.9. Alternative representations of the general network topology of the protein interaction network indicating the existence of highly connected hubs.

The scale-free character of the protein interaction networks is precisely what is known to give metabolic networks their tremendous versatility and robustness in the presence of exogenous disturbances [62]. The issue of robustness on scale-free network in random attacks has received significant attention [74] given its implications for various real-life networks that possess that property. A measure for deciding the stability of the network in the presence of random attacks, that is failures of nodes at random, is quantified by estimating the average diameter of the network is often defined as the mean shortest path of the network once nodes are eliminated.

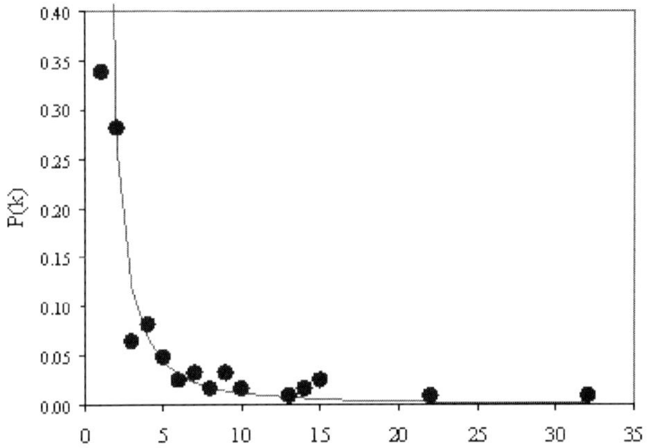

Fig. 3.10. Power-law distribution of network connectivity for experiment-specific gene interaction network based on the burn-induced inflammation data.

In order to assess the intrinsic characteristic of the protein interaction networks (Fig. 3.9) determined based on the informative genes that were identified as most responsive to the inflammatory response, we performed simulations to assess the stability of the network during failures (elimination of random nodes) and targeted attacked (elimination of nodes based on their connectivity). It has already been demonstrated with large networks exhibiting both scale-free and random (exponential) [44] the fundamental differences of the networks. Specifically, random networks have similar responses, in term of changes in network diameter, in both random failures and attacks due to the fact that no nodes exhibit higher connectivity and therefore the contributions of all nodes are equivalent. We use network diameter as a surrogate of a metric for quantifying effects on network structure. More appropriate metrics could have been adopted as well [76]. Random networks (exponential) are expected to show a similar increase in network diameter under, both, failure and attacks. However, scale-free networks are far more robust to random failures because the network can always reroute the flow of information and stabilize the change in diameter in the presence of random failures. As a result the network exhibits strong tolerance. However, targeted attacks have detrimental effects on the network. Therefore, the presence of hubs (highly connected nodes) is both an advantage, leading to robust networks, as well prone to complete failure should a critical hub be affected.

We simulated random failures and attacks by analyzing the protein interaction networks constructed based on the burn induced inflammatory response, as well as an equivalent Erdös-Renyi [13] random network with the same number of nodes, number of total links and average number of links per node. Specifically, the gene interaction network: 146 nodes, 268 connections, and the average number of connections per node was 3.67, whereas the Erdös-Renyi network: 146 nodes, 273 connections, and the average number of connections per node was 3.74. The results are depicted on Fig. 3.11. It must be realized that although many other studies analyze arbitrary networks with thousands of proteins, we concentrate on a much smaller and sensitive network which is composed of proteins very relevant to the inflammatory response resulting from severe burn injuries. However, all the important characteristics of the expected responses were recovered, which adds significant insight to the nature and structure of the generated framework. In terms of our analysis we define failure as the random elimination of a node for the network, whereas attack refers to a target removal of a specific (set) of nodes. Therefore, failures are being presented as the average of numerous (500) simulations of random eliminations of combinations of nodes from the networks. The failure results indicate the average of multiple runs. In attack, we target nodes based on their degree of connectivity, starting from the most highly connected nodes. The inherent computational property of the network that we evaluate is the average diameter of the network, which is quantified by evaluating the average shortest path length in the network. This metric assess the change in the effectiveness in communicating disturbances across the network. We therefore evaluate the structural properties of the network in propagating disturbances and the implications of such disturbances in the effective properties of the network. Our goal is to assess the fundamental differences between the two types of network, namely, scale free and random.

Based on our simulation results we observed that random (ER) networks do demonstrate very interesting characteristics. Because the distribution of connections is random, the response to random failures as well as attacks results in similar changes on the average shortest path. This is a critical property that has been previously well studied and established in ER-type of networks [44]. The deviations from the theoretical curves are mostly due to the fact that unlike simulated networks with thousands of nodes and connections, our ER network emulates the

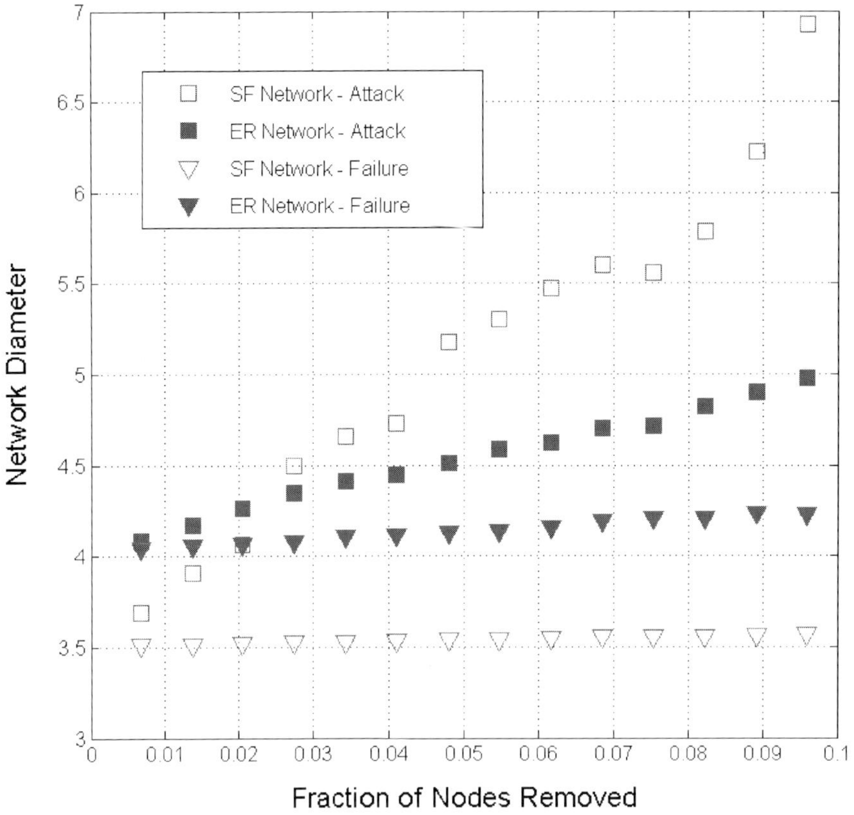

Fig. 3.11. Changes in network diameter (mean shortest path) for a scale free (sf) network based on the burn-induced inflammation data and an equivalent random (Erdös-Renyi, ER) network as a result of network attack and failure.

characteristics (size and connectivity) of the gene interaction networks emanating from the informative genes of the experiment-specific burn-induced inflammation study. Nevertheless, both random failures and targeted attacks have similar implication in the ability of the network to propagate external disturbances, and hence information, as a result of structural perturbations, such as knock-outs.

However, our computationally constructed experiment-specific scale free protein interaction network exhibits a dramatically different response. Random failures have a less pronounced effect on network diameter. The presence of the hubs guarantees and maintains the effective, average, characteristic of the network. Therefore, the mean shortest path, averaged over numerous failure simulations, does not exhibit any markedly significant changes. This property has been speculatively associated with the observed robustness of biological systems, in the sense that the central hubs provide pathways for connecting various components of the network with similar effectiveness when nodes in the network have been randomly eliminated or disabled [77]. However, elimination of the major hubs, through targeted attacks in the network, has a profound and detrimental effect on the structure of the network, as can been seen by the rapid increase in network diameter once key hubs have been removed from our network [43]. Systematic analyses of combined targeted knock-outs holds significant promise in light of the realization that controlled disturbances may have significant, albeit non detrimental effects on the networks. Thus, the effect of a major hub removal might be replicated through the compounded effect of multiple node removals, of nodes of lesser importance, without the complete network melt-down that results from eliminating a central hub [61]. The latter has not only been explored as means of analyzing networks but is also currently further analyzed as a potential method for drug target identification [78].

The analysis identified three major hubs of activity, within our protein interaction network, namely interleukin1-beta, prolactin (PRL), and mitogen activated protein kinase 14 is. Il-1B has been reported to be a dominant cytokine that acts as a central regulator of the acute inflammatory response, basically through the production of acute phase proteins [79]. This is evident in the large cascade of genes influenced through the activities of Il-1B. In addition, one specific cascade which is initiated through the activity of Il-1B, is that regulated by PRL, another of the dominant nodes we identified [80]. While Il-1B has the outcome of up-regulating a variety of genes needed in mediating the acute phase response, PRL has the inverse effect, in that it aides in the acute phase response by opposing the immunosuppressive effects of glucocorticoids and other inflammatory mediators to maintain steady-state homeostasis [81, 82]. The third hub we identified, p38MAPK, has also been established as a prominent gene involved in the acute phase response [83–86]. The p38 signaling cascade exhibits its effects following thermal injury, generally through the up-regulation of proinflammatory cytokines, such as the aforementioned Il-1B [87]. Thus, not only are these hubs capable of regulating a variety of down-stream genes, they themselves exhibit a high-degree of cross-talk, and regulate each other within the overall context of the protein interaction network. In addition, identification of these hubs provides potential therapeutic targets, to mitigate the inflammatory response observed following thermal injury.

3.4. Intervention Strategies

The idea of inhibiting multiple targets by exploring the properties of networks of interacting proteins is gaining an ever-increasing popularity [78]. Network tolerance in the presence of attacks, i.e., elimination of nodes, proves to be a promising and constructive way of determining putative intervention targets. Of particular importance is the realization that the same structural effects can be the result of not only a single node (hub) elimination which can be detrimental, but also the result of the simultaneous elimination of multiple interactions [61] with the desired structural effects yet not the lethal consequences. In fact, this is the principle behind the essential action of many multi-target drugs such as non-steroidal anti-inflammatory drugs (NSAIDs). Selecting, therefore, tentative targets for interference or elimination can be rationalized from the point of view of the resulting network interactions. Even though the importance of major hubs has been established in terms of the lethal consequences of eliminating such nodes [43] true opportunities lie in the potential for the concurrent (partial) elimination of interactions of multiple carefully selected targets.

The in silico approaches proposed in this paper, analyze interaction networks by exploring the topology of the interactions of sub-sets of maximally informative genes. Multiple approaches have been proposed for interfering with gene products in the context of altering cellular responses. Identifying a regulatory layer and the core nodes of that layer provides a mechanism to elucidate intervention points to attenuate the inflammatory process. Intervention utilizing these TF proteins could theoretically take one of three forms: 1) inhibition of TF production using knockout or silencing techniques; 2) blocking TF activity through competitive inhibition; 3) blocking TF activity through suicide inhibition. Current approaches for silencing focus on the use of siRNA techniques [88]. In this approach double-stranded RNA (dsRNA) is digested by the dsRNA-specific RNase III enzyme dicer into small interfering RNAs (siRNAs). The siRNAs then assemble with a multiprotein nuclease complex, RNA-induced silencing complex [89], which unwinds the dsRNAs and degrades target mRNAs homologous to the single stranded siRNA in a sequence-specific manner. The result of this process is the degradation of mRNA needed as a template for protein production, thereby inhibiting the production process, and depleting pools of proteins needed for specific enzymatic reactions. One specific example of siRNA utilized for intervention in inflammatory response is the application of siRNA techniques to inhibition the production of STAT-3 in order to elucidate key signaling molecules in the inflammatory response pathway [90]. A future application of this work would be to utilize the ability to block STAT-3 production to actually modulate the inflammatory response. A second plausible method used to interfere with the inflammatory response would be to intervene at the level of protein-mediated enzymatic activity. To do this, the idea of competitive inhibition could be utilized, wherein a secondary inhibitor protein would bind to the same active site as the normal enzyme substrate, without undergoing a reaction. This would then decrease the total concentration of active enzymatic protein, and decrease the overall pathway which the reaction is involved in. One example of this approach is the use of IkB Kinase beta inhibitor to block nuclear factor kappaB-mediated inflammatory response processes [91]. A third possible means of inhibiting inflammatory response at the protein level, would be to implement the idea of protein suicide inhibition. In this approach, an enzyme binds a substrate analogue and forms a complex with it during the "normal" catalysis reaction. The catalytic step will generate one of three reactive groups on the substrate analogue that will allow for the irreversible inhibition: an α- or β-haloketone, a $\beta\gamma$ unsaturated carbon, or a $\beta\gamma$ acetylenic carbon. Thus, as opposed to merely blocking the active site in a reversible manner like that in competitive inhibition, this process is irreversible, and hence makes a protein completely inactive once bound to the substrate analog. An example of this approach is the use of petrosaspongiolides which have been shown to have an in vitro and in vivo potent anti-inflammatory activity, mediated by specific inhibition of secretory phospholipase A(2) (sPLA(2) enzymes) [92].

Acknowledgements

The authors acknowledge support from the National Science Foundation under the grant NSF-BES 0519563 and the Environmental Protection Agency under the grant EPA-GAD R 832721-010. The authors would also like to acknowledge the critical input of Prof. F. Berthiaume (Center for Engineering in Medicine, Harvard Medical School) and Professors R.R. Almon, D. Dubois and W.J. Jusko (Department of Biological Sciences, SUNY Buffalo)

References

[1] V. Agoston, P. Csermely, and S. Pongor. Multiple weak hits confuse complex systems: a transcriptional regulatory network as an example. *Phys Rev E Stat Nonlin Soft Matter Phys*, 71(5 Pt 1):051909, 2005.

[2] S. T. Ahmed, A. Mayer, J. D. Ji, and L. B. Ivashkiv. Inhibition of il-6 signaling by a p38-dependent pathway occurs in the absence of new protein synthesis. *J Leukoc Biol*, 72(1):154–62, 2002.

[3] R. Albert, H. Jeong, and A. L. Barabasi. Error and attack tolerance of complex networks. *Nature*, 406(6794):378–82, 2000.

[4] B. Alberts, A. Johnson, J. Lewis, M. Raff, K. Roberts, and P. Walter. *Molecular Biology of the Cell*. Garland Science, New York, 4 edition, 2002.
[5] E. Almaas, B. Kovacs, T. Vicsek, Z. N. Oltvai, and A. L. Barabasi. Global organization of metabolic fluxes in the bacterium escherichia coli. *Nature*, 427(6977):839–43, 2004.
[6] R. R. Almon, D. C. DuBois, K. E. Pearson, D. A. Stephan, and W. J. Jusko. Gene arrays and temporal patterns of drug response: corticosteroid effects on rat liver. *Funct Integr Genomics*, 3(4):171–9, 2003.
[7] O. Alter and G. H. Golub. Integrative analysis of genome-scale data by using pseudoinverse projection predicts novel correlation between dna replication and rna transcription. *Proc Natl Acad Sci U S A*, 101(47):16577–82, 2004.
[8] M. M. Babu, N. M. Luscombe, L. Aravind, M. Gerstein, and S. A. Teichmann. Structure and evolution of transcriptional regulatory networks. *Curr Opin Struct Biol*, 14(3):283–91, 2004.
[9] G. D. Bader, M. P. Cary, and C. Sander. Pathguide: a pathway resource list. *Nucleic Acids Res*, 34(Database issue):D504–6, 2006.
[10] A. L. Barabasi, R. Albert, and H. Jeong. Scale-free characteristics of random networks: the topology of the world-wide web. *Physica A*, 281(1-4):69–77, 2000.
[11] A. L. Barabasi and Z. N. Oltvai. Network biology: understanding the cell's functional organization. *Nat Rev Genet*, 5(2):101–13, 2004.
[12] P. Blyszczuk, J. Czyz, G. Kania, M. Wagner, U. Roll, L. St-Onge, and A. M. Wobus. Expression of pax4 in embryonic stem cells promotes differentiation of nestin-positive progenitor and insulin-producing cells. *Proc Natl Acad Sci U S A*, 100(3):998–1003, 2003.
[13] B. Bollobas. *Random Graphs*. Cambridge University Press, 2001.
[14] A. L. Boulesteix and K. Strimmer. Predicting transcription factor activities from combined analysis of microarray and chip data: a partial least squares approach. *Theor Biol Med Model*, 2:23, 2005.
[15] H. J. Bussemaker, H. Li, and E. D. Siggia. Regulatory element detection using correlation with expression. *Nat Genet*, 27(2):167–71, 2001.
[16] M. P. Cary, G. D. Bader, and C. Sander. Pathway information for systems biology. *FEBS Lett*, 579(8):1815–20, 2005.
[17] X. L. Chen, Z. F. Xia, Y. X. Yu, D. Wei, C. R. Wang, and D. F. Ben. p38 mitogen-activated protein kinase inhibition attenuates burn-induced liver injury in rats. *Burns*, 31(3):320–30, 2005.
[18] A. M. Corbacho, G. Valacchi, L. Kubala, E. Olano-Martin, B. C. Schock, T. P. Kenny, and C. E. Cross. Tissue-specific gene expression of prolactin receptor in the acute-phase response induced by lipopolysaccharides. *Am J Physiol Endocrinol Metab*, 287(4):E750–7, 2004.
[19] D. L. Corcoran, E. Feingold, and P. V. Benos. Footer: a web tool for finding mammalian dna regulatory regions using phylogenetic footprinting. *Nucleic Acids Res*, 33(Web Server issue):W442–6, 2005.
[20] P. Crucitti, V. Latora, M. Marchiori, and A. Rapisarda. Error and attack tolerance of complex networks. *Physica a-Statistical Mechanics and Its Applications*, 340(1-3):388–394, 2004.
[21] P. Csermely, V. Agoston, and S. Pongor. The efficiency of multi-target drugs: the net-

work approach might help drug design. *Trends Pharmacol Sci*, 26(4):178–82, 2005.

[22] Jr. Darnell, J. E. Transcription factors as targets for cancer therapy. *Nat Rev Cancer*, 2(10):740–9, 2002.

[23] P. D'Haeseleer, S. Liang, and R. Somogyi. Genetic network inference: from co-expression clustering to reverse engineering. *Bioinformatics*, 16(8):707–26, 2000.

[24] C. Dieterich, H. Wang, K. Rateitschak, H. Luz, and M. Vingron. Corg: a database for comparative regulatory genomics. *Nucleic Acids Res*, 31(1):55–7, 2003.

[25] N. Dojer, A. Gambin, A. Mizera, B. Wilczynski, and J. Tiuryn. Applying dynamic bayesian networks to perturbed gene expression data. *BMC Bioinformatics*, 7:249, 2006.

[26] K. Dorshkind and N. D. Horseman. The roles of prolactin, growth hormone, insulin-like growth factor-i, and thyroid hormones in lymphocyte development and function: insights from genetic models of hormone and hormone receptor deficiency. *Endocr Rev*, 21(3):292–312, 2000.

[27] K. Dorshkind and N. D. Horseman. Anterior pituitary hormones, stress, and immune system homeostasis. *Bioessays*, 23(3):288–94, 2001.

[28] J. Downer, J. R. Sevinsky, N. G. Ahn, K. A. Resing, and M. D. Betterton. Incorporating expression data in metabolic modeling: a case study of lactate dehydrogenase. *J Theor Biol*, 240(3):464–74, 2006.

[29] B. L. Drees, V. Thorsson, G. W. Carter, A. W. Rives, M. Z. Raymond, I. Avila-Campillo, P. Shannon, and T. Galitski. Derivation of genetic interaction networks from quantitative phenotype data. *Genome Biol*, 6(4):R38, 2005.

[30] F. Gao, B. C. Foat, and H. J. Bussemaker. Defining transcriptional networks through integrative modeling of mrna expression and transcription factor binding data. *BMC Bioinformatics*, 5:31, 2004.

[31] B. Haefner. Can theoretical analysis of cellular regulatory networks improve drug target identification and validation? *BUSINESS BRIEFING: FUTURE Drug discovery*, pages 31–37, 2004.

[32] T. Hinoi, P. C. Lucas, R. Kuick, S. Hanash, K. R. Cho, and E. R. Fearon. Cdx2 regulates liver intestine-cadherin expression in normal and malignant colon epithelium and intestinal metaplasia. *Gastroenterology*, 123(5):1565–77, 2002.

[33] R. Hoffmann and A. Valencia. Protein interaction: same network, different hubs. *Trends Genet*, 19(12):681–3, 2003.

[34] R. Hoffmann and A. Valencia. A gene network for navigating the literature. *Nat Genet*, 36(7):664, 2004.

[35] V. R. Iyer, C. E. Horak, C. S. Scafe, D. Botstein, M. Snyder, and P. O. Brown. Genomic binding sites of the yeast cell-cycle transcription factors sbf and mbf. *Nature*, 409(6819):533–8, 2001.

[36] H. Jeong, S. P. Mason, A. L. Barabasi, and Z. N. Oltvai. Lethality and centrality in protein networks. *Nature*, 411(6833):41–2, 2001.

[37] H. Jeong, B. Tombor, R. Albert, Z. N. Oltvai, and A. L. Barabasi. The large-scale organization of metabolic networks. *Nature*, 407(6804):651–4, 2000.

[38] G. Joshi-Tope, M. Gillespie, I. Vastrik, P. D'Eustachio, E. Schmidt, B. de Bono, B. Jassal, G. R. Gopinath, G. R. Wu, L. Matthews, S. Lewis, E. Birney, and L. Stein. Reactome: a knowledgebase of biological pathways. *Nucleic Acids Res*, 33(Database issue):D428–32, 2005.

[39] G. Joshi-Tope, I. Vastrik, G. R. Gopinath, L. Matthews, E. Schmidt, M. Gillespie, P. D'Eustachio, B. Jassal, S. Lewis, G. Wu, E. Birney, and L. Stein. The genome knowledgebase: a resource for biologists and bioinformaticists. *Cold Spring Harb Symp Quant Biol*, 68:237–43, 2003.

[40] K. C. Kao, L. M. Tran, and J. C. Liao. A global regulatory role of gluconeogenic genes in escherichia coli revealed by transcriptome network analysis. *J Biol Chem*, 280(43):36079–87, 2005.

[41] K. C. Kao, Y. L. Yang, R. Boscolo, C. Sabatti, V. Roychowdhury, and J. C. Liao. Transcriptome-based determination of multiple transcription regulator activities in escherichia coli by using network component analysis. *Proc Natl Acad Sci U S A*, 101(2):641–6, 2004.

[42] K. C. Kao, Y. L. Yang, J. C. Liao, R. Boscolo, C. Sabatti, and V. Roychowdhury. Network component analysis of escherichia coli transcriptional regulation. *Abstracts of Papers of the American Chemical Society*, 227:U216–U217, 2004.

[43] M. Kato, N. Hata, N. Banerjee, B. Futcher, and M. Q. Zhang. Identifying combinatorial regulation of transcription factors and binding motifs. *Genome Biol*, 5(8):R56, 2004.

[44] S. Kauffman, C. Peterson, B. Samuelsson, and C. Troein. Random boolean network models and the yeast transcriptional network. *Proc Natl Acad Sci U S A*, 100(25):14796–9, 2003.

[45] E. Keogh, J. Lin, and A Fu. Hot sax: Efficiently finding the most unusual time series subsequences. *5th IEEE International Conference on Data Mining*, 2005.

[46] H. Kitano. Biological robustness. *Nat Rev Genet*, 5(11):826–37, 2004.

[47] D. E. Knuth. *The Art of Computer Programming*. Addison-Wesley, Reading, Massachusetts, 3 edition, 1997.

[48] J. C. Lee, A. J. Lusis, and P. Pajukanta. Familial combined hyperlipidemia: upstream transcription factor 1 and beyond. *Curr Opin Lipidol*, 17(2):101–9, 2006.

[49] K. Lee, F. Berthiaume, G. N. Stephanopoulos, and M. L. Yarmush. Profiling of dynamic changes in hypermetabolic livers. *Biotechnol Bioeng*, 83(4):400–15, 2003.

[50] L. K. Lee and C. M. Roth. Antisense technology in molecular and cellular bioengineering. *Curr Opin Biotechnol*, 14(5):505–11, 2003.

[51] T. I. Lee, N. J. Rinaldi, F. Robert, D. T. Odom, Z. Bar-Joseph, G. K. Gerber, N. M. Hannett, C. T. Harbison, C. M. Thompson, I. Simon, J. Zeitlinger, E. G. Jennings, H. L. Murray, D. B. Gordon, B. Ren, J. J. Wyrick, J. B. Tagne, T. L. Volkert, E. Fraenkel, D. K. Gifford, and R. A. Young. Transcriptional regulatory networks in saccharomyces cerevisiae. *Science*, 298(5594):799–804, 2002.

[52] D. Lejeune, L. Dumoutier, S. Constantinescu, W. Kruijer, J. J. Schuringa, and J. C. Renauld. Interleukin-22 (il-22) activates the jak/stat, erk, jnk, and p38 map kinase pathways in a rat hepatoma cell line. pathways that are shared with and distinct from il-10. *J Biol Chem*, 277(37):33676–82, 2002.

[53] D. E. Levy and Jr. Darnell, J. E. Stats: transcriptional control and biological impact. *Nat Rev Mol Cell Biol*, 3(9):651–62, 2002.

[54] Lun Li, David Alderson, Reiko Tanaka, John C. Doyle, and Walter Willinger. Towards a theory of scale-free graphs: Definition, properties, and implications (extended version). http://arxiv.org/abs/cond-mat/0501169, 2005.

[55] J. C. Liao, R. Boscolo, Y. L. Yang, L. M. Tran, C. Sabatti, and V. P. Roychowd-

hury. Network component analysis: reconstruction of regulatory signals in biological systems. *Proc Natl Acad Sci U S A*, 100(26):15522–7, 2003.

[56] J. Lim, T. Hao, C. Shaw, A. J. Patel, G. Szabo, J. F. Rual, C. J. Fisk, N. Li, A. Smolyar, D. E. Hill, A. L. Barabasi, M. Vidal, and H. Y. Zoghbi. A protein-protein interaction network for human inherited ataxias and disorders of purkinje cell degeneration. *Cell*, 125(4):801–14, 2006.

[57] J. Locker, D. Ghosh, P. V. Luc, and J. Zheng. Definition and prediction of the full range of transcription factor binding sites–the hepatocyte nuclear factor 1 dimeric site. *Nucleic Acids Res*, 30(17):3809–17, 2002.

[58] N. M. Luscombe, M. M. Babu, H. Yu, M. Snyder, S. A. Teichmann, and M. Gerstein. Genomic analysis of regulatory network dynamics reveals large topological changes. *Nature*, 431(7006):308–12, 2004.

[59] A. M. McGuire and G. M. Church. Predicting regulons and their cis-regulatory motifs by comparative genomics. *Nucleic Acids Res*, 28(22):4523–30, 2000.

[60] M. C. Monti, A. Casapullo, R. Riccio, and L. Gomez-Paloma. Further insights on the structural aspects of pla(2) inhibition by gamma-hydroxybutenolide-containing natural products: a comparative study on petrosaspongiolides m-r. *Bioorg Med Chem*, 12(6):1467–74, 2004.

[61] E. A. Nelson, S. R. Walker, J. V. Alvarez, and D. A. Frank. Isolation of unique stat5 targets by chromatin immunoprecipitation-based gene identification. *J Biol Chem*, 279(52):54724–30, 2004.

[62] A. Ng, B. Bursteinas, Q. Gao, E. Mollison, and M. Zvelebil. pstiing: a 'systems' approach towards integrating signalling pathways, interaction and transcriptional regulatory networks in inflammation and cancer. *Nucleic Acids Res*, 34(Database issue):D527–34, 2006.

[63] A. Nikitin, S. Egorov, N. Daraselia, and I. Mazo. Pathway studio–the analysis and navigation of molecular networks. *Bioinformatics*, 19(16):2155–7, 2003.

[64] P. Patel, E. Keogh, J. Lin, and S. Lonardi. Mining motifs in massive time series databases. *ICDM 2002*, pages 370–377, 2002.

[65] J. Pradines, L. Rudolph-Owen, J. Hunter, P. Leroy, M. Cary, R. Coopersmith, V. Dancik, Y. Eltsefon, V. Farutin, C. Leroy, J. Rees, D. Rose, S. Rowley, A. Ruttenberg, P. Wieghardt, C. Sander, and C. Reich. Detection of activity centers in cellular pathways using transcript profiling. *J Biopharm Stat*, 14(3):701–21, 2004.

[66] D. N. Rassokhin and D. K. Agrafiotis. Kolmogorov-smirnov statistic and its application in library design. *J Mol Graph Model*, 18(4-5):368–82, 2000.

[67] E. Ravasz, A. L. Somera, D. A. Mongru, Z. N. Oltvai, and A. L. Barabasi. Hierarchical organization of modularity in metabolic networks. *Science*, 297(5586):1551–5, 2002.

[68] A. W. Rives and T. Galitski. Modular organization of cellular networks. *Proc Natl Acad Sci U S A*, 100(3):1128–33, 2003.

[69] P. Ruminy, C. Gangneux, S. Claeyssens, M. Scotte, M. Daveau, and J. P. Salier. Gene transcription in hepatocytes during the acute phase of a systemic inflammation: from transcription factors to target genes. *Inflamm Res*, 50(8):383–90, 2001.

[70] A. Sandelin, W. W. Wasserman, and B. Lenhard. Consite: web-based prediction of regulatory elements using cross-species comparison. *Nucleic Acids Res*, 32(Web Server issue):W249–52, 2004.

[71] G. L. Semenza. Transcription factors and human disease. *Oxford Monographs on Medical Genetics*, 37, 1999.
[72] V. Spirin and L. A. Mirny. Protein complexes and functional modules in molecular networks. *Proc Natl Acad Sci U S A*, 100(21):12123–8, 2003.
[73] N. Sun, R. J. Carroll, and H. Zhao. Bayesian error analysis model for reconstructing transcriptional regulatory networks. *Proc Natl Acad Sci U S A*, 103(21):7988–93, 2006.
[74] T. Takai-Igarashi and R. Mizoguchi. Cell signaling networks ontology. *In Silico Biol*, 4(1):81–7, 2004.
[75] T. Takai-Igarashi and R. Mizoguchi. Ontological integration of data models for cell signaling pathways by defining a factor of causality called 'signal'. *Genome Inform*, 15(2):255–65, 2004.
[76] C. Tellez and M. Bar-Eli. Role and regulation of the thrombin receptor (par-1) in human melanoma. *Oncogene*, 22(20):3130–7, 2003.
[77] L. M. Tran, M. P. Brynildsen, K. C. Kao, J. K. Suen, and J. C. Liao. gnca: A framework for determining transcription factor activity based on transcriptome: identifiability and numerical implementation. *Metabolic Engineering*, 7(2):128–141, 2005.
[78] B. van Steensel, J. Delrow, and H. J. Bussemaker. Genomewide analysis of drosophila gaga factor target genes reveals context-dependent dna binding. *Proc Natl Acad Sci U S A*, 100(5):2580–5, 2003.
[79] M. Vemula, F. Berthiaume, A. Jayaraman, and M. L. Yarmush. Expression profiling analysis of the metabolic and inflammatory changes following burn injury in rats. *Physiol Genomics*, 18(1):87–98, 2004.
[80] M. V. Verga Falzacappa, M. Vujic Spasic, R. Kessler, J. Stolte, M. W. Hentze, and M. U. Muckenthaler. Stat-3 mediates hepatic hepcidin expression and its inflammatory stimulation. *Blood*, 2006.
[81] W. Wang, J. M. Cherry, D. Botstein, and H. Li. A systematic approach to reconstructing transcription networks in saccharomycescerevisiae. *Proc Natl Acad Sci U S A*, 99(26):16893–8, 2002.
[82] W. W. Wasserman and A. Sandelin. Applied bioinformatics for the identification of regulatory elements. *Nat Rev Genet*, 5(4):276–87, 2004.
[83] D. Wen, Y. Nong, J. G. Morgan, P. Gangurde, A. Bielecki, J. Dasilva, M. Keaveney, H. Cheng, C. Fraser, L. Schopf, M. Hepperle, G. Harriman, B. D. Jaffee, T. D. Ocain, and Y. Xu. A selective small molecule ikappab kinase beta inhibitor blocks nuclear factor kappab-mediated inflammatory responses in human fibroblast-like synoviocytes, chondrocytes, and mast cells. *J Pharmacol Exp Ther*, 317(3):989–1001, 2006.
[84] J. Westra, J. Bijzet, B. Doornbos-van der Meer, M. H. van Rijswijk, and P. C. Limburg. Differential influence of p38 mitogen activated protein kinase (mapk) inhibition on acute phase protein synthesis in human hepatoma cell lines. *Ann Rheum Dis*, 65(7):929–35, 2006.
[85] M. S. Wu, S. Kwon, J. F. Driscoll, and G. M. Faeth. Preferential diffusion effects on the surface structure of turbulent premixed hydrogen/air flames. *Combust. Sci. Tech.*, 78:69–96, 1991.
[86] E. Yang, T. Maguire, M L Yarmush, F Berthiaume, and I. P. Androulakis. Bioinformatics analysis of the early inflammatory response in a rat thermal injury model. *BMC Bioinformatics*, accepted for publication, 2006.

[87] E. Yang, T. Maguire, M. L. Yarmush, F. Berthiaume, and I. P. Androulakis. Bioinformatics analysis of the early inflammatory response in a rat thermal injury model. *BMC Bioinformatics*, 8(1):10, 2007.

[88] M. K. Yeung, J. Tegner, and J. J. Collins. Reverse engineering gene networks using singular value decomposition and robust regression. *Proc Natl Acad Sci U S A*, 99(9):6163–8, 2002.

[89] S. H. Yook, Z. N. Oltvai, and A. L. Barabasi. Functional and topological characterization of protein interaction networks. *Proteomics*, 4(4):928–42, 2004.

[90] A. Yuryev, Z. Mulyukov, E. Kotelnikova, S. Maslov, S. Egorov, A. Nikitin, N. Daraselia, and I. Mazo. Automatic pathway building in biological association networks. *BMC Bioinformatics*, 7:171, 2006.

[91] A. Zauberman, D. Zipori, M. Krupsky, and R. Ben-Levy. Stress activated protein kinase p38 is involved in il-6 induced transcriptional activation of stat3. *Oncogene*, 18(26):3886–93, 1999.

[92] T. L. Zimmerman, S. Thevananther, R. Ghose, A. R. Burns, and S. J. Karpen. Nuclear export of retinoid x receptor alpha in response to interleukin-1beta-mediated cell signaling: roles for jnk and ser260. *J Biol Chem*, 281(22):15434–40, 2006.

Chapter 4

Graph-based Approaches for Motif Discovery

Elena Zaslavsky

Department of Computer Science
Princeton University, Princeton, NJ 08544, USA
elenaz@cs.princeton.edu

Sequence motif finding is a very important and long-studied problem in computational molecular biology. While various motif representations and discovery methods exist, a recent development of graph-based algorithms has allowed practical concerns, such as positional correlations within motifs, to be taken into account. This survey provides an overview of the multi-partite graph formulation of motif finding, and focuses on algorithmic aspects of various motif discovery methodologies.

Motif finding has been recast as a number of different graph substructure identification problems. First we review a formulation as a maximum-weight clique finding problem, and examine two different integer linear programs to model it. The motif finding algorithms use graph pruning techniques and a cutting planes approach in conjunction with linear programming relaxations. Secondly, we discuss a formulation of motif discovery as that of maximum density subgraph finding, and review a maximum flow based algorithm in an appropriately augmented flow network. Finally, we mention the 'subtle' motifs formulation, and define its corresponding graph problem of maximal clique identification. We discuss two different approaches to tackle this problem, one based on winnowing spurious edges and the other on divide-and-conquer sub-clique finding.

4.1. Introduction

With the advent of the post-genomic era when scores of genomes have been sequenced and many protein structures solved, functionally important elements can be mined from vast amounts of genomic data through identification of common patterns or *motifs*. Existence of such shared sequence or structure elements allows one to draw links between various genes or proteins in an organism, creating a network of biological associations. The ability to analyze such networks has shown great potential in uncovering new insight in a variety of biological systems.

A problem of great interest that explores non-coding regions of DNA for sequence motifs is one of transcription factor binding site identification. Transcription factors are cellular proteins that take part in regulating gene expression, serving as activators or inhibitors of transcription. Uncovering their binding targets is a crucial step towards understanding the mechanisms of transcriptional activity in the cell. Biological approaches, which include both in-vivo [20] and in-vitro [31] experimental designs, to identifying DNA binding sites are still time-consuming, costly, sensitive to perturbation and often imprecise in pinpointing the exact locations of binding sites. It is clear that computational methods are needed to address this very important problem.

Computational discovery of transcription factor binding sites is typically cast as the problem of finding mutually similar substrings in unaligned sequence data. We refer to this task as the *motif finding* or *motif discovery* problem. Motif finding algorithms operate on sets of sequences that are presumed to possess a common motif. Two orthogonal approaches exist: one attempts to identify binding sites among a set of regulatory regions of orthologous genes across genomes of varying phylogenetic distance [3, 7, 16, 29, 30], referred to as *phylogenetic footprinting*, and the other analyzes regulatory sequences for sets of genes from a single genome assumed to be controlled by a common transcription factor. While in the first case data are collected via gene orthology determination, in the second case data are made available through DNA microarray studies [40, 43], chromatin immunoprecipitation (ChIP-chip) experiments [20] and protein binding microarrays [31]. In DNA microarray studies normalized gene expression levels of many genes are analyzed to reveal similar patterns in expression. Under the assumption that co-expressed genes are likely co-regulated, the regulatory regions of such co-expressed genes can be subjected to motif finding. In the latter approaches the binding of a regulatory protein to DNA is recognized directly via molecular methods. The group of DNA sequences, to which the protein was bound, can be input to a motif finding algorithm to identify the binding sites precisely.

There are two broad categories of motif finding algorithms, and they are fundamentally linked to the choice of the underlying motif representation [33, 41]. One class of algorithms is based on the *consensus* motif model, and the search strategies focus on finding word-based patterns via various approaches including enumerative [12, 15, 27, 34, 39, 44] and clustering [5, 35] methods. The probabilistic algorithms, based on the position-specific scoring matrix model (*PSSM*), use greedy strategies [11] or parameter estimation techniques, such as Expectation Maximization or Gibbs Sampling [14, 18, 19, 24, 38] to maximize information content of the sought motif.

A comprehensive comparison study by Tompa and coauthors [45] showed that the performance of these two broad groups of methods seem to be complementary in many cases, with a slight performance advantage demonstrated by representative methods of the combinatorial class (e.g., *Weeder* [34]). However, both categories of motif finding algorithms have their limitations. Many combinatorial methods enumerate every possible pattern, and are thus limited in the length of the motifs they can search for. While this may be less of an issue in eukaryotic genomes, where transcriptional regulation is mediated combinatorially by a large number of transcription factors with relatively short binding sites, substantially longer motifs are found when considering either DNA binding sites in prokaryotic genomes (e.g., for helix-turn-helix binding domains of transcription factors [37]) or protein motifs [4]. Limitations exist in methods based on the PSSM model as well. These methods typically employ heuristic search techniques that don't guarantee convergence to the globally optimal motif (though some recent progress has been made in this direction [21]). Additionally, a fundamental problem exists with the PSSM representation. Position-specific scoring matrices assume independence between motif positions, and though this assumption leads to satisfactory results in many applications [2], recent work [6, 26] has shown evidence of dependencies between positions in transcription factor binding sites.

While the positional independence assumption in PSSMs can be relaxed by using richer models [1, 48], risk of overfitting exists. More importantly, other analyses [32] find that the distribution of motif instances in many eukaryotic motifs exhibit complex dependencies that are not easily modeled probabilistically. It has been shown [33] that modeling these dependencies using the sequence pattern-based approaches leads to improved performance in representing and searching for binding sites; a similar statistically significant improvement is not observed with PSSMs. In fact, Naughton and colleagues [32] report that simulated motifs generated by PSSM models display a distribution very different from instances of real biological motifs. They show that biological motifs exhibit a significantly higher degree of clustering as measured by sequence similarity. As such, it is better to evaluate a candidate motif instance while considering a number of its nearest neighbors, rather than a model of the entire motif.

These observations naturally lead to considering graph-based approaches to motif finding, in which graph vertices correspond to substrings in the input, edges denote the degree of similarity between their DNA sequences, and highly connected graph sub-structures exhibit the clustering behavior necessary to identify motifs. There exist graph-based formulations and algorithms that possess an advantage over enumerative methods in that they are not exponential in the problem input size and thus are not limited by the length of the sought motif. Likewise, exploiting the graph structure of the problem and looking for highly-weighted clique-like substructures allows for implicit modeling of complex positional dependencies within motifs, otherwise not possible with PSSMs.

We explore a number of formulations and accompanying algorithms in graph-based motif discovery that have been proposed and applied to find biologically interesting motifs. The ability of the graph-based approaches to formulate the problem in such a way that optimal solutions are attainable while overcoming many of the limitations of the standard approaches demonstrates the power of these search procedures. Though we don't focus on experimental results in this article, it should be noted that the performance of these methods in uncovering known motifs illustrates their utility for novel sequence motif discovery.

The rest of this review focuses on various methodologies in graph-based algorithms, and does not encompass the broader motif finding problem. Recent surveys [10, 22, 25] and comparative studies [13, 45] can serve as a good practical guide when choosing a method to be applied in practice.

4.2. Graph-Theoretic Formulation

The motif finding problem, with the input consisting of a length parameter l of the putative motif, and N sequences, each of a possibly different length, to be searched for motif occurrences, can be formulated in graph-theoretic terms [46]. The problem is recast as a complete, weighted N-partite graph $G = (V, E)$, with a graph part corresponding to each input sequence. Denoting the set of nodes in the part corresponding to sequence i by V_i, the full vertex set is $V = V_1 \cup \cdots \cup V_N$. In part V_i there is a node for every substring of length l in sequence i. Thus, for a sequence i of length L there are $L - l + 1$ nodes in V_i (see Fig. 4.1).

Each pair of nodes u and v in different graph parts is joined by an edge $(u, v) \in E$. Letting *seq(u)* denote the input substring corresponding to node u, the weight w_{uv} on edge (u, v) is a function of the similarity between *seq(u)* and *seq(v)* as well as the background nucleotide distribution. Thus, the weight of an edge connecting pairs of substrings that are unlikely to appear in the background, is increased. A motif in this formulation is a selection of vertices, corresponding to substrings collectively exhibiting a higher degree of similarity than would be expected based on the background distribution. It is also possible to relax the specification of allowing only one motif instance to occur in every sequence, thus modeling a biologically more realistic version of the problem. The graph construction is altered to permit edges between vertices in the same graph part.

A related formulation of motif finding that has been recast similarly in graph-theoretic terms by many methods, is that of 'subtle' motifs, introduced by Pevzner

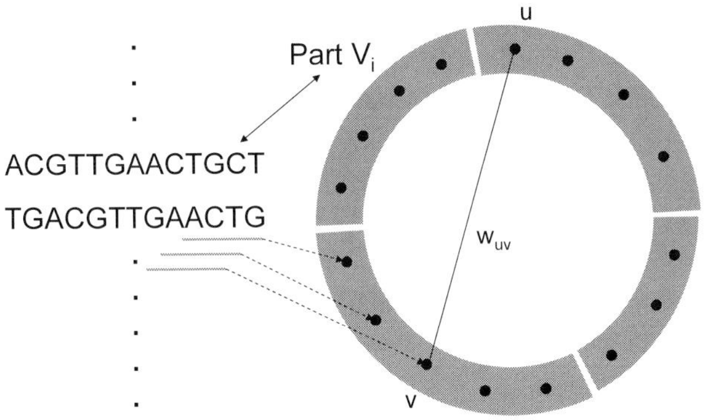

Fig. 4.1. Graph representation for the motif finding problem. A graph part V_i corresponds to every sequence, and a vertex corresponds to every possible motif position. If the motif length l is four, the picture matches up the last few substrings to their graph vertices. Every pair of vertices u and v in distinct parts are connected by an edge w_{uv}.

and Sze [35]. In this formulation an unknown pattern of a given length is inserted with modifications into each of the input sequences. The positions of insertion are unknown, as are the modifications in the instances of the pattern. Pevzner and Sze focus on what they call the (l, d)–signal version, in which the pattern is a string of length l and each instance differs from the pattern in exactly d positions. The mutations are allowed to occur anywhere in the pattern, and thus any two instances of the pattern may differ from one another in as many as $2d$ positions. On the other hand, if two substrings differ in more than $2d$ positions, they cannot simultaneously be instances of the implanted pattern. The graph version of the problem remains largely the same as above, except that it is no longer a complete N–partite graph. By definition, vertices that correspond to substrings separated by a Hamming distance that is greater than $2d$, are not connected by an edge, as such a vertex pair can not be an instantiation of a single pattern. The weights on the edges are distance rather than similarity-based and are computed by considering the number of matches and mismatches between the substrings.

We first address the algorithms designed for the standard motif finding problem, and then review some of the subtle motif methods.

4.3. Linear Programming-based Algorithms

Utilizing a linear programming (LP) framework for motif finding [17, 47] allows for a very flexible formulation of the problem that is applicable to a number of variants. This includes the standard motif finding formulation as well as the phylogenetic footprinting and subtle motifs problems. Underlying the approach, motif discovery is cast as the problem of finding the best gapless local multiple sequence alignment using the sum-of-pairs (SP) scoring scheme, which seeks a l–long substring from every input sequence such that the sum of all their pair-wise similarities is maximized. In the graph $G = (V, E)$ the equivalent is a selection of vertices, one from each graph part, such that the weight of the induced clique is maximized. We refer to this problem as finding the highest weight N–clique.

4.3.1. *Edge-Modeling Formulation*

For a graph $G = (V, E)$, where $V = V_1 \cup \ldots \cup V_N$ and $E = \{(u, v) : u \in V_i, v \in V_j, i \neq j\}$, Zaslavsky and Singh [47] introduce a formulation that explicitly models both vertices and edges in the graph. They define a binary decision variable X_u for every vertex u, and a binary decision variable Y_{uv} for every edge (u, v). Setting X_u to 1 corresponds to selecting vertex u for the N–clique and thus choosing the sequence position corresponding to u in the alignment. Setting variable Y_{uv} to 1 corresponds to choosing the edge (u, v) for the N–clique.

The following integer linear program (ILP) models the motif finding problem formulated above:

$$\begin{array}{lll} \text{Maximize} & \sum_{(u,v)\in E} w_{uv} \cdot Y_{uv} & \\ \text{subject to} & & \\ \quad \sum_{u\in V_j} X_u = 1 & & \text{for } 1 \leq j \leq N \\ \quad \sum_{u\in V_j} Y_{uv} = X_v & & \text{for } 1 \leq j \leq N, v \in V \setminus V_j \\ \quad X_u, Y_{uv} \in \{0, 1\} & & \text{for } u \in V, (u, v) \in E \end{array}$$

The first set of constraints ensures that exactly one vertex is picked from every graph part, corresponding to one position being chosen from every input sequence. The second set of constraints relates vertex variables to edge variables, allowing the objective function to be expressed in terms of finding a maximum edge-weight clique. An edge is chosen only if it connects two chosen vertices.

ILP itself is NP-hard, but replacing the integrality constraints on the X and Y variables with $0 \leq X_u, Y_{uv} \leq 1$ allows for a polynomial-time heuristic for the problem. It is important to note that should a linear programming solution happen to be integral, it is guaranteed to be optimal for the original ILP and motif finding problem. Non-integral solutions, on the other hand, are not feasible for the ILP and do not translate to a selection of positions for the motif finding problem. Those instances need to be solved by other means, such as using an ILP solver. Zaslavsky and Singh [47], who proposed this ILP formulation, couple the linear programming heuristic with a suite of graph pruning techniques, which enable a drastic reduction in graph sizes, and lead to effective and efficient application of the LP/ILP solvers to much smaller graphs.

4.3.1.1. *Graph Pruning*

The authors [47] introduce number of successively more powerful optimality-preserving dead-end elimination (DEE) techniques for pruning graphs corresponding to motif finding problems. The basic idea is to discard vertices and/or edges that cannot possibly be part of the optimal solution.

Here we mention the simplest of the techniques. Suppose there exists an N-clique of weight C^* in G. Then a vertex u, whose participation in any possible N-clique in G reduces the weight of that clique below C^*, is incompatible with the optimal alignment and can be safely eliminated.

More formally, for vertex $u \in V_i$ define $star(u)$ to be a selection of vertices from every graph part other than V_i. Let F_u be the value induced by the edge weights for the set of vertices in the $star(u)$ that form best pairwise alignments with u:

$$F_u = \sum_{j \neq i} \max_{v \in V_j} w_{uv} \tag{4.1}$$

If u were to participate in any N-clique in G, F_u is the maximum it can contribute to the weight of the clique. Similarly, let F_i^* be the value of the best possible $star(u)$ among all $u \in V_i$:

$$F_i^* = \max_{u \in V_i} F_u \tag{4.2}$$

F_i^* is an upper bound on what any vertex in V_i can contribute to any alignment.

If F_z, the most a vertex $z \in V_k$ can contribute to a clique, assuming the best possible contributions from all other graph parts, is insufficient compared to the value C^* of an existing clique, i.e. if

$$F_z < 2 \times C^* - \sum_{i \neq k} F_i^*, \tag{4.3}$$

z can be discarded. The clique value C^* is used with a factor of 2 since two edges are accounted for between every pair of graph parts in the above inequality.

In fact, the values of F_i^* are further constrained by requiring a connection to z when z is under consideration. That is, when considering a node $z \in V_k$ to eliminate, and calculating F_i^* according to Equation 4.2 among all possible $u \in V_i$, the F_u of Equation 4.1 is instead computed as:

$$F_u = w_{zu} + \sum_{j \neq i,k} \max_{v \in V_j} w_{uv} \tag{4.4}$$

The value of C^* can be computed from any "good" alignment, such as the weight of the clique imposed by the best overall *star*.

More complex graph-pruning techniques, based on three-way, as opposed to pairwise alignments, and divide-and-conquer graph decomposition approaches [47], are more effective in reducing graph size, but are also significantly less efficient. The overall algorithm applies the various pruning techniques in order of increasing complexity, and stops when the resulting graph has reached the desired size, and can be modeled and successfully solved as a linear program.

Interestingly, the vast majority of motif finding instances are not only effectively pruned by the optimality-preserving DEE methods, but also lead to linear programs whose optimal solutions are integral, and do not require the invocation of an ILP solver. These two conditions together guarantee optimality of the final solution for the original SP-based motif finding problem, making the method a polynomial-time algorithm for many practical instances of the problem.

4.3.2. *Cost-Aggregating Formulation*

In the context of a metric, such as the Hamming distance or 1/0 match/mismatch similarity score, which, when imposed on pairs of substrings allows for only a discrete set of possible values, Kingsford and coauthors [17] introduce an alternative integer linear program that better exploits the structure of the graph problem by allowing aggregation of edges of the same weight. The previous ILP formulation modeled edges in the graph $G = (V, E)$, where $V = V_1 \cup \ldots \cup V_N$ explicitly, while they are only used to ensure that if two nodes u and v are chosen in the optimal solution then the edge weight w_{uv} is added to the cost of the clique. In this formulation no edge variables exist; instead, in addition to the node variables X_u, there's a variable Y_{ujc} for each node u, each graph part j such that $u \notin V_j$, and each edge weight c. The intuition is that Y_{ujc} is 1 if node u and some node $v \in V_j$ are chosen such that the edge weight $w_{uv} = c$. These Y variables model groupings of the edges by weight into *cost bins*, as shown in Fig. 4.2, and selection of a cost bin requires the "payment" of the appropriate cost by the objective function.

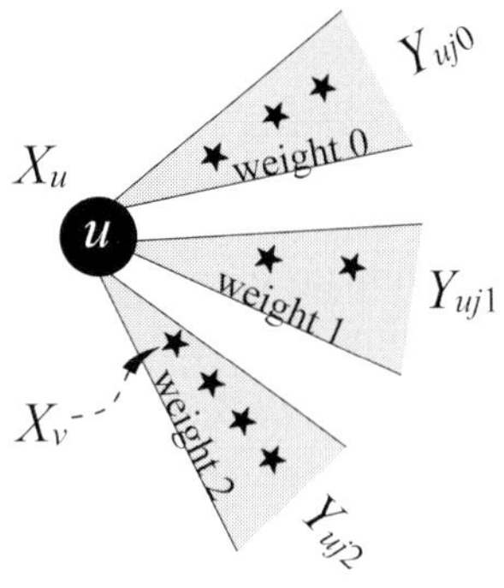

Fig. 4.2. Schematic of the cost-aggregating formulation. Adjacent to a node $u \in V_i$ there are at most $|D|$ (where D be the set of possible edge weights), cost bins for each part $j > i$, each associated with a variable Y_{ujc}. For each cost c there are the nodes $v \in V_j$ for which $w_{uv} = c$ (stars).

Formally, let D be the set of possible edge weights (costs) and let $W = \{(u, j, c) : c \in D, u \in V, j \in 1, \ldots N$ and $u \notin V_j\}$ be the set of triples over which the Y_{ujc} variables are indexed, and let $\mathbf{part}(u) = i$ if $u \in V_i$. Then the following ILP models the motif-finding graph problem (note that in the paper [17] a distance-based metric is used and the formulation follows a minimization problem):

$$\begin{array}{ll} \text{Maximize} & \sum_{(u,j,c)\in W:\mathbf{part}(u)<j} c \cdot Y_{ujc} \\ \text{subject to} & \\ \quad \sum_{u\in V_i} X_u = 1 & \text{for } i = 1, \ldots, N \\ \quad \sum_{c\in D} Y_{ujc} = X_u & \text{for } j \in 1, \ldots, N \text{ and } u \in V \setminus V_j \\ \quad \sum_{v\in V_j : w_{uv}=c} X_v \geq Y_{ujc} & \text{for } (u, j, c) \in W \text{ s.t. } u \in V_i \text{ and } i < j \\ \quad X_u, Y_{ujc} \in \{0, 1\} & \end{array} \tag{4.5}$$

As above, the first set of constraints forces a single node to be chosen in each graph part. The second set of constraints makes certain that if a node u is chosen, then for every other part j, one of its "adjacent" cost bins must also be chosen (Fig. 4.2). The third set of constraints ensures that Y_{ujc} can be selected only if some node $v \in V_j$, such that $w_{uv} = c$, is also selected. Figure 4.2 gives a schematic drawing of these constraints.

As shown by Kingsford and coauthors [17], this ILP formulation correctly models the SP motif finding problem. When considering a linear programming relaxation of the cost-aggregating formulation in order to develop an polynomial-time algorithm for the problem, the authors find that while it is weaker than the alternative LP formulation for motif finding [47], an exponentially-sized class of constraints can be added to make the two LP formulations equivalent. However, they then show that it is not necessary to explicitly add all these constraints by giving a separation algorithm, based on the Max-flow min-cut theorem. In the presence of a separation algorithm the ellipsoid method [9] can be used to find a solution to the tightened LP in time polynomial in the number of variables of the mathematical programming problem. In practice, the authors use a polynomially-sized subset of constraints in a cutting-planes approach to find solutions to the LP problem. Finally, it is interesting to note that the LP relaxations often have integral optimal solutions, making solving the LP sufficient in many cases for solving the original ILP, similar to what was observed above with the edge-modeling LP formulation.

4.4. Maximum Density Subgraph-based Algorithm

Looking for sequence motifs by identifying maximum density subgraphs (MDS) was suggested by Matsuda [28], as applied to the problem of detection of conserved domains in protein sequences. Fratkin and coauthors [8] employ maximum density subgraphs at the core of their MotifCut algorithm when searching for DNA motifs using motif finding graphs. Given the construction of such graphs in that the edges with higher weights correspond to pairs of substrings that are more similar and more likely to constitute motif instances, graph structures with greatest edge-weight density are likely to represent motif occurrences. Note that the MotifCut algorithm operates on motif finding graphs such that all pairs of vertices are connected by edges.

While the notion of subgraph density can be characterized in a number of ways, the authors use the most common tractable definition, and define density for a graph $G = (V, E)$ as $||E||/||V||$, where $||V||$ is the number of vertices and $||E||$ is the total weight of all the edges. They also provide some empirical evidence that their choice is a good one for biological data by evaluating which definitions lead to densities of motifs being most different from densities of subgraphs in the background. Building on this definition, [8] formulate motif finding as the problem of search for the maximum density subgraph $G^* = argmax_{G' \subseteq G}(||E'||/||V'||)$.

The solution to the maximum density subgraph problem is based on the Max-flow min-cut theorem. The graph is augmented by a source and sink nodes to become a flow network, with appropriate capacities connecting the source/sink to the rest of the graph, and the maximum flow is computed while also finding the minimum cut. If the maximum flow in the network satisfies certain conditions relating average degree of vertices and current graph density, then one of the subgraphs induced by the minimum cut is guaranteed to have higher density than the original graph. A new flow network is constructed from the subgraph with higher density and the process is repeated, converging to the maximum density subgraph after a polynomial number of iterations.

To achieve a speed-up over the *push/relabel* algorithm for the maximum flow computations, Fratkin and colleagues [8] propose a heuristic that looks for the maximum density subgraph in local, high average degree neighborhoods around vertices. Finally, the authors use a refinement step to eliminate the high number of false positives in the obtained set of candidate motif instances by greedily finding a subset of minimum entropy. The overall approach is interesting and novel in that it poses motif finding as an optimization problem that is solved by a polynomial-time algorithm and seems to scale well with longer input sequences.

4.5. Subtle Motif Algorithms

The subtle motif finding formulation of Pevzner and Sze [35] has generated a host of algorithms to address the (15, 4) challenge problem issued in the paper, which was to find implanted motifs of length 15 with 4 random mutations, and many of the subsequent algorithms solve this problem successfully (see [36] for comparative analysis). Additionally, they also address more difficult variants, in which a higher percentage of motif positions is corrupted and the motif is implanted into longer background sequences. Here we discuss a few of these methods, which focus on looking for structure in the subtle motif finding graphs $G = (V, E)$. Note that these differ from the standard motif finding graphs in that they aren't complete, as some pairs of substrings are not close enough in terms of their Hamming distances and can't simultaneously be instances of the same implanted motif. Finding subtle motifs reduces to identifying cliques of size N in these N-partite graphs.

First, we note that the linear programming formulations and graph pruning approaches above are also applicable to the subtle motifs problem, and indeed solve many variants successfully with excellent performance statistics in identifying motifs. The LP formulations remain largely unchanged except that the variables corresponding to non-existent edges are removed, and summations in the corresponding constraints are taken only over existing edges.

4.5.1. *Winnowing Techniques*

In their paper proposing the subtle motifs problem, Pevzner and Sze [35] make the observation that the vast majority of edges in G appear due to spurious similarities between pairs of substrings in the input that are not instances of the implanted motif. They suggest a winnowing approach, which attempts to identify these *spurious* edges and remove them from the graph, so that the remaining edges are *signal* edges, connecting the instances of the motif.

The *Winnower* algorithm utilizes the notion of an *extendable* clique. This notion captures the ability to grow a clique of size smaller than N, removing spurious edges along the way. More formally, let vertex u be a *neighbor* of a clique $C = \{v_1, \ldots, v_k\}$ if $\{v_1, \ldots, v_k, u\}$ is also a clique in the graph. A clique with vertices in graph parts $V_1, \ldots, V_k$ wlog is called *extendable* if it has at least one neighbor vertex in every other part of the graph. Certainly, edges that are not part of any extendable cliques can be removed, and these are the edges defined as *spurious*. For $k = 1$, a vertex u is a neighbor of clique $C = \{v\}$ of size one, if u and v are connected by an edge in the graph. The winnowing condition amounts to removing vertices that do not have a connection to every graph part. For $k = 2$, a vertex u is a neighbor of clique $C = \{v, w\}$ if $\{u, v, w\}$ form a triangle in the graph. The winnowing strategy for this case deletes edges that do not form three-way connections to every other part of the graph. For $k \geq 3$, the authors make a stronger observation by noting that not only does an edge that is part of a maximal N-clique have to belong to one extendable clique, but rather it has to belong to at least $\binom{N-2}{k-2}$ extendable cliques of size k. For $k = 3$ this condition removed edges that belong to fewer than $N - 2$ extendable cliques.

The general complexity of the winnowing approach for cliques of size k is $O(n^{k+1})$, where n is the total number of vertices in the motif finding graph. The authors demonstrate in a probabilistic analysis, based on the probability of an edge existing between two random vertices, that the expected runtime of the algorithm is of lower complexity, making it practical for many instances of the problem.

In follow-up work, Liang and coauthors [23] improve sensitivity of the method, enabling it to detect weaker motifs present in longer input sequences.

4.5.2. *Clique Finding with Consensus Constraint*

Sze and colleagues [42] proposed to formulate motif finding as the problem of looking for large cliques in motif finding graphs with the additional constraint that there must exist a string s that is close to every motif instance. The existence of such a string ensures that motif instances are derived from a single pattern. The authors observe that the maximal clique in an N-partite graph is much smaller than the size of the graph, and use a divide-and-conquer approach to solve the problem. They apply a branch-and-bound algorithm to each subproblem, checking along the way whether a string, close to the current set of possible motif instances, exists.

The main idea in the clique-finding part of the algorithm is to subdivide the problem into a number of independent N_0-partite subgraphs with $N_0 < N$, identifying all the cliques in each of the smaller subgraphs, and combine the results. Cliques, found in the smaller subgraphs, form vertices in a new graph, with two cliques from the same subproblem corresponding to vertices in the same part of the new graph. Two vertices in different parts are connected by and edge if their corresponding cliques combine to form a larger clique in the original graph. The authors show how to choose an appropriate subdivision parameter N_0, so that the number of cliques in each subproblem does not grow too high, and the second graph problem based on sub-cliques is not more complex than the original one.

Deciding whether a string, close to a given set of strings of the same length, exists is an NP-hard problem. Sze and coauthors [42] circumvent this difficulty by noting that the close-string problem need not be solved every time a clique of size j is expanded to size $j + 1$, but rather when a clique potentially larger than before is found. Weaker necessary conditions are used instead during other clique expansion steps to prune impossible branches. These conditions constrain existence of a close string based on pair-wise distances between strings in the set.

The authors further generalize their approach for problems such that not every input sequence contains a motif instance, and demonstrate an optimized branch-and-bound algorithm for clique finding that is faster than the standard methods and is applicable to relatively large graphs.

4.6. Discussion

We have presented an overview of graph-based methods for motif discovery. The authors of the respective methods, the LP-based methods and MotifCut in particular, thoroughly test the performance of the algorithms and show improved results in biological and/or simulated data over a number of standard and well-regarded approaches. The methods are based on motif models substantially different from the common PSSM model, and allow for the relaxation of the positional independence assumption within motifs. This line of research provides a successful and powerful alternative to traditional stochastic and enumerative techniques.

Acknowledgements

The author thanks Mona Singh and the anonymous referees for their helpful suggestions on the manuscript. E.Z. is supported by the NSF IGERT award DGE-9972930 and the NIH award HHSN266200500021C.

References

[1] Y. Barash, G. Elidan, N. Friedman, and T. Kaplan. Modeling dependencies in protein-dna binding sites. In *Proceedings of the Seventh Annual International Conference on Research in Computational Molecular Biology*, pages 28–37. ACM Press, 2003.
[2] P. V. Benos, M. L. Bulyk, and G. D. Stormo. Additivity in protein-dna interactions: how good an approximation is it? *Nucleic Acids Research*, 30(20):4442–4451, 2002.
[3] M. Blanchette and M. Tompa. Discovery of regulatory elements by a computational method for phylogenetic footprinting. *Genome Res.*, 12:739–748, 2002.
[4] B. Boeckmann, A. Bairoch, R. Apweiler, M.-C. Blatter, A. Estreicher, E. Gasteiger, M.J. Martin, K. Michoud, C. O'Donovan, I. Phan, S. Pilbout, and M. Schneider. The swiss-prot protein knowledgebase and its supplement trembl in 2003. *Nucleic Acids Res.*, 31:365–370, 2003.
[5] J. Buhler and M. Tompa. Finding motifs using random projections. *J. Comput. Biol.*, 9(2):225–242, 2002.
[6] M. L. Bulyk, P. L. Johnson, and G. M. Church. Nucleotides of transcription factor binding sites exert interdependent effects on the binding affinities of transcription factors. *Nucleic Acids Research*, 30(5):1255—1261, 2002.
[7] P. Cliften, P. Sundarsanam, A. Desikan, L. Fulton, B. Fulton, J. Majors, R. Waterston, B.A. Cohen, and M. Johnston. Finding functional features in Saccharomyces genomes by phylogenetic footprinting. *Science*, 301(5629):71–76, 2003.
[8] E. Fratkin, B. T. Naughton, D. L. Brutlag, and S. Batzoglou. Motifcut: regulatory motifs finding with maximum density subgraphs. *Bioinformatics*, 22(14):e150–e157, 2006.
[9] M. Grötschel, L. Lovász, and A. Schrijver. *Geometric Algorithms and Combinatorial Optimization*. Springer-Verlag, Berlin, Germany, 2nd edition, 1993.
[10] D. GuhaThakurta. Computational identification of transcriptional regulatory elements in dna sequence. *Nucleic Acids Res.*, 34(12):3585–3598, 2006.
[11] G. Hertz and G. Stormo. Identifying dna and protein patterns with statistically significant alignments of multiple sequences. *Bioinformatics*, 15:563–577, 1999.
[12] L. S. Hon and A. N. Jain. A deterministic motif finding algorithm with application to the human genome. *Bioinformatics*, 22(9):1047–1054, 2006.
[13] J. Hu, B. Li, and D. Kihara. Limitations and potentials of current motif discovery algorithms. *Nucleic Acids Res.*, 33(15):4899–4913, 2005.
[14] J. Hughes, P. Estep, S. Tavazoie, and G. Church. Computational identification of cis-regulatory elements associated with groups of functionally related genes in *S. cerevisiae*. *J. Mol. Biol.*, 296:1205–1214, 2000.
[15] U. Keich and P. Pevzner. Finding motifs in the twilight zone. *Bioinformatics*, 18:1374–1381, 2002.

[16] M. Kellis, N. Patterson, M. Endrizzi, B. Birren, and E. Lander. Sequencing and comparison of yeast species to identify genes and regulatory elements. *Nature*, 423:241–254, 2003.

[17] C. Kingsford, E. Zaslavsky, and M. Singh. A compact mathematical programming formulation for dna motif finding. In *Proceedings of the Seventeenth Annual Symposium on Combinatorial Pattern Matching (CPM), Barcelona, Spain*, pages 233–245. Springer, 2006.

[18] C. Lawrence, S. Altschul, M. Boguski, J. Liu, A. Neuwald, and J. Wootton. Detecting subtle sequence signals: a gibbs sampling strategy for multiple alignment. *Science*, 262:208–214, 1993.

[19] C. Lawrence and A. Reilly. An expectation maximization (em) algorithm for the identification and characterization of common sites in unaligned biopolymer sequences. *Proteins: Structure, Function, and Genetics*, 7:41–51, 1990.

[20] T.I. Lee, N.J. Rinaldi, F. Robert, D.T. Odom, Z. Bar-Joseph, G.K. Gerber, N.M. Hannett, C.T. Harbison, C.M. Thompson, I. Simon, J. Zeitlinger, E.G. Jennings, H.L. Murray, D.B. Gordon, B. Ren, J.J. Wyrick, J.B. Tagne, T.L. Volkert, E. Fraenkel, D.K. Gifford, and R.A. Young. Transcriptional regulatory networks in saccharomyces cerevisiae. *Science*, 298(5594):799–804, 2002.

[21] H.C. Leung and F.Y. Chin. Finding exact optimal motifs in matrix representation by partitioning. *Bioinformatics*, 21 (Suppl. 2):ii86–ii92, 2005.

[22] N. Li and M. Tompa. Analysis of computational approaches for motif discovery. *Algorithms for Molecular Biology*, 1:8, 2006.

[23] S. Liang, M.P. Samanta, and B.A. Biegel. Cwinnower algorithm for finding fuzzy DNA motifs. *J. Bioinform. Comput. Biol.*, 2(1):47—60, 2004.

[24] X. Liu, D.L. Brutlag, and J.S. Liu. Bioprospector: discovering conserved dna motifs in upstream regulatory regions of co-expressed genes. In *Proceedings of the Sixth Pacific Symposium on Biocomputing*, pages 127–138. International Society for Computational Biology, 2001.

[25] K. MacIsaac and E. Fraenkel. Practical strategies for discovering regulatory dna sequence motifs. *PLoS Computational Biology*, 2(4):e36, 2006.

[26] T. K. Man and G. D. Stormo. Non-independence of mnt repressor-operator interaction determined by a new quantitative multiple fluorescence relative affinity (qumfra) assay. *Nucl. Acids Res.*, 29:2471–2478, 2001.

[27] L. Marsan and M. F. Sagot. Algorithms for extracting structured motifs using a suffix tree with an application to promoter and regulatory site consensus identification. *J. Comput. Biol.*, 7:345–362, 2000.

[28] H. Matsuda. Detection of conserved domains in protein sequences using a maximum-density subgraph algorithm. *IEICE Trans. Fund. Elec. Comm. Comp. Sci.*, E83-A(4):713–721, 2000.

[29] L. McCue, W. Thompson, C. Carmack, M. Ryan, J. Liu, V. Derbyshire, and C. Lawrence. Phylogenetic footprinting of transcription factor binding sites in proteobacterial genomes. *Nucleic Acids Res.*, 29(3):774–782, 2001.

[30] A. McGuire, J. Hughes, and G. Church. Conservation of dna regulatory motifs and discovery of new motifs in microbial genomes. *Genome Res.*, 10(6):744–757, 2000.

[31] S. Mukherjee, M.F. Berger, G. Jona, X.S. Wang, D. Muzzey, M. Snyder, R.A. Young, and M.L. Bulyk. Rapid analysis of the dna-binding specificities of transcription fac-

tors with dna microarrays. *Nature Genetics*, 36(12):1331–1339, 2004.

[32] B. T. Naughton, E. Fratkin, S. Batzoglou, and D. L. Brutlag. A graph-based motif detection algorithm models complex nucleotide dependencies in transcription factor binding sites. *Nucleic Acids Res.*, 34:5730–5739, 2006.

[33] R. Osada, E. Zaslavsky, and M. Singh. Comparative analysis of methods for representing and searching for transcription factor binding sites. *Bioinformatics*, 20(18):3516–3525, 2004.

[34] G. Pavesi, P. Mereghetti, G. Mauri, and G. Pesole. Weeder web: discovery of transcription factor binding sites in a set of sequences from co-regulated genes. *Nucleic Acids Res.*, 32:W199–W203, 2004.

[35] P. Pevzner and S. Sze. Combinatorial approaches to finding subtle signals in dna sequences. In *Proceedings of the Eighth International Conference on Intelligent Systems for Molecular Biology*, pages 269–278. International Society for Computational Biology, AAAI Press, 2000.

[36] A. Price, S. Ramabhadran, and P. Pevzner. Finding subtle motifs by branching from sample strings. *Bioinformatics*, 19 (Suppl 2):ii149–ii155, 2003.

[37] K. Robison, A. M. McGuire, and G. M. Church. A comprehensive library of dna-binding site matrices for 55 proteins applied to the complete *Escherichia coli* k-12 genome. *J. Mol. Biol.*, 284:241–254, 1998.

[38] R. Siddharthan, E. D. Siggia, and E. van Nimwegen. Phylogibbs: a gibbs sampling motif finder that incorporates phylogeny. *PLoS Computational Biology*, 1(7):e67, 2005.

[39] S. Sinha and M. Tompa. A program for discovery of novel transcription factor binding sites by statistical overrepresentation. *Nucleic Acids Res.*, 31(13):3586–3588, 2003.

[40] P. T. Spellman, G. Sherlock, M. Q. Zhang, V. R. Iyer, K. Anders, M. B. Eisen, P. O. Brown, D. Botstein, and B. Futcher. Comprehensive identification of cell cycle-regulated genes of the yeast saccharomyces cerevisiae by microarray hybridization. *Molecular Biology of the Cell*, 9(12):3273–3297, 1998.

[41] G. D. Stormo. Dna binding sites: representation and discovery. *Bioinformatics*, 16:16–23, 2000.

[42] S. H. Sze, S. Lu, and J. Chen. Integrating sample-driven and pattern-driven approaches in motif finding. In *WABI*, pages 438–449, 2004.

[43] S. Tavazoie, J. D Hughes, M. J. Campbell, R. J. Cho, and G.M. Church. Systematic determination of genetic network architecture. *Nature Genetics*, 22(3):281–285, 1999.

[44] M. Tompa. An exact method for finding short motifs in sequences, with application to the ribosome binding site problem. In *Proceedings of the Seventh International Conference on Intelligent Systems for Molecular Biology*, pages 262–271. International Society for Computational Biology, AAAI Press, 1999.

[45] M. Tompa, N. Li, T. L. Bailey, G. M. Church, B. De Moor, E. Eskin, A. V. Favorov, M. C. Frith, Y. Fu, W. J. Kent, V. J. Makeev, A. A. Mironov, W. S. Noble, G. Pavesi, G. Pesole, M. Regnier, N. Simonis, S. Sinha, G. Thijs, J. van Helden, M. Vandenbogaert, Z. Weng, C. Workman, C. Ye, and Z. Zhu. Assessing computational tools for the discovery of transcription factor binding sites. *Nature Biotech.*, 23(1):137–144, 2005.

[46] M. Vingron and P. Pevzner. Multiple sequence comparison and consistency on mul-

tipartite graphs. *Advances in Applied Mathematics*, 16:1–22, 1995.

[47] E. Zaslavsky and M. Singh. A combinatorial optimization approach for diverse motif finding applications. *Algorithms for Molecular Biology*, 1:13, 2006.

[48] Q. Zhou and J. S. Liu. Modeling within-motif dependence for transcription factor binding site predictions. *Bioinformatics*, 20:909—916, 2004.

Chapter 5

Statistical Clustering Analysis: An Introduction

Hang Zhang

Department of Industrial Engineering
Arizona State University
Tempe, Arizona 85287-5906, USA
hang.zhang@asu.edu

Clustering analysis is to segment objects in a dataset into meaningful subsets such that objects with high similarity are segmented into the same subset, and objects with low similarity are segmented into different subsets. This chapter introduces three fundamental but core topics in clustering analysis: the definition of similarity and dissimilarity measure, the clustering algorithm, and determining the number of clusters. For each topic, we introduce the ones that are most popularly used, and emphasize their statistical backgrounds.

5.1. Introduction

Clustering analysis is to group objects in a dataset into subsets such that objects with high similarity are segmented into the same subset and objects with low similarity are segmented into different subsets. The grouping results, subsets, are called clusters.

A dataset to be clustered consists of a collection of objects. An object may be characterized by a vector of feature values. For example, in a dataset of fish, an object is just an observation of fish represented by a vector of features such as its weight, length, color, etc. We name clustering these objects as observation clustering. An object may also be characterized by a sequence of observations, e.g., the time series of a stock price in one year. If we want to find the segmentation such that stocks having high dependency are grouped into the same cluster, and stocks with low dependency into different clusters, we take each sequence as an object. Specifically, we call clustering these objects (sequences) as variable clustering.

One question comes up with the definition of clustering analysis: what is a cluster. The answer to this question varies in different applications of clustering analysis. For example, in image segmentation, clusters are regions in the image, each of which is considered to "be homogeneous with respect to some image property of interests such as intensity, color or texture" [16]. In variable clustering, usually a cluster is a group of sequences that are associated with each other. For instance, clustering the spike train data, sequences of the spiking time stamps, of multiple brain neurons identifies the associations among brain neurons. In some other applications, a cluster may be considered as a sample from an underlying probabilistic distribution. In the example of fish data, objects in a cluster can be considered as a random sample from a multivariate distribution.

The goal of clustering analysis also varies with applications. In image processing, the purposes of clustering analysis mostly include detecting edges of objects [28], and image segmentation. Image segmentation is a common problem in image processing. It involves taking an image and identifying particular features, such as the figure of human beings or a vehicle, for further purpose such as movement tracking. If properly implemented, clustering analysis can automatically divide an image into similar regions. In some other applications, clustering analysis may be to refer the underlying distributions generating the clusters, such as the number of underlying distributions and the parameters of each distribution.

The readers should be noted about the difference between clustering and classification. Classification is also called supervised learning. Given a collection of labeled objects, we derive the discrimination model which is later used to label a new object without a class label. Clustering, also called unsupervised learning, is to group a collection of unlabeled objects into meaningful clusters. After clustering, objects in the same cluster are given the same labels. Objects in different clusters are labeled differently.

In this chapter, we introduce clustering analysis mostly from the perspective of multivariate statistics. For the convenience of the readers, we also introduce some heuristic methods in case the readers may need them in some applications where it is not proper to assume the multivariate probability distribution. We focus on two basic aspects of clustering analysis: clustering and determining the number of clusters.

As in the definition of clustering analysis, measure of similarity (or dissimilarity) plays an important role. Before we go into those two topics, we describe the measures of similarity (or dissimilarity) between two objects.

Before moving forward, we first give the notations which will be used in the remainder of this chapter. We denote the dataset to be clustered as $\mathbf{X}'$, which is an $N \times P$ matrix where P is the number of features (variables), and N is the number of observations. Here, $\mathbf{X}'$ stands for the transpose of $\mathbf{X}$. In observation clustering, observation i is characterized by the ith row of $\mathbf{X}'$, denoted as $\mathbf{x}_i'$, which is a $1 \times P$ row vector. In variable clustering, each object is the sequence of a variable. So, object j is represented as the jth column of matrix $\mathbf{X}'$, denoted as $\mathbf{x}_j$ .We also denote the similarity and dissimilarity between two objects $\mathbf{x}_i$ and $\mathbf{x}_j$ as $s(\mathbf{x}_i, \mathbf{x}_j)$ and $d(\mathbf{x}_i, \mathbf{x}_j)$, respectively. The readers should be notified that in observation clustering, $\mathbf{x}_i'$ is the ith row of $\mathbf{X}'$. However, in variable clustering, $\mathbf{x}_j$ is the jth column of $\mathbf{X}'$.

5.2. Similarity (Dissimilarity) Measures

In different applications of clustering analysis, we need to choose the proper similarity (dissimilarity) measures. In observation clustering analysis, commonly used measures include three dissimilarity measures (Euclidean distance, Minkowski distance and Mahalanobis distance) and one similarity measure, the cosine. In variable clustering analysis, the commonly used measures include Pearson's correlation coefficients, mutual information, and cross intensity for point processes. We want the objects with high similarity (low dissimilarity) to be grouped into the same cluster, and objects with low similarity (high dissimilarity) to be grouped into different clusters.

5.2.1. *Measures for Observation Clustering*

5.2.1.1. *Euclidean Distance and Minkowski Distance*

Euclidean distance is the most popular measure of dissimilarity, denoted by $d_2(\mathbf{x}_i, \mathbf{x}_j)$:

$$d_2(\mathbf{x}_i, \mathbf{x}_j) = [\sum_{k=1}^{p}(x_{ik} - x_{jk})^2]^{1/2} = \| \mathbf{x}_i - \mathbf{x}_j \|_2 \tag{5.1}$$

where $\| \cdot \|_2$ is called *Norm 2*. More generally, Minkowski distance is defined as:

$$d_m(\mathbf{x}_i, \mathbf{x}_j) = [\sum_{k=1}^{p} | x_{ik} - x_{jk} |^m]^{1/m} \tag{5.2}$$

Clearly, Euclidean distance is just a special case of Minkowski distance when m=2.

Taking Euclidean distance as the dissimilarity measure is intuitive. It works well when each cluster is a hyper-sphere in space. From the perspective of statistics, it means that each cluster is generated by a multivariate normal distribution with mean μ_i and variance-covariance matrix $\Sigma_i = \sigma^2\mathbf{I}$, $i = 1, 2, \ldots, K$, where

μ_i and μ_j are significantly different when $i \neq j$, $\mathbf{I}$ is an identity matrix of dimension p, and K is the number of underlying probability distributions generating the dataset.

When clusters are not hyper-spheres but hyper-ellipsoids, which means the variance-covariance matrix can not be expressed by $\Sigma_i = \sigma^2\mathbf{I}$, taking Euclidean distance as the dissimilarity measure has poor performance. Figure 5.1 shows two points clustered incorrectly because of using Euclidean distance as the dissimilarity measure. In Fig. 5.1, we cluster a point into a subset whose center has the minimal Euclidean distance to the point (how to find the centers is described in K-means clustering algorithm in Sec. 5.3). Obviously in Fig. 5.1, there are two ellipsoid clusters. The centers of these two clusters are denoted as C_1 and C_2 on Fig. 5.1. Consider two points P_1 and P_2. Obviously, if we take into account the ellipsoid nature of clusters, P_1 should be clustered to subset centered at C_1, and P_2 should be assigned to subset centered at C_2. However, since P_1 has smaller distance to C_2, if we take Euclidean distance as the dissimilarity measure, P_1 is clustered incorrectly into subset centered at C_2. Similarly, P_2 is clustered incorrectly into subset centered at C_1.

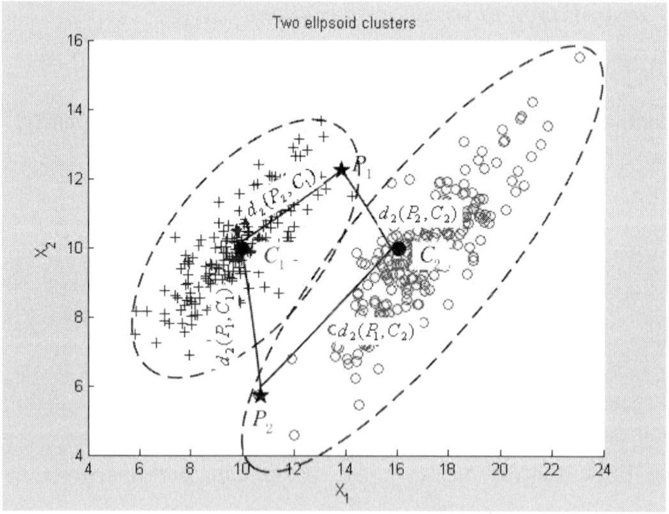

Fig. 5.1. Poor performance of Euclidean distance when clusters are ellipsoids.

5.2.1.2. *Mahalanobis Distance*

Mahalanobis distance takes the different variances in different features and the possible correlation between any two features into consideration. The Mahalanobis distance between any two objects $\mathbf{x}_i$ and $\mathbf{x}_j$ is:

$$d_M(\mathbf{x}_i, \mathbf{x}_j) = [(\mathbf{x}_i - \mathbf{x}_j)'\Sigma^{-1}(\mathbf{x}_i - \mathbf{x}_j)]^{1/2} \quad (5.3)$$

where Σ is the sample variance-covariance matrix. For example, in Fig. 5.1, $d_M(\mathbf{x}_{P_1}, \mathbf{x}_{C_1}) = [(\mathbf{x}_{P_1} - \mathbf{x}_{C_1})'\Sigma_{C_1}^{-1}(\mathbf{x}_{P_1} - \mathbf{x}_{C_1})]^{1/2}$, $d_M(\mathbf{x}_{P_1}, \mathbf{x}_{C_2}) = [(\mathbf{x}_{P_1} - \mathbf{x}_{C_2})'\Sigma_{C_2}^{-1}(\mathbf{x}_{P_1} - \mathbf{x}_{C_2})]^{1/2}$, Matrix Σ_{C_j} is the sample variance-covariance matrix of objects assigned to cluster centered at $\mathbf{x}_{C_j}$, j=1 and 2. The distance $d_M(\mathbf{x}_{P_1}, \mathbf{x}_{C_1}) < d_M(\mathbf{x}_{P_1}, \mathbf{x}_{C_2})$, so point P_1 is assigned to cluster centered at C_1 correctly. Similarly, point P_2 is assigned to cluster centered at C_2 correctly.

Mahalanobis distance is very analogous to Hotelling's T^2 statistic. Hotelling's T^2 statistic is a very popularly used multivariate statistic to measure the weighted distance between a high-dimension point to a population center [23]. We introduce Hotelling's T^2 statistic here briefly to help understand the Mahalanobis distance. Hotelling's T^2 statistic is calculated as:

$$T^2 = [(\mathbf{x}_i - \overline{\mathbf{x}})'\mathbf{S}^{-1}(\mathbf{x}_i - \overline{\mathbf{x}})] \quad (5.4)$$

where $\overline{\mathbf{x}}$ is the sample mean and $\mathbf{S}$ is the sample variance-covariance matrix. If point $\mathbf{x}_i$ has high T^2 statistic, it means with low probability, $\mathbf{x}_i$ is generated from an underlying population whose probability density function (pdf) has sample mean $\overline{\mathbf{x}}$ and sample variance-covariance matrix $\mathbf{S}$. The readers can find the analogy between Eqs. 5.3 and 5.4. Assigning a point to a cluster whose center has the minimum Mahalanobis distance to the point is just assigning a point to a population where the point has the minimum Hotelling's T^2 statistic, i.e., the highest probability that the point is generated by that population.

Users can also find the similarity between Mahalanobis distance and the likelihood value of the observation under the assumption that the observation is a sample from a multivariate normal distribution. For instance, in Fig. 5.1, the likelihood value of an observation to a multivariate normal distribution, with sample mean $\mathbf{x}_{C_j}$ and sample variance-covariance matrix Σ_{C_j} is: $L_j(\mathbf{x}) = \frac{1}{(2\pi)^{p/2}|\Sigma_{C_j}|^{1/2}} exp(-\frac{1}{2}(\mathbf{x} - \mathbf{x}_{C_j})'\Sigma_{C_j}^{-1}(\mathbf{x} - \mathbf{x}_{C_j}))$, j=1, 2. Taking Eq. 5.3 into consideration, we get

$$L_j(\mathbf{x}) = \frac{1}{(2\pi)^{p/2} \mid \Sigma_{C_j} \mid^{1/2}} exp(-\frac{1}{2} d_M(\mathbf{x}, \mathbf{x}_{C_j})), \quad (5.5)$$

Term $\mid \Sigma_{C_j} \mid$ in the right hand side of Eq. 5.5 is the determinant of matrix Σ_{C_j}. It is also called generalized variance [17]. From Eqs. 5.3 and 5.5, we can see that assigning a point to a cluster according to the minimal Mahalanobis distance with the cluster center is equivalent to assigning it to a cluster according to the maximum likelihood value, as long as the distributions of clusters have similar general variances.

5.2.1.3. *Cosine*

Cosine is widely used as a similarity measure in text clustering [27], which is:

$$s_C(\mathbf{x}_i, \mathbf{x}_j) = \frac{\mathbf{x}_i \cdot \mathbf{x}_j}{\| \mathbf{x}_i \| \cdot \| \mathbf{x}_j \|} \tag{5.6}$$

where $\mathbf{x}_i \cdot \mathbf{x}_j = \mathbf{x}_i'\mathbf{x}_j$, the inner product of two vectors. In text clustering, usually texts are coded according the presence (code 1) or absence (code 0) of the interested words or sentences. For instance, we are interested in five words (features) A, B, C, D and E. Two texts are coded as $\mathbf{x}_1' = [1, 0, 0, 0, 0]$ and $\mathbf{x}_2' = [0, 0, 0, 0, 1]$, which means in $\mathbf{x}_1$, only word A is present, and in $\mathbf{x}_2$, only word E is present. If we use Euclidean distance to measure their dissimilarity, $d_2(\mathbf{x}_1, \mathbf{x}_2) = \sqrt{2}$. Now, we consider another two texts $\mathbf{x}_3' = [1, 1, 1, 1, 0]'$ and $\mathbf{x}_4' = [0, 1, 1, 1, 1]$. Their Euclidean distance $d_2(\mathbf{x}_3, \mathbf{x}_4) = \sqrt{2}$. Clearly, texts $\mathbf{x}_1$ and $\mathbf{x}_2$ have no word in common, but $\mathbf{x}_3$ and $\mathbf{x}_4$ have 3 out of 5 words in common. Texts $\mathbf{x}_3$ and $\mathbf{x}_4$ should have lower dissimilarity than $\mathbf{x}_1$ and $\mathbf{x}_2$. However, Euclidean distance measures their dissimilarities the same.

Cosine solves this problem. The cosine of texts $\mathbf{x}_1$ and $\mathbf{x}_2$ is $s_C(\mathbf{x}_1, \mathbf{x}_2) = 0$, and that of texts $\mathbf{x}_3$ and $\mathbf{x}_4$ is $s_C(\mathbf{x}_3, \mathbf{x}_4) = 3/4$. It means that texts $\mathbf{x}_3$ and $\mathbf{x}_4$ have higher similarity than $\mathbf{x}_1$ and $\mathbf{x}_2$.

5.2.2. *Measures for Variable Clustering*

Variable clustering is very important in identifying the dependency among variables, causal analysis, and selecting variables to reduce the dimension of data. For instance, in stock market place, it is of significant importance to understand which stocks are inter-dependent, the causal/result relationship among these inter-dependent stocks, and which stocks are affecting the stocks of interest. In neuroscience, in order to understand how neurons are cooperating with each other from the neural activity data, one can cluster the neurons by calculating the similarity (dissimilarity) measures among the spike train data (sequences) of neurons *in vivo*.

In this subsection, we introduce two commonly association measures: Pearson's correlation coefficient and mutual information.

5.2.2.1. *Pearson's Correlation Coefficient*

Pearson's correlation coefficient of sequences $\mathbf{x}_i$ and $\mathbf{x}_j$ is calculated as:

$$\rho_{ij} = \frac{S_{ij}}{\sqrt{S_{ii}S_{jj}}} \tag{5.7}$$

where $S_{ij} = \frac{1}{N-1}\sum_{k=1}^{N}(x_{ki} - \bar{x}_i)(x_{kj} - \bar{x}_j)$, and $\bar{x}_i = \frac{1}{N}\sum_{k=1}^{N} x_{ki}$. Usually we call Pearson's correlation coefficient correlation for simplification.

Readers should be noted that correlation is ranged from -1 to 1. When one variable is just a linear function of the other one, $X_2 = aX_1 + b$, $\rho = 1$, if $a > 0$, and $\rho = -1$ if $a < 0$. So, if one is only concerned with the extent of the correlation but not the direction, one can use the square or the absolute value of ρ as the similarity measure.

Correlation captures the linear dependency between two sequences, as illustrated in Fig. 5.2(a) and (b). However, it does not capture the nonlinear dependency, as shown in Fig. 5.2(c), where $X_2 = sin(X_1) + \epsilon$, and ϵ is a normally distributed noise term.

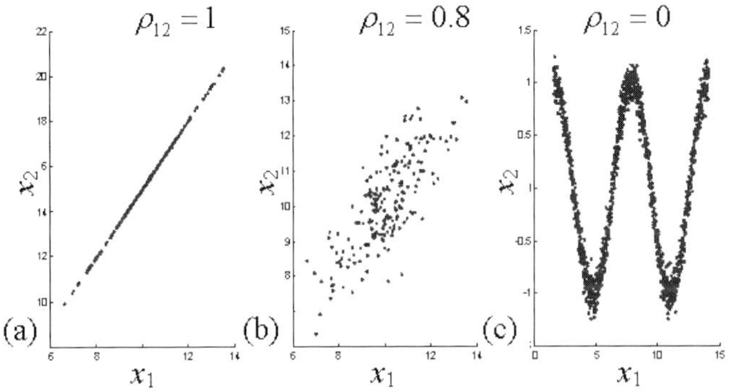

Fig. 5.2. Correlation of two variables: (a) and (b) linear dependence; (c) nonlinear dependence.

5.2.2.2. *Mutual Information*

In addition to linear dependency as shown in Fig. 5.2(a) and (b), mutual information is also capable of capturing nonlinear dependency such as the one shown in Fig. 5.2(c). Mutual information comes from information theory and is based on Shannon entropy. The mutual information between two sequences $\mathbf{x}_i$ and $\mathbf{x}_j$ is calculated as:

$$\boldsymbol{I}(\mathbf{x}_i, \mathbf{x}_j) = \frac{1}{N} \sum_{k=1}^{N} log(\frac{\hat{p}_{ij}([x_{ki}, x_{kj}])}{\hat{p}_i(x_{ki})\hat{p}_j(x_{kj})}) \tag{5.8}$$

where $\hat{p}_{ij}([x_{ki}, x_{kj}])$ is the estimated likelihood of the sample $[x_{ki}, x_{kj}]$ under the joint distribution of sequences $\mathbf{x}_i$ and $\mathbf{x}_j$, and $\hat{p}_i(x_{ki})$ is the estimated likelihoods of sample x_{ki} under the marginal distribution of $\mathbf{x}_i$. The estimated likelihood value is calculated as follows:

$$\begin{aligned} \hat{p}_{ij}([x_i, x_j]) &= \tfrac{1}{Nh^r} \textstyle\sum_{k=1}^{N} g(\tfrac{[x_i, x_j] - [x_{ki}, x_{kj}]}{h}) \\ \hat{p}_i(x) &= \tfrac{1}{Nh^r} \textstyle\sum_{k=1}^{N} g(\tfrac{x - x_{ki}}{h}) \end{aligned} \tag{5.9}$$

where $g(\cdot)$ is a kernel function. A frequent choice for the kernel function is the standard normal density, i.e.,

$$g(\mathbf{x}) = (2\pi)^{-r/2} exp(-\frac{1}{2}\mathbf{x}'\mathbf{x}) \tag{5.10}$$

where r is the dimension of $\mathbf{x}$. In the first equation in Eq. 5.8, r=2. Parameter h in Eq. 5.9 is given by:

$$h = [\frac{4}{(2r+1)N}]^{1/(r+4)} \tag{5.11}$$

The mutual information $\boldsymbol{I}(\mathbf{x}_i, \mathbf{x}_j)$ between sequences $\mathbf{x}_i$ and $\mathbf{x}_j$ is always non-negative, and is zero if and only if these two sequences are stochastically independent. Usually we use the normalized mutual information as the similarity measure which is ranged from 0 to 1. The transformation of mutual information to the similarity measure is as follows:

$$s(\mathbf{x}_i, \mathbf{x}_j) = \sqrt{1 - exp(-2\boldsymbol{I}(\mathbf{x}_i, \mathbf{x}_j))} \tag{5.12}$$

Kojadinovic [20] gives the details of mutual information and how to use mutual information as the similarity measure to cluster variables. Mutual information has also been used in neuroscience to identify the associations between neurons [2]. Readers should be noted that the mutual information introduced here is for continuous variables. For discrete variables, such as the sequence of spike train data of neurons, the calculation may be different.

There are some other measures to capture the association between variables, such as the cross-intensity and coherence for associations between point processes [3]. These two measures are commonly used for sequences of spike train data from neurons.

5.3. Clustering Algorithm

A big variant of clustering algorithms has been devised to cover different applications. Currently there is no one clustering algorithm that performs well on all dataset. Which clustering algorithm should be chosen depends on the definition of clusters and the size of the dataset in a specific application. For instance, in image segmentation, density-based methods are popularly used such as DBSCAN and GDBSCAN algorithms [6] [7]. Some other clustering algorithms are devised to cluster large datasets, such as CLARA and CLARANS. Wei *et al.* [31] gives a good review of these large dataset clustering algorithms. For other clustering algorithms, Han and Chamber [13] give good general references.

In this section, for the sake of space, we only introduce four basic but most popularly used clustering algorithms: K-means, E-M algorithm, hierarchical clustering algorithm and self-organization maps (SOM) algorithm. K-means algorithm is one of the most popularly used clustering algorithms. E-M algorithm is a method popularly used in multivariate statistics for missing value estimation. If we take the label of each object as the missing value in clustering analysis, E-M algorithm applies. We introduce E-M algorithm here because it is analogous to K-means algorithm and it provides statistical background for K-means algorithm. We introduce hierarchical clustering algorithm as one example of heuristic algorithms. K-means, E-M, and hierarchical clustering algorithms, in most common case, load all the data into the computer memory at the same time for clustering analysis. Contrarily in SOM algorithm, the data enters the computer memory for clustering one by one. It has advantage in handling large dataset or data streams, where either the dataset is too large to be all loaded to the computer memory at one time, or the data itself emerges sequentially.

5.3.1. *K-Means Algorithm*

The K-means algorithm assumes the number of clusters K is known. It works iteratively as follows, where we use t to denote the iteration number:

(1) Randomly select K objects as the initial centers of these K clusters, denoted as $\bar{\mathbf{x}}_1^t, \bar{\mathbf{x}}_2^t, ..., \bar{\mathbf{x}}_K^t$; $t=0$;

(2) For object i, calculate $s_*(\mathbf{x}_i, \bar{\mathbf{x}}_j^t)$ or $d_*(\mathbf{x}_i, \bar{\mathbf{x}}_j^t), j=1, 2, ..., K$. Assign cluster label to object i by

$$\begin{gathered} CL_i^t = \arg\max_j [s_*(\mathbf{x}_i, \bar{\mathbf{x}}_j^t), j = 1, 2, ..., K] \\ or \\ CL_i^t = \arg\min_j [d_*(\mathbf{x}_i, \bar{\mathbf{x}}_j^t), j = 1, 2, ..., K], i = 1, 2, ..., N \end{gathered} \tag{5.13}$$

where CL_i^t stands for cluster label of object i at iteration t;

(3) Update the centers of the K clusters by

$$\mathbf{x}_j^{t+1} = \frac{1}{N_{C_j}^t} \sum_{i=1}^{N} I(CL_i^t = j)\mathbf{x}_i, j = 1, 2, ..., K,$$

where $I(X)$ is an identity function such that $I(X) = 1$ if X is true, and $I(X) = 0$ otherwise. The denominator, $N_{C_j}^t$, is the number of objects assigned to cluster j at the tth iteration; Let t=t+1;

(4) If the convergence criterion is satisfied, algorithm stops. Otherwise go to step (2).

K-means algorithm has several variants [1]. For instance, in step (1), the initial centers of these K clusters can be randomly generated points which are evenly distributed in the area occupied by the objects in the dataset. Or we can just randomly pick up K objects in the dataset as the initial centers. In different applications, the candidates for the similarity or dissimilarity measure in step (2) can include d_2, d_m, d_M and s_C. Or, if users are clustering variables, correlation and mutual information can be the candidated similarity measures. If d_M is selected as the dissimilarity measure, in step (1), K variance-covariance matrices $\Sigma_{C_j}^0$, j=1, 2, ..., K, are needed to be initialized. Usually, we choose $\Sigma_{C_j}^0 = \mathbf{I}$, where $\mathbf{I}$ is a $p \times p$ identity matrix. In step (3), the variance-covariance matrix of cluster j at iteration t $\Sigma_{C_j}^t$ should also be updated by

$$\Sigma_{C_j}^{t+1} = \frac{1}{N_{C_j}^t - 1} \sum_{i=1}^{N} I(CL_i^t = j)(\mathbf{x}_i - \bar{\mathbf{x}}_j^t)(\mathbf{x}_i - \bar{\mathbf{x}}_j^t)'$$

i.e., the sample variance-covariance matrix of objects assigned to cluster j.

Because of the randomness of the initialization of K cluster centers in step (1), K-means cluster algorithm may give different clustering results on the same dataset in different runs. One can run the K-means clustering algorithm on the dataset for D duplicates, $D > 1$. For duplicate k (k=1, 2, ..., D), we calculate the sum of the intra-cluster dissimilarities ($SICD$) from each object to its cluster center, denoted as $SICD^k$:

$$SICD^k = \sum_{j=1}^{K} \sum_{i=1}^{N} d_2(\mathbf{x}_i, \bar{\mathbf{x}}_{kj}) I(CL_{ki} = j) \tag{5.14}$$

where $\bar{\mathbf{x}}_{kj}$ is the center of the jth cluster in the kth duplicate, CL_{ki} is the cluster label of object i in the kth duplicate. The clustering result with the smallest $SICD^k$ is the final clustering result, i.e.,

$$k^* = \arg\min_k (SICD^k, k = 1, 2, ..., D), CL_i = CL_{k^*,i}, i = 1, 2, ..., N \tag{5.15}$$

K-means algorithm is a hard clustering algorithm, i.e., each subject is clustered into a single cluster with probability 1. Its counter part, soft clustering algorithm, assigns a set of probabilities to a subject, each of which represents the probability that the subject belongs to a cluster. The probabilities assigned to a single subject are all non-negative and sum up to 1. The soft clustering counter part of K-means is fuzzy K-means algorithm. Generally speaking, the differences between fuzzy K-means and K-means are that in fuzzy K-means algorithm, each subject is assigned a set of probabilities, and in each iteration, the clusters centers are updated by weighted average of all subjects in the dataset, where the weight of each subject is the probability that this subject belongs to the cluster [9]. Other examples of soft clustering algorithms include fuzzy clustering by local approximation of memberships (FLAME) [8] and neuro-fuzzy [25] algorithms.

5.3.2. *E-M Algorithm*

E-M algorithm stands for Expectation-Maximization algorithm. It has two steps: Expectation step and Maximization step. By assuming that the dataset is generated from a mixture of K multivariate normal distribution, the E-M algorithm works as follows in clustering analysis.

Expectation step (E-step): Make expectations of the mean vector and variance-covariance matrix of cluster j, denoted as $\hat{\mu}_j$ and $\hat{\Sigma}_j$ respectively, using objects assigned to cluster j, $j = 1, 2, ..., K$. If it is in the first iteration, randomly select K different vectors $\hat{\mu}_j$'s and K $p \times p$ positive-definitive and symmetric matrices $\hat{\Sigma}_j$'s as the initial expectation of the mean vectors and variance-covariance matrices.

Maximization step (M-step): For object i, calculate the likelihood that it is generated by the jth multivariate normal distribution with mean $\hat{\mu}_j$ and variance-covariance matrix $\hat{\Sigma}_j$. Assign cluster label CL_i to object i according to the maximum likelihood, i.e.,

$$L_{ij} = \frac{1}{(2\pi)^{P/2} \mid \hat{\Sigma}_j \mid^{1/2}} exp(-\frac{1}{2}(\mathbf{x}_i - \hat{\mu}_j)'\hat{\Sigma}_j^{-1}(\mathbf{x}_i - \hat{\mu}_j)) \qquad (5.16)$$

If some stopping criterion is satisfied, the algorithm stops; otherwise go back to the E-step. The frequently used stopping criterion is that the update of $\hat{\mu}_j$'s from the previous iteration does not exceed a threshold value.

In the initial run of the E-step, $\hat{\Sigma}_j$'s are usually selected to be identity matrices.

One disadvantage of the E-M algorithm is that it is based on the assumption of the mixture of some distributions with unknown parameters, and the assumption is difficult to be justified. There is a dilemma here: to justify the assumption, we have to know the clusters first and then run the test to see whether the distribution of each cluster satisfied the distribution assumption. This means that we have to know the clustering result before running the clustering algorithm. Priebe and Marchette [26] address this problem by using a nonparametric method. Interested readers are referred to it for details.

The readers should be noted about the analogy between K-means and E-M algorithms. Their analogy is maximized when d_M is used as the dissimilarity measure between two objects in K-means algorithm. So, in the following discussions about the analogy between these two algorithms, we assume that d_M defined in Eq. 5.3 is the dissimilarity measure for the K-means algorithm.

The initialization steps of these two algorithms are similar. In K-means algorithm, randomly selecting K objects as the initial centers of the K clusters is analogous to the E-step in the first iteration of the E-M algorithm. In the E-M algorithm, K mean vectors $\hat{\mu}_j$'s are randomly selected, which is just the initialization of the K cluster centers in the K-means algorithm. We also randomly select K variance-covariance matrices $\hat{\Sigma}_j$, j=1, 2,..., K in the E-M algorithm, which is the initialization of the K variance-covariance matrices in the K-mean algorithm.

The analogy also exists in assigning each object to a cluster. In Eq. 5.13, cluster label of object i at iteration t is determined by the maximum similarity or the minimum dissimilarity with the cluster center. In Eq. 5.16, it is determined by the maximum likelihood of the object to distribution with mean $\hat{\mu}_j$ and variance-covariance matrix $\hat{\Sigma}_j$. To calculate the maximum likelihood, the term $(\mathbf{x}_i - \hat{\mu}_j)'\hat{\Sigma}_j^{-1}(\mathbf{x}_i - \hat{\mu}_j)$ in Eq. 5.16 is just the Mahalanobis distance, which is used as the dissimilarity measure in the K-means algorithm.

Another analogy between these two algorithms is the update of the cluster centers (mean vectors in E-M algorithms) and the variance-covariance matrices. In both algorithms, the cluster centers and the variance-covariance matrices are updated by the mean vectors and the variance-covariance matrices of the objects assigned to the same cluster, respectively.

Because of the similarity between K-means and E-M algorithms, the convergence of the K-means algorithm can be studied through the convergent property of the E-M algorithm. Xu and Jordan give the details of the convergence of the E-M algorithm [32].

5.3.3. *Hierarchical Clustering*

Hierarchical clustering has two directions: agglomerative and divisive hierarchical clustering. They work in two opposite directions as follows. Agglomerative hierarchical clustering initially takes each object as a single cluster. At each step, it combines two clusters with the maximum similarity or the minimum dissimilarity into one. The algorithm stops when all objects are assigned into the same cluster or some stopping criteria is satisfied. The divisive hierarchical clustering algorithm proceeds in the opposite way. It initially takes the whole dataset as a single cluster. At each step, one cluster is split into two until each cluster contains only one object or some stopping criteria is satisfied. Since agglomerative hierarchical clustering algorithm is more popularly used than the divisive algorithm, in this section, we only give the details of the agglomerative algorithm.

The agglomerative hierarchical clustering algorithm [22] is illustrated in Fig. 5.3 by an example of a 2-D dataset with 8 objects, A, B, C, ..., H, which constitute 3 clusters obviously. This algorithm yields a dendrogram which shows the combination of two clusters at each step, as shown in Fig. 5.3(b).

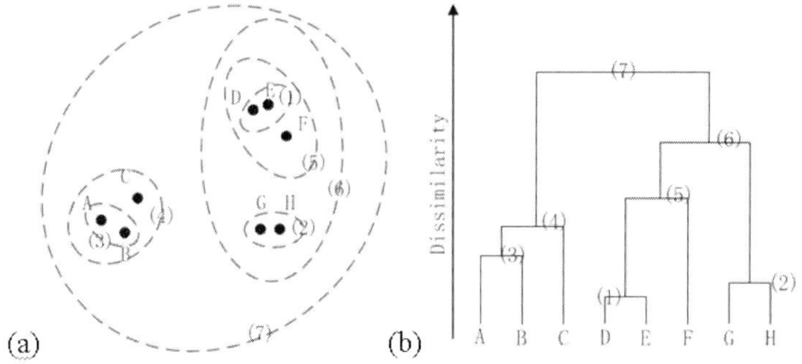

Fig. 5.3. Agglomerative hierarchical clustering algorithm: (a) 2-D example; (b) output dendrogram.

Figure 5.3(a) shows the 8 objects in the 2-D example dataset. Each slash ellipse represents a combination of two clusters. The step number when the combination happens is shown as the number in a pair of parentheses on that slash ellipse. For instance, the slash ellipse numbered (4) represents the combination of two clusters at step (4), one has objects A and B which are combined into one cluster in step (3), and the other one has a single object C.

The dendrogram shown in Fig. 5.3(b) illustrates the combinations of clusters in the procedure of the hierarchical clustering algorithm more clearly. The vertical lines represent the remained clusters. A horizontal line represents the combination of two clusters. Its two ends connect two vertical lines representing two existing clusters when the combination happens. The vertical location of each horizontal line represents the dissimilarity measure when two clusters are combined into one. For instance, in Fig. 5.3(b), clusters D and E are combined into one cluster at the first step since they have the minimum dissimilarity. So the horizontal line representing this combination is located lowest in the dendrogram, and it connects two vertical lines representing clusters D and E. Now, we have seven clusters, which are (A), (B), (C), (D, E), (F), (G) and (H). Then, clusters G and H are combined since the dissimilarity of these two clusters are the minimum among all mutual dissimilarities of remaining clusters. This procedure stops when there is only one cluster left, i.e., all objects are clustered into one cluster.

There are several variants of the agglomerative hierarchical clustering algorithm. The selection of dissimilarity measure between two clusters is one source of the variants. Another source of the variants is the stopping criteria. Well-known variants of hierarchical clustering methods include CURE [11], ROCK [12], and CHEMELEON [18].

The dissimilarity measure of two clusters used in the agglomerative hierarchical clustering algorithm includes single-linkage dissimilarity, complete-linkage dissimilarity, minimum-variance dissimilarity [30], and some others [15]. The first two are most popularly used measures. Figure 5.4 illustrates these two measures with a 2-D example. Single-linkage dissimilarity of two clusters is the minimum of the distances between all pairs of objects, one from a cluster, and the other one from another cluster. In Fig. 5.4, we use D_{KL} to represent the dissimilarity between clusters K and L, denoted as C_K and C_L respectively. Single-linkage dissimilarity is calculated as follows:

$$D_{KL} = \min_{\mathbf{x}_i \in C_K, \mathbf{x}_j \in C_L} d_2(\mathbf{x}_i, \mathbf{x}_j); \forall i, j. \tag{5.17}$$

Contrarily, complete-linkage dissimilarity between clusters C_K and C_L takes the maximum one as the dissimilarity measure, i.e.,

$$D_{KL} = \max_{\mathbf{x}_i \in C_K, \mathbf{x}_j \in C_L} d_2(\mathbf{x}_i, \mathbf{x}_j); \forall i, j. \tag{5.18}$$

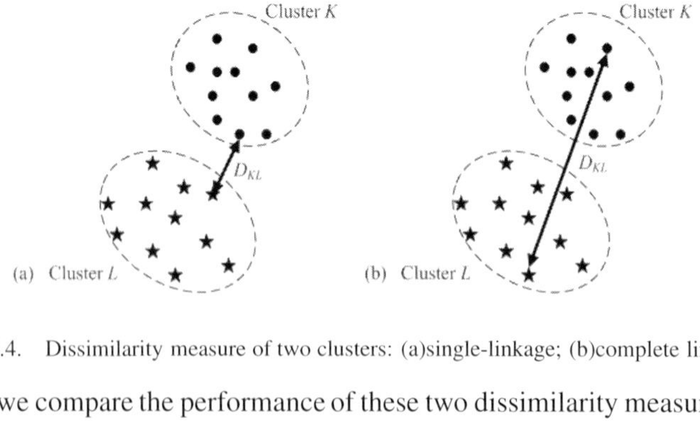

Fig. 5.4. Dissimilarity measure of two clusters: (a)single-linkage; (b)complete linkage.

Now, we compare the performance of these two dissimilarity measures in hierarchical clustering algorithms to help users to determine which measure to choose when clustering different datasets. Jain and Goel [15] compare the performance of these two dissimilarity measures and several other measures in agglomerative hierarchical clustering. They conclude that if random data is generated by uniform or normal distributions, complete-linkage dissimilarity measure performs best. Single-linkage dissimilarity measure performs more poorly than complete-linkage, but its performance tends to improve as the number of objects N in the dataset increases.

However, the single-linkage dissimilarity measure has its advantage in being more versatile in dealing with non-convex clusters. Figure 5.5 shows an example with two concentric clusters. In this figure, black dots represent objects from class 1, and circles represent objects from class 2. Hierarchical clustering algorithm using single-linkage dissimilarity measure can correctly cluster these objects, but the algorithm using complete-linkage dissimilarity measure can not.

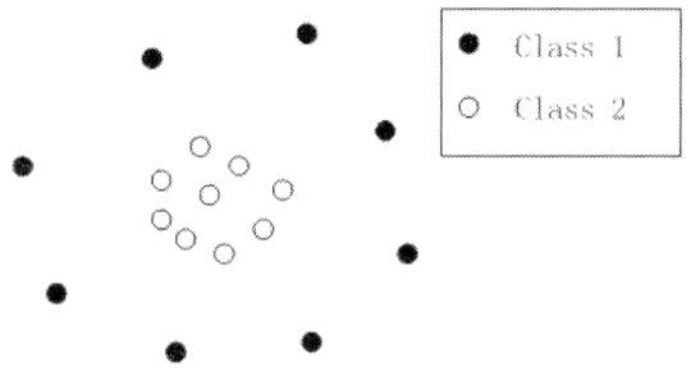

Fig. 5.5. Two concentric clusters.

The other source of variant hierarchical clustering algorithm is the stopping criteria. Some hierarchical clustering algorithms assume that the number of clusters K^* is known in advance. An algorithm stops when the number of clusters derived by the algorithm equals K^*. So, only partial dendrogram can be generated (from N clusters to K^* clusters, instead of from N clusters to 1 cluster in a full dendrogram). For instance, in Fig. 5.3, if we know that $K^* = 3$, after the fifth combination, we retrieve 3 clusters. The algorithm stops and concludes that the 3 clusters are (A, B, C), (D, E, F) and (G, H).

Some other hierarchical clustering algorithms determine K^* dynamically. The output clusters of the hierarchical clustering algorithm are just the clusters in the full dendrogram when K^* clusters are retrieved.

In the following section, we are going to introduce several methods of determining the number of clusters.

5.3.4. *Self-Organizing Map*

Self-organizing map (SOM) is a popularly used data visualization and clustering algorithm. It maps high-dimension data into a low dimension (typically 2-D or 3-D) map space. An SOM consists of components called neurons or nodes. Each neuron has the same shape, typically rectangle and hexagonal shapes. Neurons constitute the lattice of the map. In this lattice map, each neuron has its neighboring neurons, which are determined by the shape of the neuron and the structure of the map. Figure 5.6 illustrates the structure of a 2-D SOM, and the definition of neighboring neurons when the neuron has rectangle and hexagonal shapes. In Fig. 5.6, the gray neurons are the nearest neighbors, and the neurons in slash lines are the second nearest neighbors, of the neuron i shown in black, and so on. The definition of the closeness of the neighborhood to a neuron is important in the SOM algorithm, which will be shown soon.

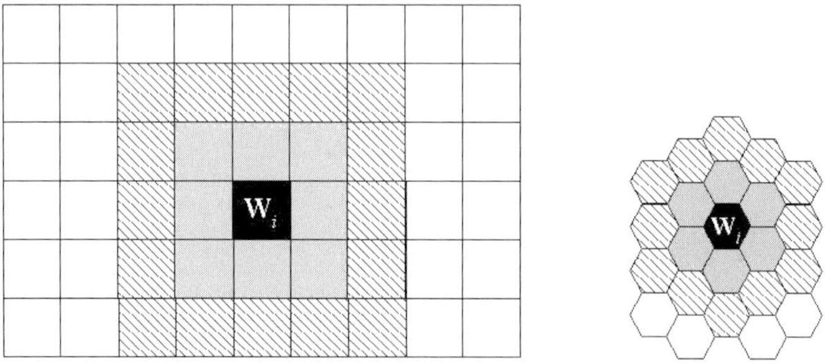

(a) (b)

Fig. 5.6. Structure of 2-D SOM with (a) rectangle (b) hexagonal neurons, the definition of neighboring neurons, and the closeness of neighborhood.

Each neuron in SOM is associated with a weight vector $\mathbf{W}$ with the same dimension as each observation. As shown in Fig. 5.6, the weight vector of neuron i is denoted as $\mathbf{W}_i$.

After the structure of the SOM, such as whether we use 2-D or 3-D SOM, the shape of the neuron, the number of neurons L, ect, are determined, the SOM algorithm for clustering analysis works in the following procedure:

(1) Initialize the weights associate with neurons by random numbers, and let the iteration step t=0,

$$\mathbf{W}_i(t) = [w_{i1}(t), w_{i2}(t), ..., w_{ip}(t)], i = 1, 2, ..., L.$$

(2) Draw a sample randomly from the dataset, denoted as $\mathbf{x}$ from the dataset $\mathbf{X}$.

(3) Find the winner neuron i^* with the associated weight vector $\mathbf{W}_{i^*}$ according to the maximum similarity (or minimal dissimilarity)

$$\begin{gathered} i^* = \arg\max_i(s_*(\mathbf{x}, \mathbf{W}_i)) \\ or \\ i^* = \arg\min_i(d_*(\mathbf{x}, \mathbf{W}_i)), i = 1, 2, ..., L \end{gathered} \tag{5.19}$$

(4) Update the weight vectors of the winner and its neighboring neurons, with the rate according to the closeness to the winner neuron, as follows:

$$\mathbf{W}_i(t+1) = \mathbf{W}_i(t) + \alpha(t)\Lambda(i,i^*)(\mathbf{x} - \mathbf{W}_i(t)), i = 1, 2, ..., L \quad (5.20)$$

where $\alpha(t)$ is the learning rate, which is a monotonically decreasing function of the iteration step t so that in the beginning iterations, the SOM has a fast learning rate, and slower rate later on. Function $\Lambda(i,i^*)$ is a neighborhood function, which is also a monotonically decreasing function of the closeness between neuron i and the winner neuron i^*. A frequently used neighborhood function is:

$$\Lambda(i,i^*) = exp(- \| r_i - r_{i^*} \|^2 /(2\sigma^2(t)) \quad (5.21)$$

Term $\| r_i - r_{i^*} \|^2$ is the closeness of neuron i to the winner neuron i^*. One example of the closeness of the neighborhood is shown in Fig. 5.6, where we can code the closeness of the neurons in black as 0, the closeness of neurons in gray as 1, the closeness of neurons in slash lines as 2, and etc. Term $\sigma^2(t)$ is a scale parameter, it is also a monotonically decreasing function of the iteration number t.

(5) Compute the amount of the weight vectors updates

$$E_t = \sum_{i=1}^{L} \| \mathbf{W}_i(t+1) - \mathbf{W}_i(t) \|_2 \quad (5.22)$$

If E_t is not greater than a threshold ε, stop; otherwise let $t = t+1$, go back to (2).

The output of the SOM algorithm will be a set of neurons, where in some regions of the map, the neighboring neurons will have weight vectors with high similarity or small dissimilarity, and have weight vectors with small similarity or high dissimilarity with neurons in other regions. Usually, we map the distance between two neighboring neurons into a gray scale or a color map, and the output can be visualized as in Fig. 5.7.

Figure 5.7 can be interpreted in the following way. In Fig. 5.7(a), it clear that the whole map is divided by a bell of dark neurons into two parts. From the vertical bar in Fig. 5.7(a), we can see that the darker gray means longer distance. So, Fig. 5.7(a) shows that the whole dataset is divided into two dense areas by a sparse area shown by the dark colored neurons. Similarly, Fig. 5.7(b) tells us that the whole dataset has three clusters since the light grayed neurons divide the whole map into three dark grayed regions.

(a) (b)

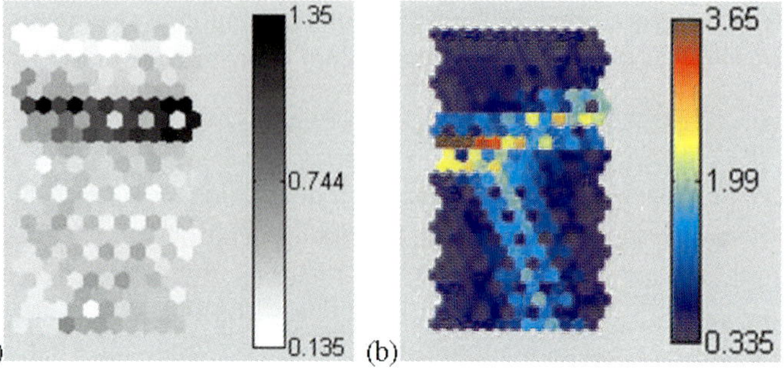

Fig. 5.7. Output of SOM where distances between neurons are mapped to (a) gray scale; (b) color map (Kohonen [19]).

For clustering analysis, what is left is to assign a cluster label to each observation in matrix **X**. After visualizing the output of SOM, on the map, one can label the regions considered to be clusters with different numbers. For instance, in Fig. 5.7(b), we can label the top region, the left-bottom region, and the right-bottom region as 1, 2, and 3, respectively. Then for each observation, the neuron on the map with the highest similarity (or lowest dissimilarity) with the observation is identified. The observation is assigned a cluster label according to the label of the region where the identified neuron falls in. If the identified neuron falls in the areas separating the regions of clusters, the observation is identified as an outlier.

The advantages of SOM exist in the following folds. First, it does not require the number of cluster as the input. It completes the clustering and identifying the number of clusters at the same time. Second, the observations enter the algorithm sequentially, which means we do not have to load the whole dataset into the memory for clustering analysis. It is very helpful when the dataset is too large for the computer to load all into the memory at the same time. It is also very helpful in the case that the whole dataset is not available but that the observations come sequentially. Third, SOM is a distance preserving data visualization method. It maps the high dimensional dataset into a 2-D map, where the distance between observations is preserved in the distance between the weight vectors associated with neurons, and the distance is visualized by gray scale or color maps, as in Fig. 5.7. Fourth, statistically, SOM simulates the density distribution of the dataset. For example, in Fig. 5.7(b), we can see that the whole dataset has three areas with high density, and different dense regions have different density distributions.

The biggest problem of SOM is that it is subjective. Although SOM identifies the number of clusters and cluster the objects at the same time, the number of clusters is still based on subjective judgment of human beings. For instance, in

Fig. 5.7(b), one can consider the two bottom dark regions as one single cluster because they are connected by light grayed neurons. It totally depends on the threshold of color or gray scale one chooses to separate regions. When there is only one cluster in the dataset, the visualized output map usually misleads users to identify clusters more than 1.

For detailed information of SOM, readers are referred to Kohonen [19].

5.4. Determining the Number of Clusters

Determining the number of clusters is a difficult and unresolved problem in clustering analysis. The reason partially comes from the fact that the number of clusters is closely subject to the definition of clusters. In image segmentation, if we define clusters as regions in the image, each of which is considered to be homogeneous with respect to intensity, the dataset shown in Fig. 5.8 should be considered consisting of only one cluster. Contrarily, if we define clusters as samples of multivariate normal probability distributions, two clusters should be determined in Fig. 5.8: one for the open circles, and the other one for the solid circles.

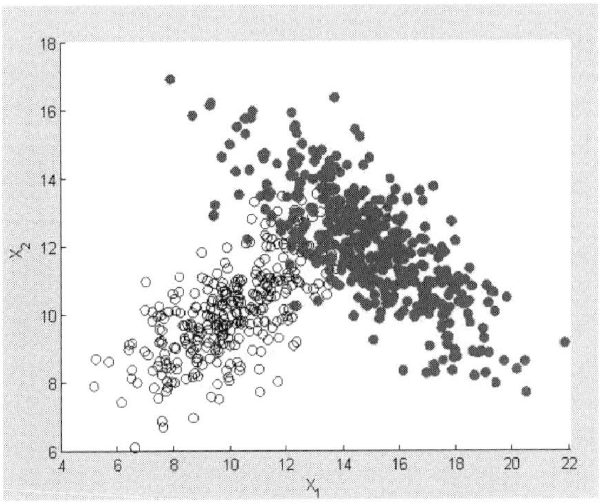

Fig. 5.8. Two-dimensional dataset where the determination of the number of clusters varies in different applications.

In this section, we introduce two categories of methods to determine the number of clusters in a dataset: model-based method and scale-based method. Readers interested in the applications of image segmentation should refer to the density-based method, such as the shared nearest neighbors (SNN) method [5].

5.4.1. *Model-based Method*

Model-based method assumes that the dataset consists of samples from a mixture of populations, and each population has a determined form of probability distribution with unknown parameters. The pdf of the mixture model is:

$$f(\mathbf{x}) = \sum_{j=1}^{K} f(\mathbf{x}|\mathbf{x} \in C_j)p(\mathbf{x} \in C_j) \tag{5.23}$$

where $f(\mathbf{x}|\mathbf{x} \in C_j)$ is the pdf of vector $\mathbf{x}$ conditioning on vector $\mathbf{x}$ is generated by class j, denoted by C_j , $p(\mathbf{x} \in C_j)$ is the probability that vector $\mathbf{x}$ is generated by C_j, and K is the number of assumed mixed probability populations.

Model-based method to determine the number of clusters works as follows:

(1) Assume the form of pdf $f(\mathbf{x}|\mathbf{x} \in C_j)$. Usually, we take the same form of function for $f(\mathbf{x}|\mathbf{x} \in C_j), j = 1, 2, ..., K$. We denote the model parameters of $f(\mathbf{x}|\mathbf{x} \in C_j)$ as θ_j. For $f(\mathbf{x}|\mathbf{x} \in C_i)$ and $f(\mathbf{x}|\mathbf{x} \in C_j)$, if $i \neq j$, $\theta_i \neq \theta_j$.

(2) Let K takes value from 1 to $\bar{K}$, where $\bar{K}$ is a large integer. For each K, we cluster the dataset into K clusters. For cluster j, j=1 to K, we calculate the maximum likelihood estimates of θ_j, denoted as $\hat{\theta}_j$. Probability $p(\mathbf{x} \in C_j)$ is calculated by:

$$p(\mathbf{x} \in C_j) = \frac{\sum_{i=1}^{N} I(\mathbf{x}_i \in C_j)}{N} \tag{5.24}$$

It is just the percent of the observations assigned into cluster j.

(3) For each K, calculate the adjusted log-likelihood value by:

$$l(K) = 2\sum_{i=1}^{N} log(\sum_{j=1}^{K} f(\mathbf{x}_i|\mathbf{x}_i \in C_j, \hat{\theta}_j)p(\mathbf{x}_i \in C_j)) - g(K) \tag{5.25}$$

where $f(\mathbf{x}_i|\mathbf{x}_i \in C_j, \hat{\theta}_j)$ is the pdf of $\mathbf{x}_i$ with conditions that $\mathbf{x}_i$ is generated by class C_j and model parameters are $\hat{\theta}_j$. Function $g(K)$ is a penalty function. It is a monotonously increasing function of K. Fraley and Raftery [10] suggest $g(K) = m_K log(N)$, where m_K is the number of independent parameters to be estimated in the mixture model defined in Eq. 5.23.

(4) Choose the K corresponding to the largest $l(K)$ as the number of clusters, denoted as K^* , i.e.,

$$K^* = \arg\max_K (l(K), K = 1, 2, ..., \bar{K}) \tag{5.26}$$

Let us review Eq. 5.25 in step (3). We can find that Eq. 5.25 is very similar to the Bayesian Information Criterion (BIC) when we determine the best model in multivariate regression analysis. To determine the best model, models are fitted with all possible combinations of the explanatory variables. For each fitted model, the BIC is calculated as:

$$BIC = nln(\frac{RSS}{n}) + kln(n) \tag{5.27}$$

where RSS is the residual sum of squares of the fitted model, n is the number of observations in the training dataset, and k is the number of explanatory variables in the model. The model with the smallest BIC is selected as the best model.

The similarity between the model-based method to determine the number of clusters and the BIC method to choose the best model exists in three sides. First, Eq. 5.25 is very similar with Eq. 5.27; Second, the second component of the right hand side of Eq. 5.25 is the penalty to prevent too many clusters. In Eq. 5.27, $kln(n)$ is also a penalty to prevent too many explanatory variables, which leads to overfitting the data. Third, in BIC, we choose the model with the smallest BIC as the best model. Contrarily in model-based method to determine the number of clusters, we choose the number of clusters as the one corresponding to the largest adjusted loglikelihood value.

The disadvantage of the model-based method is that one has to assume the form of the underlying probability density function before we apply this method to determine the number of clusters. Usually, users assume multivariate normal distributions. This implies that all clusters are convex. The convexity of clusters is difficult to justify, especially when the dimension of the dataset is high. Figure 5.9 illustrates this disadvantage with a two-dimension dataset which obviously has three clusters, one of which is non-convex. The model-based method with the assumption of bi-variate normal distribution detects 6 clusters, instead of the true value 3.

5.4.2. *Scale-based Method*

The determination of the number of clusters is not only subject to the definition of clusters, as stated in the introduction of this section, but also subject to the resolution level we choose when we view the clusters. Figure 5.10 shows a case where different resolution may give different determination of the number of clusters. In Fig. 5.10, if we use a high resolution, we can conclude that there are 3 clusters, as in Fig. 5.10(a). Contrarily, if we use a low resolution, we can conclude with 2 clusters; see Fig. 5.10(b).

The readers should be noted that different number of clusters caused by using different resolution is also subject to the definition of clusters. As shown in Fig. 5.10(b), if we use a low resolution, we get two clusters. Points in cluster 1

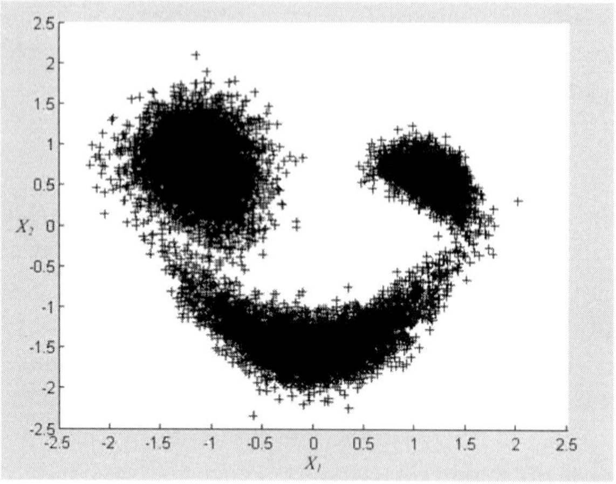

Fig. 5.9. Two-dimensional dataset with one non-convex cluster.

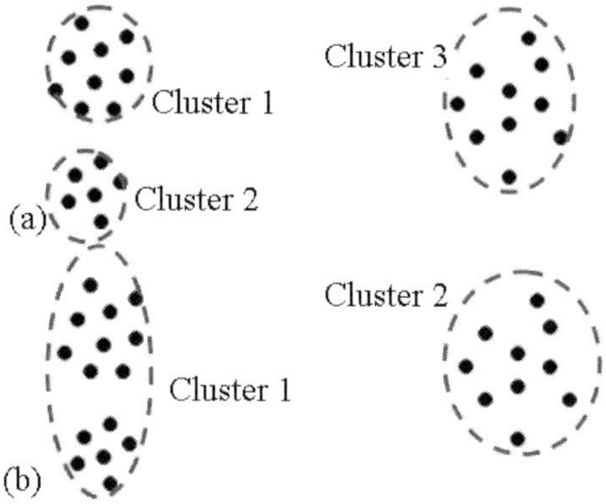

Fig. 5.10. Different determinations of the number of clusters with different resolution levels: (a) 3 clusters with high resolution; (b) 2 clusters with low resolution.

constitute a single cluster only if this cluster satisfies our definition of cluster. Otherwise, we should choose other resolution such that each cluster found satisfies the definition, as in Fig. 5.10(a).

Scale-based method is a method which gives the numbers of clusters under different resolution levels. The winning number of clusters is the number survives in the largest range of resolution levels.

Scale-based method works as follows:

(0) Standardize the dataset such that each dimension has mean 0 and standard deviation 1, denoted as $\mathbf{X}'_{Std}$. Let t=0, scale parameter λ_t starts from a small

number such as 0.2, i.e., $\lambda_t = 0.2$. Let the step size of the scale parameter to be a small number such as 0.1, i.e., $\Delta\lambda = 0.1$. Assume the number of clusters K_t to be larger than the reasonable maximum number of clusters in the dataset. For instance, if we think that the number of clusters in the dataset can not exceeds 8, we let $K_t = 9$;

(1) Cluster the dataset into K_t clusters. Users can use any clustering algorithm which utilizes the number of clusters as an input parameter, such as K-means and hierarchical clustering algorithms. Calculate cluster centers;
(2) If any two cluster centers are closer than the scale parameter λ_t, combine these two clusters into one cluster.
(3) Increase λ_t by one step size, i.e., $\lambda_{t+1} = \lambda_t + \Delta\lambda$. Let $t = t + 1$. Update K_t with the number of remaining clusters.
(4) If $K_t = 1$, stop; otherwise go to (1).

After we run the above algorithm, we can plot K_t against λ_t. Figure 5.11 is an example of this plot. We choose the K_t (not including K_0) surviving in the longest range of λ_t as the number of clusters K^*. In Fig. 5.11, we choose $K^* = 3$.

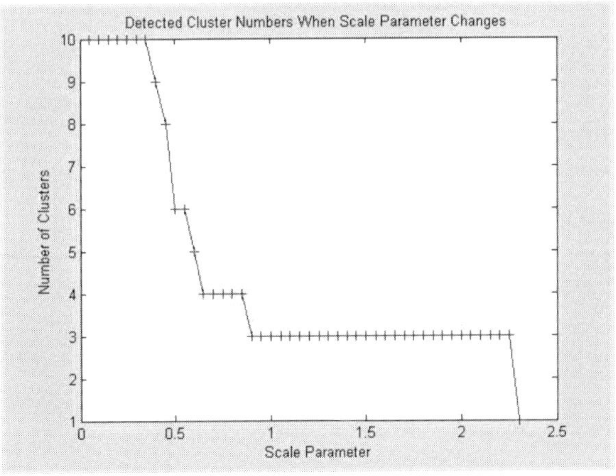

Fig. 5.11. K_t vs. λ_t plot of scale-based method.

One advantage of the scale-based method over the model-based method is that it is capable of giving the correct number of clusters when non-convex clusters exist. For example, the scale based method can correctly identify 3 clusters in the dataset shown in Fig. 5.9.

The disadvantage of the scale-based method is that the algorithm stops when one cluster is reached, as shown in step (4). Thus, it always concludes on a number greater than 1, which means it can not give correct answer when there is only one cluster in the dataset.

This problem is solved by adding a dummy dimension to the original standardized dataset and clone the original dataset in the space with the dummy dimension [33] such that the augmented dataset has at least two clusters. The augmented dataset is denoted as $\mathbf{X}_{Std}^{D}$, which is:

$$\mathbf{X}_{Std}^{D} = \begin{pmatrix} \mathbf{X}'_{Std}, & \mathbf{0} \\ \mathbf{X}'_{Std}, & \mathbf{d} \end{pmatrix} \tag{5.28}$$

where $\mathbf{0}$ is an $N \times 1$ zero column vector, and $\mathbf{d}$ is another $N \times 1$ vector with all elements d. They call this method as scale-based with dummy dimension (SBDD) method.

We can apply the scale-based method on this augmented dataset $\mathbf{X}_{Std}^{D}$. In this way, the augmented dataset has at least two clusters. So, we can compare whether the number of clusters 2 survives in the longest range of the scale parameter or any other number does. The number of clusters in the original dataset $\mathbf{X}'_{Std}$ is just the number of clusters identified in $\mathbf{X}_{Std}^{D}$ divided by 2.

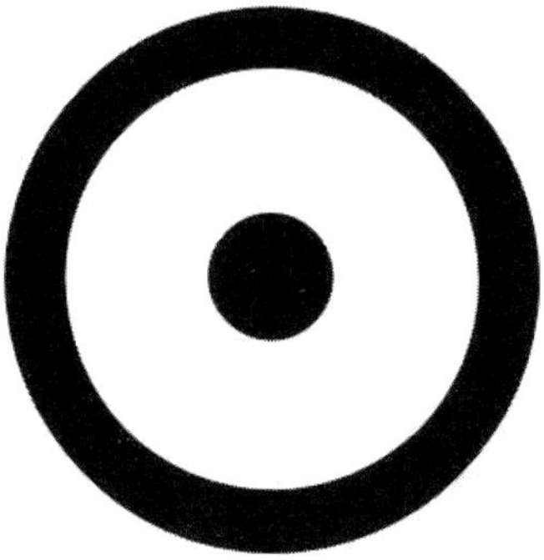

Fig. 5.12. One cluster encircled by another one.

There is one user-specified parameter d in the SBDD method. The value of d is suggested to start with a small value of, such as d=2. With each value of d, the augmented dataset is constructed by Eq. 5.28. Scale-based method is applied on $\mathbf{X}_{Std}^{D}$. If the scale-based method identifies clusters whose centers have the following pattern:

$$\begin{pmatrix} (\bar{\mathbf{x}}'_1, 0), (\bar{\mathbf{x}}'_2, 0), ..., (\bar{\mathbf{x}}'_K, 0) \\ (\bar{\mathbf{x}}'_1, d), (\bar{\mathbf{x}}'_2, d), ..., (\bar{\mathbf{x}}'_K, d) \end{pmatrix} \tag{5.29}$$

the SBDD algorithm stops. Otherwise, increase d by a step size such as $\Delta d = 0.5$.

The readers should be notified that although scale-based methods (also the SBDD method) can handle slightly or moderate non-convex clusters, they are not capable of dealing with those extremely non-convex clusters such as one cluster is encircled by another one, as shown in Fig. 5.12. For this type of extremely non-convex clusters, SNN method may be a good choice.

Kothari and Pitts [21] and Zhang and Albin [33] give more details of the scale-based method and the SBDD method, respectively. Readers are also referred to some other variants of the scale-based method to determine the number of clusters, such as the multi-scale clustering [24], influence zones [14], SOM [4] and kernel density estimation [29].

References

[1] M. Anderberg. *Cluster analysis for applications*. Academic Press, New York, 1973.

[2] A. Borst and F. Theunissen. Information theory and neural coding. *Nature Neuroscience*, 2: 947–957, 1999.

[3] D. Brillinger. Nerve cell spike train data analysis: a progression of technique. *Journal of the American Statistical Association*, 87: 260–271, 1992.

[4] J. Costa and M. Netto. Estimating the number of clusters in multivariate data by self-organizing maps. *International Journal of Neural Systems*, 9(3): 195–202, 1999.

[5] M. Daszykowski, B. Walczak, and D. Massart. Looking for natural patterns in data - Part 1. Density-based approach. *Chemometrics and Intelligent Laboratory Systems*, 56: 83–92, 2001.

[6] M. Ester, H.-P. Kriegel, J. Sander, and X. Xu. A density-based algorithm for discovering clusters in large spatial databases with noise. In *Proc. of the 2nd Int. Conf. on Knowledge Discovery and Data Mining (KDD-96)*, pages 226–231, 1996.

[7] M. Ester, H.-P. Kriegel, J. Sander, and X. Xu. Density-based clustering in spatial databases: the algorithm GDBSCAN and its applications. *Data Mining and Knowledge Discovery, An International Journal*, 2(2): 169-194, 1998.

[8] L Fu and E. Medico. FLAME, a novel fuzzy clustering method for the analysis of DNA microarray data. *BMC Bioinformatics*, 8: 3, 2007.

[9] J. C. Bezdek. *Pattern Recogniztion with Fuzzy Objective Function Algorithms*. Plenum press, New York, 1981.

[10] C. Fraley and A. E. Raftery. How many clusters? Which clustering method? Answers via model-based cluster analysis. *Computer Journal*, 41: 578–588, 1998.

[11] S. Guha, R. Rastogi, and K. Shim. CURE: an efficient clustering algorithm for large databases. *Information Systems*, 26: 35-58, 2001.

[12] S. Guha, R. Rastogi, and K. Shim. ROCK: a robust clustering algorithm for categorical attributes. In *Proceedings of the 15th international conference on data engineering*, page 512, 1999.

[13] J. Han and M. Kamber. *Data Mining: Concepts and Techniques*. Morgan Kaufmann Publishers, 2005.

[14] M. Herbin, N. Bonnet, and P. Vautrot. Estimation of the number of clusters and influ-

ence zones. *Pattern Recognition Letters*, 22: 1557-1568, 2001.

[15] N. C. Jain, A. Indrayan, and L. R. Goel. Monte Carlo comparison of six hierarchical clustering on random data. *Pattern Recognition*, 19(1): 95–99, 1986.

[16] A. K. Jain, M. N. Murty, and P. J. Flynn. Data clustering: a review. *ACM Computing Surveys*, 31(3): 264–323, 1999.

[17] R. Johnson and D. Wichern. *Applied multivariate statistical analysis*. Prentice Hall, New Jersey, 1998.

[18] G. Karypis, E.-H. Han, and V. Kumar. CHAMELEON: A hierarchical clustering algorithm using dynamic modeling *Computer*, 32(8): 68–75, 1999.

[19] T. Kohonen. *Self-Organizing Maps*. Springer, New York, 2001.

[20] I. Kojadinovic. Agglomerative hierarchical clustering of continuous variables based on mutual information. *Comp. Stat. and Data Analysis*, 46; 269–294, 2004.

[21] R. Kothari and D. Pitts. On finding the number of clusters. *Pattern Recognition Letters*, 20: 405–416, 1999.

[22] A. Lukasova. Hierarchical agglomerative clustering procedure. *Pattern Recognition*, 11: 365–381, 1979.

[23] D. Montgomery and G. Runger. *Applied statistics and probability for engineers*. John Wiley and Sons, New Jersey, 2007.

[24] E. Nakamura and N. Kehtarnavaz. Determining number of clusters and prototype locations via multi-scale clustering. *Pattern Recognition Letters*, 19: 1265–1283, 1998.

[25] Y. Jin. Fuzzy modeling of high-dimensional systems: complexity reduction and interpretability improvement. *IEEE Transactions on Fuzzy Systems*, 8(2): 212-221, 2000.

[26] C. Priebe and D. Marchette. Adaptive mixture density estimation. *Pattern Recognition*, 26: 771-785, 1993.

[27] A. Strehl, J. Ghosh, and R. Mooney Impact of similarity measures on web-page clustering. In *AAAI-2000: Workshop of Artificial Intelligence for Web Search*, pages 58–64, 2000.

[28] H. Tao and T. Huang. Color image edge detection using cluster analysis. In *Proceedings of International Conference on Image Processing*, 1, pages 834–836, 1997.

[29] W.-J. Wang, Y.-X. Tan, J.-H. Jiang, J.-Z. Lu, G.-L. Shen, and R.-Q. Yu. Clustering based on kernel density estimation: nearest local maximum searching algorithm. *Chemometrics and Intelligent Laboratory Systems*, 72: 1–8, 2004.

[30] J. H. Ward, Jr. Hierarchical grouping to optimize an objective function. *Journal of the American Statistical Association*, 58: 236–244, 1963.

[31] C.-P. Wei, Y.-H. Lee, and C.-M. Hsu. Empirical comparison of fast clustering algorithms for large data sets. In *Proceedings of the 33rd Hawaii International Conference on System Sciences*, 2000.

[32] L. Xu and M. I. Jordan. On convergence properties of the EM algorithm for Gaussian mixtures. *Neural Computation*, 8(1): 129–151, 1996.

[33] H. Zhang and S. Albin. Determining the number of operational modes in baseline multivariate spc data. *IIE Transactions*, 39(12): 1103–1110, 2007.

PART 2

New Methods and Applications

Chapter 6

Diversity Graphs

P. Blain

Swarthmore College Mathematics, Swarthmore, PA, USA
pblain1@swarthmore.edu

C. Davis

University of Utah Mathematics, Salt Lake, UT, USA
davis@math.utah.edu

A. Holder

Rose-Hulman Institute of Technology, Mathematics, Terre Haute, IN, USA
aholder@rose-hulman.edu *

J. Silva

University of Colorado Applied Mathematics, Denver, CO, USA
jsilva2105@msn.com

C. Vinzant

Oberlin College Mathematics, Oberlin, OH, USA
cvinzant@oberlin.edu

Bipartite graphs have long been used to study and model matching problems, and in this paper we introduce the bipartite graphs that explain a recent matching problem in computational biology. The problem is to match haplotypes to genotypes in a way that minimizes the number of haplotypes, a problem called the Pure Parsimony problem. The goal of this work is not to address the computational or biological issues but rather to explore the mathematical structure through a study of the underlying graph theory.

---

*Research conducted at Trinity University, San Antonio, TX, with partial support of the National Science Foundation, grant DMS-0353488.

6.1. Introduction

The burgeoning field of computational biology is advancing the science of genetics and transforming traditional 'wet lab' research into computational efforts. The preponderance of the current research emphasizes computational aspects, which have made significant strides in projects such as the human genome project. These advances have the potential of redefining standard medical practice and have already proven to be a significant contribution to mankind.

One of the problems currently receiving attention is that of describing how genetic diversity propagates from one generation to the next. Such problems are called *haplotyping* problems, and a brief biological description is warranted. Genes are sequences of DNA that code for specific traits, with the vast majority of DNA being common among all individuals. The locations on the genome where diversity occurs are called single nucleotide polymorphisms (SNPs). Diploid organisms like humans have two distinct copies of each gene, one from each parent, which together describe a trait. A collection of SNPs in a single copy of a gene is called a *haplotype*, and a pair of haplotypes forms a *genotype*. Each SNP of a haplotype is in one of two states, denoted by -1 or 1, that corresponds to the two distinct nucleotide base pairs of the DNA. Each SNP of a genotype is in one of three states, -2, 0, or 2, where the SNP is -2 (resp, 2) if and only if each of the haplotypes that form the genotype have a -1 (resp. 1) at that SNP, and the SNP is 0 if and only if one of the haplotypes has a -1 and the other a 1 at that SNP.

Biologists are capable of efficiently determining an individual's genotype, but it is difficult and costly to determine the haplotypes. However, haplotypes are more valuable to biologists, and a haplotyping problem is to calculate the haplotypes knowing only the genotypes. In particular, finding small collections of haplotypes that explain the genotypes is biologically relevant. The problem of finding a smallest collection of haplotypes is called the *Pure Parsimony* problem and empirical evidence suggests that these minimum solutions naturally occur. The initial investigations into haplotyping were undertaken by Clark in [5], and since then there has been flourish of activity addressing computational issues [2–4, 6–19]. Theoretically we know that the parsimony problem is APX-hard and that practically it is difficult to solve on large data sets. Our goal is not directly computational, and instead we address the underlying structure of the problem through graph theory. We do provide closed form solutions in a some instances.

6.2. Notation, Definitions and Preliminary Results

The results of this paper are graph theoretical, and a basic understanding of bipartite graph theory is expected. We point readers to [1] for a thorough development.

The degree and neighborhood of node v are denoted by $\deg(v)$ and $N(v)$, respectively. The vector of ones is denoted by e, where length is decided by the context of its use. If x is a vector, then diag(x) is the symmetric matrix whose diagonal elements correspond to x and whose off-diagonal elements are zero. So, diag(e) is the identity matrix. For any real number C we define $C_+ = \max\{C, 0\}$.

We assume that haplotypes are of length n and that SNPs are indexed by $i = 1, 2, \ldots, n$. The set of all possible haplotypes of length n is the collection of sequences $\mathcal{H} = \{-1, 1\}^n$. Similarly, the collection of all genotypes of length n is $\{-2, 0, 2\}^n$. The arithmetic of *mating* haplotypes to form a genotype is simply coordinate-wise addition. So, if the maternal haplotype is $(-1, 1, -1, 1)$ and the fraternal haplotype is $(1, 1, -1, -1)$, the genotype is

$$(-1, 1, -1, 1) + (1, 1, -1, -1) = (0, 2, -2, 0). \tag{6.1}$$

We extend this coordinatewise addition so that we can generally add elements in $\{-2, -1, 0, 1, 2\}^n$. Let $a, b \in \{-2, -1, 0, 1, 2\}$ and define $\oplus$ so that

$$a \oplus b = \begin{cases} -2, & a < 0, b < 0 \\ 2, & a > 0, b > 0 \\ 0, & \text{otherwise.} \end{cases}$$

This binary operator is commutative, and to add elements of $\{-2, -1, 0, 1, 2\}^n$ we perform the operation componentwise. For example,

$$(-1, 1, -1, 1) \oplus (1, 0, -2, 1) \oplus (1, -1, -1, 1) = (0, 0, -2, 2).$$

Notice that $\oplus$ reduces to the typical addition in (6.1) if the terms on the left are in $\{-1, 1\}^4$. Unfortunately, $\oplus$ does not generally satisfy a cancellation rule since $a \oplus 0 = b \oplus 0$ does not mean that $a = b$. For much of the paper simple addition as described in (6.1) is sufficient, and to distinguish $\oplus$ from $+$ we use $+$ in all instances where it appropriate.

SNP values of 0 are called *ambiguous* because the orientation of the 1 and -1 in the parental donations could be reversed. The question we address begins with a collection of genotypes and asks us to construct a collection of haplotypes that form the genotypes under this arithmetic. If there were no ambiguous SNPs, this process would be trivial, and hence, we assume that each genotype has at least one ambiguous SNP. The subset of $\{-2, 0, 2\}^n$ with this property is denoted $\mathcal{G}$. For any $\mathcal{G}' \subseteq \mathcal{G}$, we say $\mathcal{H}' \subseteq \mathcal{H}$ is a *solution* to $\mathcal{G}'$ if, for all $\mathbf{g} \in \mathcal{G}'$, there exist $\mathbf{h}', \mathbf{h}'' \in \mathcal{H}$ such that $\mathbf{g} = \mathbf{h}' + \mathbf{h}''$. A solution $\mathcal{H}$ to $\mathcal{G}'$ is *minimal* if $\mathcal{H} \backslash \{\mathbf{h}\}$ is not a solution to $\mathcal{G}'$, for all $\mathbf{h} \in \mathcal{H}$. We say $\mathcal{H}$ is a *minimum solution* if there exists no solution $\mathcal{H}'$ to $\mathcal{G}'$ such that $|\mathcal{H}'| < |\mathcal{H}|$.

Our intent is to study the underlying graph theory of finding solutions, and we introduce the concept of a *Diversity Graph*. Informally, a diversity graph is a labeled (or colored) bipartite graph with one set of nodes representing genotypes, the other set representing haplotypes, and edges representing the possible relationships between them.

Definition 6.1. For $\mathcal{H}' \subset \mathcal{H}$ and $\mathcal{G}' \subset \mathcal{G}$, a bipartite graph $(\mathcal{V}, \mathcal{W}, \mathcal{E})$, and functions $\eta : \mathcal{V} \to \mathcal{H}'$ and $\gamma : \mathcal{W} \to \mathcal{G}'$, we say $(\mathcal{V}, \mathcal{W}, \mathcal{E}, \eta, \gamma)$ is a *diversity graph* on n SNPs if

(1) η and γ are one-to-one,
(2) for each $w \in \mathcal{W}$, there exists some $v \in \mathcal{V}$ such that $(v, w) \in \mathcal{E}$, and
(3) $\mathcal{E}$ has the property that if $(v', w) \in \mathcal{E}$, there exists some $v'' \in \mathcal{V}\backslash\{v'\}$ such that $(v'', w) \in \mathcal{E}$ and $\mathbf{h}' + \mathbf{h}'' = \mathbf{g}$, where $\mathbf{h}' = \eta(v')$, $\mathbf{h}'' = \eta(v'')$, and $\mathbf{g} = \gamma(w)$.

The requirement that η and γ be one-to-one ensures that each haplotype and genotype are represented by exactly one node. The rest of the definition guarantees that $\mathcal{H}'$ is a solution to $\mathcal{G}'$. If $\eta(v') + \eta(v'') = \gamma(w)$, we say that v' and v'' or $\eta(v')$ and $\eta(v'')$ are *mates* for w or $\gamma(w)$. Notice that a diversity graph is a labeled bipartite graph, and we make the distinction between the structure represented by the graph and the biology represented by the labeling. The elements of $\mathcal{H}$ and $\mathcal{G}$ are denoted by $\mathbf{h}$ and $\mathbf{g}$ or by $\eta(v)$ and $\gamma(w)$, where v and w are elements of $\mathcal{V}$ and $\mathcal{W}$. Different elements of $\mathcal{H}$ and $\mathcal{G}$ are indicated with superscripts and SNP locations are indicated with subscripts. We say that a bipartite graph *supports diversity* if there are sets $\mathcal{H}' \subseteq \mathcal{H}$ and $\mathcal{G}' \subseteq \mathcal{G}$, and functions η and γ that fulfill the definition.

An important observation is that the pure parsimony problem as stated assumes that $\mathcal{G}'$ is known, that $\eta(V) = \mathcal{H}$ and that $\mathcal{E}$ is as large as possible. However, the parsimony problem makes sense on other graphs, and in general we address the problem of starting with a diversity graph $(\mathcal{V}, \mathcal{W}, \mathcal{E}, \eta, \gamma)$ and finding a smallest subset of $\mathcal{V}$, say $\mathcal{V}'$, such that

- $\eta(\mathcal{V}')$ is a solution to $\gamma(\mathcal{W}) = \mathcal{G}'$, and
- if v' and v'' are in $\mathcal{V}'$, then they are allowed to mate and form w if and only if (v', w) and (v'', w) are in $\mathcal{E}$.

So, when we say that $\mathcal{H}$ is a minimal or minimum solution we mean that it is a solution with respect to a diversity graph. If $\mathcal{V}$ and η are such that $\eta(\mathcal{V}) = \mathcal{H}$ and $\mathcal{E}$ is as large as possible, then we are considering the original parsimony problem.

Before continuing with an investigation into the bipartite graphs that support diversity, we establish some general results about diversity graphs. The first of these results shows how to order the elements of $\mathcal{H}$ so that we can conveniently pair them to form the genotype $(0, 0, \ldots, 0)$. The imposed ordering is *lexico-*

graphic, meaning that $(\mathbf{h}_1, \mathbf{h}_2, \ldots, \mathbf{h}_n) < (\mathbf{h}'_1, \mathbf{h}'_2, \ldots, \mathbf{h}'_n)$ if the first component with different values satisfies $\mathbf{h}_i < \mathbf{h}'_i$. The proof of this lemma is simple and omitted.

Lemma 6.1. *If the elements of $\mathcal{H}$ are ordered lexicographically, then for unique i and j between 1 and 2^n we have that $h^i + h^{i+1} \neq h^j + h^{j+1}$ and that $h^j + h^{(2^n-j+1)} = (0, 0, \ldots, 0)$.*

Since haplotypes mate in unique pairs to form a genotype, the degree of every node in $\mathcal{V}$ is even. An immediate consequence of this observation is that not every bipartite graph supports diversity. Moreover, even if every node of a bipartite graph has even degree it does not mean the graph supports diversity. As an example, the complete bipartite graph $K_{2,2}$ does not support diversity because any η and γ that satisfies the third and fourth conditions of the definition violates the fact that γ is one-to-one. Theorem 6.1 does not characterize the graphs that support diversity, but it does establish a necessary condition.

Theorem 6.1. *Let $(\mathcal{V}, \mathcal{W}, \mathcal{E}, \eta, \gamma)$ be a diversity graph. Let w^1 and w^2 be in $\mathcal{W}$, with $w^1 \neq w^2$. Then,*

$$\max\{\deg(w^1), \deg(w^2)\} \geq 2|N(w^1) \cap N(w^2)|.$$

Proof. Let $\mathcal{H}' = \{\eta(v) : v \in N(w^1) \cap N(w^2)\}$ and assume to the contrary that

$$\max\{\deg(w^1), \deg(w^2)\} < 2|\mathcal{H}'|.$$

Since haplotypes mate in unique pairs, there must be fewer than $|\mathcal{H}'|$ pairs of haplotypes mating to form each of $\gamma(w^1) = \mathbf{g}^1$ and $\gamma(w^2) = \mathbf{g}^2$. It follows that there exists $\mathbf{h}^1, \mathbf{h}^2, \mathbf{h}^3$ and $\mathbf{h}^4$ in $\mathcal{H}'$ such that $\mathbf{h}^1 + \mathbf{h}^2 = \mathbf{g}^1$ and $\mathbf{h}^3 + \mathbf{h}^4 = \mathbf{g}^2$. Suppose that $\mathbf{g}^1_j = 2$, which means $\mathbf{h}_j = 1$ for all $\mathbf{h} \in \mathcal{H}'$. It follows that $\mathbf{g}^2_j = 2$. Similarly, if $\mathbf{g}^1_j = -2$, we have that $\mathbf{g}^2_j = -2$. Suppose that $\mathbf{g}^1_j = 0$. Then $\mathbf{g}^2_j$ can not be 2 or -2 because if so, the same argument would guarantee that $\mathbf{g}^1_j$ is 2 or -2, respectively. We conclude that $\mathbf{g}^2_j = 0$. Since j was arbitrary, we have the contradiction that $\mathbf{g}^1 = \mathbf{g}^2$. □

We now turn our direction to a matrix equation that is satisfied by every diversity graph. Consider the diversity graph $(\mathcal{V}, \mathcal{W}, \mathcal{E}, \eta, \gamma)$, where $|\mathcal{V}| = m$ and $|\mathcal{W}| = k$. List the elements of $\mathcal{V}$ as $v^1, v^2, \ldots, v^m$ and $\mathcal{W}$ as $w^1, w^2, \ldots, w^k$, and assume that $\eta(v^i) = \mathbf{h}^i$ and $\gamma(w^i) = \mathbf{g}^i$. Let H be the $m \times n$ matrix so that $H_{(i,j)} = h^i_j$, and let G be the $k \times n$ matrix defined by $G_{(i,j)} = g^i_j$. Also let E be the $m \times k$ biadjacency matrix —i.e. $E_{(i,j)} = 1$ if $(v^i, w^j) \in \mathcal{E}$ and $E_{(i,j)} = 0$ otherwise. Note that the column sums of E must be even from the definition of a diversity graph.

The $k \times n$ matrix product $E^T H$ aggregates the mating structure for each genotype. Without loss of generality, let $E_{(1,i)} = E_{(2,i)} = \ldots = E_{(t,i)} = 1$ and $E_{(t+1,i)} = E_{(t+2,i)} = \ldots = E_{(m,i)} = 0$. Then the ith row of $E^T H$ is $\eta(v^1) + \eta(v^2) + \ldots + \eta(v^t)$. From the definition of a diversity graph we know there are $t/2$ disjoint pairs, (v^p, v^q), with p and q no greater than t, such that $\eta(v^p) + \eta(v^q) = \gamma(w^i)$. This means that the ith row of $E^T H$ is $(t/2)\gamma(w^i)$. We have just established the following result.

Theorem 6.2. *If $(\mathcal{V}, \mathcal{W}, \mathcal{E}, \eta, \gamma)$ is a diversity graph, then $E^T H = \text{diag}\left(\frac{1}{2}E^T e\right) G$.*

The matrix equation in Theorem 6.2 succinctly separates the structure of the graph, explained by E, from the labeling of the graph, explained by H and G. Unfortunately, satisfying the matrix equation does not guarantee the graph is a diversity graph because the aggregated information ignores the need of a mating structure. As an example

$$\begin{aligned} E^T H &= (1\ 1\ 1\ 1) \begin{pmatrix} 1 & 1 & -1 & -1 \\ -1 & 1 & 1 & -1 \\ -1 & -1 & -1 & 1 \\ 1 & -1 & 1 & 1 \end{pmatrix} \\ &= (2)\,(0\ 0\ 0\ 0) \\ &= \text{diag}\left(\frac{1}{2}E^T e\right) G. \end{aligned}$$

This labeling of $K_{4,1}$ does not lead to a diversity graph since no pair of haplotypes (no two rows of H) add to form the single genotype (the row of G).

We conclude this section with a discussion of a logical operator that helps address the failure of Theorem 6.2 to characterize graphs with the stated matrix equation. The *logical join* of a sequence of matrices is determined by the logical operator "or" over each component of these matrices. The component-wise logical join is defined so that $0 \vee 0 = 0$, $0 \vee 1 = 1$, and $1 \vee 1 = 1$. The set $\{A^1, A^2, \ldots, A^s\}$ is a *logical decomposition* of A if A is the logical join of the matrices in this set, denoted:

$$\bigvee_{1 \le i \le s} A^i = A^1 \vee A^2 \vee \ldots \vee A^s = A,$$

where we assume that all matrix elements are 0 or 1. For example, the matrices on the left are a logical decomposition of the matrix on the right,

$$\begin{pmatrix} 1\,1\,0\,0 \\ 1\,0\,1\,0 \end{pmatrix} \vee \begin{pmatrix} 1\,1\,0\,0 \\ 0\,1\,0\,1 \end{pmatrix} = \begin{pmatrix} 1\,1\,0\,0 \\ 1\,1\,1\,1 \end{pmatrix}.$$

Such decompositions are used in the next section to characterize the graphs that support diversity.

6.3. Graphs That Support Diversity

A natural goal is to characterize the bipartite graphs that support diversity, and the first result of this section does this for complete bipartite graphs.

Theorem 6.3. *The complete graph $K_{p,q}$ supports diversity if and only if p is even and $q = 1$.*

Proof. Assume that $(\mathcal{V}, \mathcal{W}, \mathcal{E}, \eta, \gamma)$ is a complete diversity graph. Suppose that $|\mathcal{W}| > 1$. Then, from Theorem 6.1 we have for any w^1 and w^2 in $\mathcal{W}$ that

$$\max\{\deg(w^1), \deg(w^2)\} \geq 2|N(w^1) \cap N(w^2)| = 2|\mathcal{V}|,$$

which is a contradiction. So, $|\mathcal{W}| = 1$. The fact that genotypes need pairs of haplotypes guarantees that $|\mathcal{V}|$ is even.

Now assume that p is even and $q = 1$. Select n so that $|\mathcal{V}| < 2^n$ and let γ be such that $\gamma(w) = (0, 0, \ldots, 0)$. There are 2^{n-1} disjoint pairs of haplotypes that can mate to form $\gamma(w)$. Pick $|\mathcal{V}|/2$ of these pairs and let $\mathcal{H}'$ be the set of haplotypes in these pairs. Allowing $\eta : \mathcal{V} \to \mathcal{H}'$ to be a bijection, we see that $(\mathcal{V}, \mathcal{W}, \mathcal{E}, \eta, \gamma)$ is a diversity graph. □

From Theorem 6.3 we see that the probability of generating a complete bipartite graph of the form $K_{p,1}$ that supports diversity is one half (assuming that even and odd values of p are equally likely). In some ways, the next result extends this idea by showing that the probability of generating a random bipartite graph that supports diversity is low. The result decomposes the biadjacency matrix into matrices whose rows sums are all 2, which guarantees a mating structure.

Theorem 6.4. *The bipartite graph $(\mathcal{V}, \mathcal{W}, \mathcal{E})$ supports diversity if and only if the biadjacency matrix E has a logical decomposition $E^1, E^2, \ldots, E^s$ so that*

- $e^T E^k = 2e^T$ *for all* $1 \leq k \leq s$,
- *there exists an* $H \in \{-1, 1\}^{|\mathcal{V}| \times n}$ *with distinct rows and the property that* $(E^1)^T H = (E^2)^T H = \ldots = (E^s)^T H$, *and*
- *the rows of* $(E^k)^T H$ *are distinct.*

Proof. Assume that $(\mathcal{V}, \mathcal{W}, \mathcal{E}, \eta, \gamma)$ is a diversity graph, and let $s = (1/2)\max\{\deg(w) : w \in \mathcal{W}\}$. Order the elements of $\mathcal{W}$ so that $\deg(w^1) \geq \deg(w^2) \geq \ldots \geq \deg(w^{|\mathcal{W}|})$. We construct the matrices $E^1, E^2, \ldots, E^s$ that form a desired logical decomposition. The neighborhood of each w^j can be written as the disjoint union of $(1/2)\deg(w^j)$ pairs,

$$N(w^j) = \bigcup_k \{v^{k_{j'}}, v^{k_{j''}}\}, \tag{6.2}$$

where $1 \leq k \leq (1/2)\deg(w^j)$. Let every element of the first column of each E^k be a zero except for the $k_{1'}$ and $k_{1''}$ positions, which are both set to 1. If the $\deg(w^1) = \deg(w^2)$, form the second column of each E^k similarly, replacing $1'$ and $1''$ with $2'$ and $2''$. Otherwise, $\deg(w^1) > \deg(w^2)$ and this construction terminates once k reaches $(1/2)\deg(w^2)$. In this case, let the second column of E^k, for $(1/2)\deg(w^2) + 1 \leq k \leq s$, be the same as the second column of E^1. Continue in this fashion through the remaining nodes in $\mathcal{W}$, duplicating columns from E^1 as needed. From (6.2) we have that $E_{(i,j)} = 1$ if and only if $E^k_{(i,j)} = 1$ for at least one k, from which we conclude that $E = E^1 \vee E^2 \vee \ldots \vee E^s$.

Let H and G be the matrices in Theorem 6.2. Each column of E^k corresponds to an element of $\mathcal{W}$ that in turn corresponds to a genotype under γ. Moreover, each column of E^k contains two 1s that identify a pair of haplotypes (via η) that mate to form the genotype. So, each column sum of every E^k sums to 2 and $(E^k)^T H = G$, for every k. From the fact that γ is one-to-one we have that the rows of $(E^k)^T H$ are distinct.

Now assume that E has a logical decomposition, say $E^1, E^2, \ldots E^s$, that satisfies the three conditions. List the elements of $\mathcal{V}$ from 1 to $|\mathcal{V}|$ and define $\eta(v^i)$ to be the ith row of H. The fact that the rows of H are unique ensures that η is one-to-one. Similarly, list the nodes in $\mathcal{W}$ from 1 to $|\mathcal{W}|$ and let $\gamma(w^i)$ be the ith row of $(E^k)^T H$, which is common for $1 \leq k \leq s$. The assumption that the rows of $(E^k)^T H$ are distinct guarantees that γ is one-to-one. From the condition that each column sum of E^k is 2, we have that each column of E has at least two ones. This means that there are at least two elements of V that are adjacent to each element of W. The same condition together with the definition of η and γ further guarantee that if $(v', w) \in \mathcal{E}$, then there is a v'' so that $(v'', w) \in \mathcal{E}$ and $\eta(v') + \eta(v'') = \gamma(w)$. □

The logical decomposition in Theorem 6.4 extends Theorem 6.2 to characterize graphs that support diversity by adding the necessary condition that the edge structure must accommodate a mating structure. The fact that $(E^1)^T H = (E^2)^T H = \ldots = (E^s)^T H$ shows that every pairwise differences $(E^i)^T - (E^j)^T$ must share a non-trivial null space, which is restrictive and demonstrates that bipartite graphs that support diversity are rare.

The last two results indicate that the structural requirements needed to support diversity are important and that most bipartite graphs can not be labeled to represent a population. We point out that this is true even with the number of SNPs being arbitrary, which is somewhat counter intuitive because the complexity of a mating scheme can increase as the number of SNPs grows. We conclude this section by showing that we can add nodes and edges to any bipartite graph so that it does support diversity. We only consider adding nodes to $\mathcal{V}$ since in real problems the genotypic information corresponding to $\mathcal{W}$ is defined by the biological data.

For $w \in W$, define

$$T(w) = \bigcup_{w' \neq w} (N(w) \cap N(w')).$$

So, $T(w)$ is the collection of nodes in the neighborhood of w that are also in the neighborhood of another node in W. We extend the neighborhood of each w so that the number of points in $N(w) \backslash T(w)$ plus the number of points in the extension is at least the number of points in $T(w)$. Let $\hat{\mathcal{V}}(w)$ be a collection of nodes whose cardinality is either $(2|T(w)| - |N(w)|)_+$ or $1 + (2|T(w)| - |N(w)|)_+$ to ensure that $|N(w) \cup \hat{\mathcal{V}}(w)|$ is even or 0. Notice that if $2|T(w)| \leq |N(w)|$, then $N(w)$ is not extended. The extended vertex and edge sets are

$$\bar{\mathcal{V}} = \mathcal{V} \cup \left(\bigcup_{w \in W} \hat{\mathcal{V}}(w) \right) \text{ and } \bar{\mathcal{E}} = \mathcal{E} \cup \left(\bigcup_{w \in \mathcal{W}} \{(v, w) : v \in \hat{\mathcal{V}}(w)\} \right).$$

Lemma 6.2. *Any bipartite graph $(\mathcal{V}, \mathcal{W}, \mathcal{E})$ with no isolated nodes can be extended and labeled to become a diversity graph by adding no more than $\sum_{w \in \mathcal{W}} |\hat{\mathcal{V}}(w)|$ nodes to $\mathcal{V}$. In particular, $(\bar{\mathcal{V}}, \mathcal{W}, \bar{\mathcal{E}})$ is an extension of $(\mathcal{V}, \mathcal{W}, \mathcal{E})$ that adds this number of nodes to $\mathcal{V}$ that supports diversity.*

Proof. The proof follows by induction on $|\mathcal{W}|$. Let $(\mathcal{V}, \mathcal{W}, \mathcal{E})$ be a bipartite graph with no isolated nodes such that $|\mathcal{W}| = 1$. Let $W = \{w\}$ and notice that $T(w) = \emptyset$. Hence, $(2|T(w)| - |N(w)|)_+ = 0$, and we add a single node to $\mathcal{V}$ if and only if $|\mathcal{V}|$ is odd. The resulting $\bar{\mathcal{V}}$ has an even number of nodes, and from Theorem 6.3 we know that that this graph supports diversity.

Assume the result holds if $|\mathcal{W}| \leq k$. Let $(\mathcal{V}, \mathcal{W}, \mathcal{E})$ be a bipartite graph with no isolated nodes such that $|\mathcal{W}| = k + 1$. Select $w^1 \in W$, and let $(\mathcal{V}', \mathcal{W}', \mathcal{E}')$ be the subgraph of $(\mathcal{V}, \mathcal{W}, \mathcal{E})$ with the vertices in $\{w^1\} \cup N(w^1) \backslash T(w^1)$ and edges incident to w^1 removed. Extend the subgraph so that $(\bar{\mathcal{V}}', \mathcal{W}', \bar{\mathcal{E}}', \eta', \gamma')$ is a diversity graph, where the image sets of η' and γ' are in $\{-1, 1\}^{n'}$ and $\{-2, 0, 2\}^{n'}$, respectively. The functional descriptions of η and γ below depend on η' and γ', and a slight abuse of notation is used to describe this dependence. As an example, if $\eta'(v) = (1, -1, 1)$, we assume that $(\eta'(v), 1, 1, 1) = (1, -1, 1, 1, 1, 1)$, which allows us to embed η' into a larger collection of haplotypes.

The argument is established in 2 cases, each of which constructs η and γ so that $(\bar{\mathcal{V}}, \mathcal{W}, \bar{\mathcal{E}}, \eta, \gamma)$ is a diversity graph. Notice that the number of nodes added to $\mathcal{V}$ is additive over $\mathcal{W}$, and hence,

$$\sum_{w \in \mathcal{W}} |\hat{\mathcal{V}}(w)| = |\hat{\mathcal{V}}(w^1)| + \sum_{w \in \mathcal{W}'} |\hat{\mathcal{V}}(w)|.$$

This fact guarantees that the constructions below add the maximum number of vertices allowed by the result.

Case 1: Suppose that $T(w^1) = \emptyset$. Then, $(2|T(w^1)| - |N(w^1)|)_+ = 0$, and $|\hat{\mathcal{V}}(w^1)|$ is either 0 or 1 depending on whether $|N(w^1)|$ is even or odd, respectively. If $|N(w^1)|$ is even (odd), then no nodes are (a single node is) added to $(\mathcal{V}', \mathcal{W}, \mathcal{E}')$. Let $|N(w^1) \cup \hat{\mathcal{V}}(w^1)| = 2p$ for the natural number p. Let k be a natural number so that $2^k > 2p$. List the elements of $\{-1, 1\}^k$ lexicographically as $\mathbf{h}^1, \mathbf{h}^2, \ldots, \mathbf{h}^{2^k}$. Then, denoting the elements of $N(w^1) \cup \hat{\mathcal{V}}(w^1)$ as v^i for $i = 1, 2, \ldots, 2p$, we define η and γ by

$$\eta : \bar{\mathcal{V}} \to \{-1, 1\}^{n'+k} : v \mapsto \begin{cases} (\eta'(v), 1, 1, \ldots, 1), & v \in \mathcal{V}' \\ (1, 1, \ldots, 1, \mathbf{h}^{i+1}), & v = v^i, \\ & i = 1, 2, \ldots p \\ (1, 1, \ldots, 1, \mathbf{h}^{2^k - i + p}), & v = v^i, \\ & i = p + 1, \ldots 2p \end{cases}$$

and

$$\gamma : \mathcal{W} \to \{-2, 0, 2\}^{n'+k} : w \mapsto \begin{cases} (\gamma'(v), 2, 2, \ldots, 2), & w \in \mathcal{W}' \\ (2, 2, \ldots 2, 0, 0, \ldots 0), & w = w^1, \end{cases}$$

where $\gamma(w^1)$ has k zeros. We mention that Lemma 6.1 is used to guarantee that the last k elements of $\eta(v^i)$, for $i = 1, 2, \ldots, 2p$, can be paired to satisfy the definition of a diversity graph.

Case 2: Suppose $T(w^1) \neq \emptyset$. The difficulty with this case lies in the fact that $\eta'(v)$ is defined for $v \in T(w^1)$. Notice that

$$\begin{aligned} &(|N(w^1)| - 2|T(w^1)|) + (2|T(w^1)| - |N(w^1)|)_+ \\ &= (|N(w^1)| - 2|T(w^1)|)_+ \geq 0, \end{aligned}$$

which guarantees that there are enough nodes in $(N(w^1) \cup \hat{\mathcal{V}}(w^1)) \backslash T(w^1)$ to be uniquely paired with the nodes in $T(w^1)$. Let $\{Z, Z^C\}$ be a two set partition of $(N(w^1) \cup \hat{\mathcal{V}}(w^1)) \backslash T(w^1)$ so that $|Z| = |T(w^1)|$. Notice that the definition of $\hat{\mathcal{V}}(w^1)$ guarantees that both $|T(w^1) \cup Z|$ and $|Z^C|$ are even.

List the elements of $T(w^1)$, Z and Z^C so that

$$T(w^1) = \{v^1, v^2, \ldots, v^{|T(w^1)|}\}, \tag{6.3}$$

$$Z = \{v^{|T(w^1)|+1}, v^{|T(w^1)|+2}, \ldots, v^{2|T(w^1)|}\}, \text{ and} \tag{6.4}$$

$$Z^C = \{v^{2|T(w^1)|+1}, v^{2|T(w^1)|+2}, \ldots, v^{|\hat{\mathcal{V}}(w^1)|}\}. \tag{6.5}$$

Let $|N(w^1) \cup \hat{\mathcal{V}}(w^1)| = 2p$ for the natural number p and let k be such that $2^k > p$. Label $\{-1, 1\}^k$ lexicographically as $\mathbf{h}^1, \mathbf{h}^2, \ldots, \mathbf{h}^{2^k}$. Define $\eta : \bar{\mathcal{V}} \to \{-1, 1\}^{n'+k}$ so that

$$v \mapsto \begin{cases} (\eta'(v), 1, 1, \ldots, 1), & v \in \mathcal{V}' \backslash T(w^1) \\ (\eta'(v^i), \mathbf{h}^{i+1}), & v = v^i, 1 \le i \le |T(w^1)| \\ (-\eta'(v^{i-|T(w^1)|}), \mathbf{h}^{2^k - i + |T(w^1)|}), & v = v^i, \\ & |T(w^1)| + 1 \le i \le 2|T(w^1)| \\ (\eta'(v^i), \mathbf{h}^{i-|T(w^1)|}), & v = v^i, \\ & 2|T(w^1)| + 1 \le i \le |\hat{\mathcal{V}}(w^1)|, \\ & i \text{ odd} \\ (-\eta'(v^{i-1}), \mathbf{h}^{2^k - i + 2|T(w^1)|}), & v = v^i, \\ & 2|T(w^1)| + 1 \le i \le |\hat{\mathcal{V}}(w^1)|, \\ & i \text{ even} \end{cases}$$

and $\gamma : \mathcal{W} \to \{-2, 0, 2\}^{n'+k}$ so that

$$w \mapsto \begin{cases} (\gamma'(w), 2, 2, \ldots, 2), & w \in \mathcal{W}' \\ (0, 0, \ldots, 0), & w = w^1. \end{cases}$$

□

We conclude this section by showing that any bipartite graph, including those with isolated nodes, can be extended and labeled to become a diversity graph. The result is an extension of Lemma 6.2, but the edges added between isolated nodes are handled outside the definition of $\bar{\mathcal{E}}$.

Theorem 6.5. *Any bipartite graph $(\mathcal{V}, \mathcal{W}, \mathcal{E})$ can be extended and labeled to become a diversity graph by adding no more than*

$$\sum_{w \in \mathcal{W}} |\hat{\mathcal{V}}(w)| + (2M_{\mathcal{W}} - M_{\mathcal{V}})_+ + M_{\mathcal{V}} (\bmod 2)$$

nodes to $\mathcal{V}$, where $M_{\mathcal{V}}$ and $M_{\mathcal{W}}$ are the number of isolated nodes in $\mathcal{V}$ and $\mathcal{W}$, respectively.

Proof. Let $\mathcal{V}_I$ and $\mathcal{W}_I$ be the isolated nodes in $\mathcal{V}$ and $\mathcal{W}$ and let $(\mathcal{V}', \mathcal{W}', \mathcal{E}')$ be the subgraph of $(\mathcal{V}, \mathcal{W}, \mathcal{E})$ with these nodes removed. Extend $(\mathcal{V}', \mathcal{W}', \mathcal{E}')$ as in Lemma6.2 so that $(\bar{\mathcal{V}}', \mathcal{W}', \bar{\mathcal{E}}', \eta', \gamma')$ is a diversity graph. Since $N(w) = T(w) = \emptyset$ for all $w \in \mathcal{W}_I$, we have that

$$\sum_{w \in \mathcal{W}'} |\hat{\mathcal{V}}(w)| = \sum_{w \in \mathcal{W}} |\hat{\mathcal{V}}(w)|,$$

and we conclude that the extension of $(\mathcal{V}', \mathcal{W}', \mathcal{E}')$ to $(\bar{\mathcal{V}}', \mathcal{W}', \mathcal{E}')$ adds $\sum_{w \in \mathcal{W}} |\hat{\mathcal{V}}(w)|$ nodes to $(\mathcal{V}, \mathcal{W}, \mathcal{E})$. Let n' be such that the images of η' and γ' are in $\{-1, 1\}^{n'}$ and $\{-2, 0, 2\}^{n'}$. Let k be a natural number so that $2^k > |\mathcal{W}_I|$.

Let $\hat{\mathcal{V}}_I$ be a set of nodes of size $(2M_{\mathcal{W}} - M_{\mathcal{V}})_+ + M_{\mathcal{V}}(\text{mod } 2)$, which guarantees that $|\mathcal{V}_I \cup \hat{\mathcal{V}}_I|$ is even and at least twice the size of $\mathcal{W}_I$. List the elements in $\mathcal{W}_I$ as $w^1, w^2, \ldots, w^{M_W}$ and the elements in $\mathcal{V}_I \cup \hat{\mathcal{V}}_I$ as $v^1, v^2, \ldots, v^q$, where $q = M_V + (2M_{\mathcal{W}} - M_{\mathcal{V}})_+ + M_{\mathcal{V}}(\text{mod } 2)$. For $i = 1, 2, \ldots, M_W$, add (v^{2i-1}, w^i) and (v^{2i}, w^i) to $\bar{\mathcal{E}}'$. Notice that this may leave some isolated nodes in $\mathcal{V}_I$, which is allowed by the definition. Let $\mathbf{h}^1, \mathbf{h}^2, \ldots, \mathbf{h}^{2^k}$ be a lexicographic ordering of $\{-1, 1\}^k$. Define $\eta : \bar{\mathcal{V}}' \cup \mathcal{V}_I \cup \hat{\mathcal{V}}_I \to \{-1, 1\}^{n'+k}$ by

$$v \mapsto \begin{cases} (\eta'(v), 1, 1, \ldots, 1), & v \in \bar{\mathcal{V}}' \\ (1, 1, \ldots, 1, \mathbf{h}^{i+1}), & v = v^i, i = 1, 2, \ldots, q/2 \\ (1, 1, \ldots, 1, \mathbf{h}^{2^k - i + (q/2)}), & v = v^i, i = q/2 + 1, q/2 + 2, \ldots, q \end{cases}$$

and $\gamma : \mathcal{W} \to \{-2, 0, 2\}^{n'+k}$ so that

$$w \mapsto \begin{cases} (\gamma', 2, 2, \ldots, 2), & w \in \mathcal{W}' \\ (2, 2, \ldots, 2, \eta(v^{2i-1}) + \eta(v^{2i})), & w = w^i, i = 1, 2, \ldots, M_W \in \mathcal{W}_I, \end{cases}$$

where the uniqueness of $\eta(v^{2i-1}) + \eta(v^{2i})$ follows from Lemma 6.1. □

6.4. Algorithms and Solutions for the Pure Parsimony Problem

Although the pure parsimony is generally difficult, there are cases where a closed form solution exists. Throughout this section we assume that $(\mathcal{V}, \mathcal{W}, \mathcal{E}, \eta, \gamma)$ is a diversity graph with the property that γ maps $\mathcal{W}$ onto $\mathcal{G}'$. We also assume that $\mathcal{H}^* \subseteq \eta(\mathcal{V})$ is a minimal solution relative to $(\mathcal{V}, \mathcal{W}, \mathcal{E}, \eta, \gamma)$.

We begin by establishing the intuitive result that a minimal solution has cardinality $2|\mathcal{G}'|$ if and only if the neighborhoods of the elements in $\mathcal{W}$ are disjoint. Although this fact is nearly obvious, we include a proof for completeness. The following lemma supports the result.

Lemma 6.3. *Suppose that $T(w) \neq \emptyset$ for some $w \in \mathcal{W}$. Then, $\mathcal{H}^*$ contains an element of $\bigcup_{w \in \mathcal{W}} \eta(T(w))$.*

Proof. Assume that $T(w) \neq \emptyset$ for some $w \in \mathcal{W}$ and suppose that $\mathcal{H}^*$ does not contain an element of $\bigcup_{w \in \mathcal{W}} \eta(T(w))$. Then for each w there is a v' and v'' in $N(w) \backslash \bigcup_{w \in \mathcal{W}} \eta(T(w))$ so that $\eta(v') + \eta(v'') = \gamma(w)$. This implies that $|\mathcal{H}^*| = 2|\mathcal{G}'|$. However, we know that $T(w)$ is nonempty for some w, which means there exists w^1 and w^2 such that $\eta(v^1) + \eta(v^2) = \gamma(w^1)$ and $\eta(v^1) + \eta(v^3) = \gamma(w^2)$ for some v^1, v^2, and v^3 in $\mathcal{V}$. This means that

$$\left(\mathcal{H}^* \backslash \{\eta(v) : \eta(v) \in \mathcal{H}^*, \{(v, w^1), (v, w^2)\} \cap \mathcal{E} \neq \emptyset\}\right) \cup \{\eta(v^1), \eta(v^2), \eta(v^3)\}$$

is a solution to $\mathcal{G}'$ whose cardinality is at most $|\mathcal{H}^*| - 1$, which is a contradiction. □

Theorem 6.6. *We have that* $|\mathcal{H}^*| = 2|\mathcal{G}'|$ *if and only if* $N(w') \cup N(w'') = \emptyset$ *for all* w' *and* w'' *in* $\mathcal{W}$.

Proof. The fact that $|\mathcal{H}^*| = 2|\mathcal{G}'|$ if $N(w') \cup N(w'') = \emptyset$ for all w' and w'' in $\mathcal{W}$ is clear. Assume that $|\mathcal{H}^*| = 2|\mathcal{G}'|$, and suppose for the sake of obtaining a contradiction that $T(w) \neq \emptyset$ for some $w \in \mathcal{W}$. From Lemma 6.3 we have that $\mathcal{H}^*$ contains an element in $\bigcup_{w \in \mathcal{W}} \eta(T(w))$. Let w^1 and w^2 be such that $\eta(v^1) + \eta(v^2) = \gamma(w^1)$ and $\eta(v^1) + \eta(v^3) = \gamma(w^2)$, for some v^1, v^2, and v^3. Let $\mathcal{W}' = \mathcal{W} \backslash \{w^1, w^2\}$, and let $\mathcal{V}' = \cup_{w \in \mathcal{W}'} N(w)$. Furthermore, let $(\mathcal{V}')^*$ be such that $\eta((\mathcal{V}')^*)$ is a minimum solution to $\gamma(\mathcal{W}')$ with respect to $(\mathcal{V}', \mathcal{W}', \mathcal{E}')$, where $\mathcal{E}'$ is $\mathcal{E}$ with the edges incident to w^1 and w^2 removed. Then, $|\eta((\mathcal{V}')^*)| \leq 2|\mathcal{G}'|$. We know that we can resolve $\mathcal{G}'$ by including v^1, v^2, and v^3 in $(H')^*$. Since all three haplotypes might not be required, we have that $2|\mathcal{G}'| = |\mathcal{H}^*| \leq |(\mathcal{H}')^*| + 3$. So,

$$2|\mathcal{G}'| = |\mathcal{H}^*| \leq |(\mathcal{V}')^*| + 3 \leq 2|\mathcal{W}'| + 3 = 2(|\mathcal{W}| - 2) + 3 = 2|\mathcal{G}'| - 1.$$

Since this is a contradiction, we have that $T(w) = \emptyset$ for all w, and consequently, $N(w') \cap N(w'') = \emptyset$, for all $w' \neq w''$. □

We continue our investigation by exploring the effects of restricting the number of times a haplotype can be used to form a genotype. This makes sense realistically since in many populations the mating structure is not random. For example, many species have a unique mate for life, which means their haplotypes are only used in conjunction with the haplotypes of another individual. To make this precise, we reduce the edge set of the initial graph. For any $\mathcal{E}' \subseteq \mathcal{E}$ we define the degree of v with respect to $\mathcal{E}'$ to be $\deg_{\mathcal{E}'}(v) = |\{(v, w) : (v, w) \in \mathcal{E}'\}|$. For the diversity graph $(\mathcal{V}, \mathcal{W}, \mathcal{E}, \eta, \gamma)$ we let $\mathcal{V}_m^* \subseteq \mathcal{V}$ be any solution to

$$\begin{aligned} &\min\{|\mathcal{V}'| : \mathcal{V}' \subseteq \mathcal{V}, \eta(\mathcal{V}') \text{ solves } \gamma(\mathcal{W}), \\ &\quad \max\{\deg_{\mathcal{E}'}(v) \leq m : v \in \mathcal{V}'\} \text{ for some } \mathcal{E}' \subseteq \mathcal{E}\}. \end{aligned} \tag{6.6}$$

The value of this optimization problem is denoted $\phi(m) = |\mathcal{V}_m^*|$, and if the problem is infeasible, we let $\phi(m) = \infty$. As an example, consider the graph in Fig. 6.1, which is easily seen to support diversity. Since $\deg(w^i) = 2$ for all i except 3, the only solution is $\eta(\{v^i : i = 1, 2, \ldots, 9\})$. If $m = 1$, then each v can be adjacent to at most one w with respect to $\mathcal{E}'$. This means we must be able to associate a unique pair in $\mathcal{V}$ with each element of $\mathcal{W}$. Biologically this means that each parent can donate one of its two haplotypes to a unique child. Since this is impossible for this graph, we have that $\phi(1) = \infty$. Notice that in general we have $\phi(1)$ is either $2|\mathcal{W}|$ or ∞ depending on whether or not (6.6) is feasible. The situation is more complex if $m > 1$, and one of the main goals of this section is to show that $\phi(2)$ can be calculated by decomposing an acyclic diversity graph into longest paths.

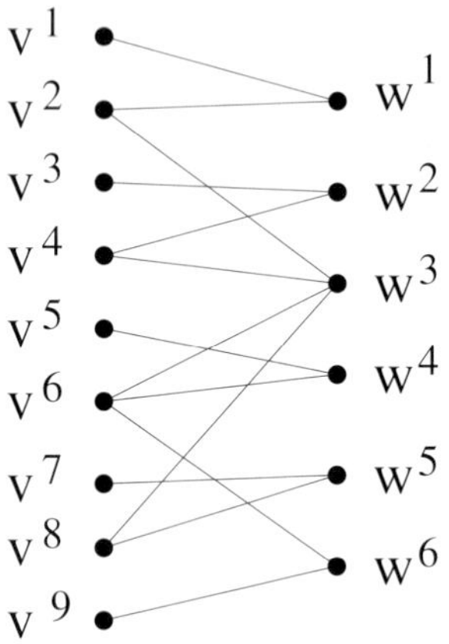

Fig. 6.1. A graph for which $\mathcal{V}_1^* = \emptyset$, $\phi(1) = \infty$, $\mathcal{V}_2^* = \mathcal{V}$, $\phi(2) = 9$, and $m^* = 2$.

At some threshold, increasing m does not change the cardinality of $\mathcal{V}_m^*$. For instance, if a haplotype is not compatible with more than m genotypes, then allowing it to mate with $m + 1$ haplotypes provides no additional benefit. Hence, for some m, $\phi(m) = \phi(m + k)$ for every natural number k. Moreover, increasing the number of possible mates that any haplotype is allowed never causes an increase in $\phi(m)$, and hence, ϕ is non-increasing. The smallest m such that $\phi(m) = \phi(m + k)$, for all $k \in \mathbb{N}$, is denoted by m^*. Clearly we have that

$$m^* \leq \max\{\deg(v) : v \in \mathcal{V}\} \leq |\mathcal{W}|.$$

An important observation is that $\phi(m^*)$ is the solution to the pure parsimony problem. So, if we knew how ϕ grew as m increased and how to bound m^*, then we could estimate the size of a biologically relevant collection of haplotypes. Unfortunately, we do not know the answer to either of these questions at this point, but these and related questions have future research promise. We initiate the investigation by studying $\phi(2)$ and $\phi(|\mathcal{W}|)$ if $m^* = |\mathcal{W}|$, the latter of which is addressed below.

Theorem 6.7. *If* $m^* = |\mathcal{W}|$, *then* $\phi(m^*) = |\mathcal{W}| + 1$.

Proof. Let $m^* = |\mathcal{W}|$. Then, there exists $v' \in \mathcal{V}_m^*$ such that for each $w^i \in \mathcal{W}$ there is a unique $v^i \in \mathcal{V}_m^* \setminus \{v'\}$ that satisfies $\eta(v') + \eta(v^i) = \eta(w^i)$. This means that $\phi(m^*) \geq |\mathcal{W}|+1$, and since $\eta(\{v', v^1, v^2, \ldots, v^{|\mathcal{W}|}\})$ solves $\mathcal{G}'$, we conclude that $\phi(m^*) = |\mathcal{W}| + 1$. □

Our next goal is to calculate $\phi(2)$ for acyclic graphs. The key observation in this case is that the most complicated subgraphs induced by a solution are paths. To motivate this intuition, consider the diversity graph in Fig. 6.2. Notice that both $\eta(\{v^1, v^3\})$ and $\eta(\{v^2, v^4\})$ are solutions to $\gamma(\{w^1\})$ but that the path v^1, w^1, v^3 has the advantage over v^2, w^1, v^4 since the single node v^5 can be appended to the path so that $\eta(\{v^1, v^3, v^5\})$ solves $\gamma(\{w^1, w^2\})$. If we had instead selected v^2, w^1, v^4, then both v^3 and v^5 would have been needed so that $\eta(\{v^2, v^3, v^4, v^5\})$ solved $\gamma(\{w^1, w^2\})$. It is clear in this example that $\phi(2) = 3$ and that $m^* = 2$. The intuition is that we want to decompose the graph into longest paths, a process explained by Algorithm 6.1. The fact that this technique minimizes the number of paths is established in Theorem 6.8. The proof is by induction on $|\mathcal{W}|$ and relates $\phi(2)$ as defined on $(\mathcal{V}, \mathcal{W}, \mathcal{E})$ to $\phi(2)$ as defined on one of its subgraphs.

Algorithm 6.1. Theorem 6.8 shows that this algorithm calculates $\phi(2)$. The removal of the path in Step 4 means that all nodes and edges in P_k are removed.

An Algorithm to Decompose the acyclic bipartite graph $(\mathcal{V}, \mathcal{W}, \mathcal{E})$ into the Fewest Paths

Step 1: Set $k = 0$ and $(\mathcal{V}_k, \mathcal{W}_k, \mathcal{E}_k) = (\mathcal{V}, \mathcal{W}, \mathcal{E})$.

Step 2: Find the longest path in $(\mathcal{V}_k, \mathcal{W}_k, \mathcal{E}_k)$, say P_k. If no path exists, set $P_k = \emptyset$.

Step 3: If $P_k = \emptyset$, stop.

Step 4: Set $(\mathcal{V}_{k+1}, \mathcal{W}_{k+1}, \mathcal{E}_{k+1}) = (\mathcal{V}_k, \mathcal{W}_k, \mathcal{E}_k) \backslash P_v$.

Step 5: Increase k by 1.

Step 6: Go to Step 2.

Theorem 6.8. *Let $(\mathcal{V}, \mathcal{W}, \mathcal{E}, \eta, \gamma)$ be an acyclic diversity graph. Then Algorithm 6.1 calculates $\phi(2)$, and in particular, if k is the number of paths found by the algorithm, then $\phi(2) = |\mathcal{W}| + k$.*

Proof. If $|\mathcal{W}| = 1$, Algorithm 6.1 clearly finds an optimal solution. Assume the result is true as long as $|\mathcal{W}| \leq q$. Let $(\mathcal{V}, \mathcal{W}, \mathcal{E}, \eta, \gamma)$ be an acyclic diversity graph with $|\mathcal{W}| = q + 1$. Apply Algorithm 6.1 to $(\mathcal{V}, \mathcal{W}, \mathcal{E})$ and let $P_1, P_2, ..., P_k$ be the paths in non-increasing length found by the algorithm. Denote the last path as $P_k = v^1, w^1, v^2, w^2, \ldots, w^r, v^{r+1}$. Let $\mathcal{W}' = \mathcal{W} \backslash \{w^r\}$ and $\mathcal{E}' = \mathcal{E} \backslash \{(w^r, v) : v \in N(w^r)\}$.

Case 1: Suppose $P_k \neq v^1, w^1, v^2$. Then, the algorithm applied to $(\mathcal{V}, \mathcal{W}', \mathcal{E}')$ finds the paths $P_1, P_2, ..., P_k'$, where $P_k' = v^1, w^1, v^2, w^2, \ldots, w^{r-1}, v^r$ —i.e. the last path is missing w^r and v^{r+1}. In this case, the algorithm terminates with k paths for both $(\mathcal{V}, \mathcal{W}, \mathcal{E})$ and $(\mathcal{V}, \mathcal{W}', \mathcal{E}')$. From the induction hypothesis we have that $\phi(2) = |\mathcal{W}'| + k$ for $(\mathcal{V}, \mathcal{W}', \mathcal{E}')$. Let $\hat{\mathcal{V}}_2^*$ be a solution to (6.6) for the subgraph $(\mathcal{V}, \mathcal{W}', \mathcal{E}')$. Then, $\hat{\mathcal{V}}_2^* \subseteq \mathcal{V}$, $|\hat{\mathcal{V}}_2^*| = |\mathcal{W}'| + k$, $\eta(\hat{\mathcal{V}}_2^*)$ solves $\gamma(\mathcal{W}')$, and $|N(w) \cap \hat{\mathcal{V}}_2^*| \leq 2$. Since $(\mathcal{V}, \mathcal{W}, \mathcal{E})$ is acyclic we know that (v^1, w^r) and (v^r, w^r) are not both in $\mathcal{E}$. Moreover, w^r cannot be adjacent to any of the terminal nodes $P_1, P_2, \ldots, P_{v-1}$ since this would violate the fact that each of these is a longest path. We conclude that adding w^r back to $\mathcal{W}'$ forces $\phi(2)$ for $(\mathcal{V}, \mathcal{W}, \mathcal{E})$ to be at least one greater than $\phi(2)$ for $(\mathcal{V}, \mathcal{W}', \mathcal{E}')$. Notice that $\eta(\hat{\mathcal{V}}_2^* \cup \{v^{r+1}\})$ is a solution to $\gamma(\mathcal{W})$ that is feasible to (6.6) for $(\mathcal{V}, \mathcal{W}, \mathcal{E})$. Since

$$|\hat{\mathcal{V}}_w^* \cup \{v^{r+1}\}| = |\mathcal{W}'| + k + 1 = |\mathcal{W}| + k,$$

we have that $\phi(2)$ for $(\mathcal{V}, \mathcal{W}, \mathcal{E})$ is $|\mathcal{W}| + k$.

Case 2: Suppose $P_k = v^1, w^1, v^2$ —i.e. $r = 1$. Then the algorithm applied to $(\mathcal{V}, \mathcal{W}', \mathcal{E}')$ produces the paths $P_1, P_2, \ldots, P_{k-1}$, and we have that $\phi(2) = |\mathcal{W}'| + k - 1$ for $(\mathcal{V}, \mathcal{W}', \mathcal{E}')$. Let $\hat{\mathcal{V}}_2^*$ be as in Case 1 with the cardinality condition replaced by $|\hat{\mathcal{V}}_2^*| = |\mathcal{W}'| + k - 1$. Notice that w^1 cannot be adjacent to any of the terminal nodes of $P_1, P_2, \ldots, P_{k-1}$, as this would violate the fact that these are longest paths. So, adding w^1 back to $\mathcal{W}'$ forces $\phi(2)$ for $(\mathcal{V}, \mathcal{W}, \mathcal{E})$ to be at least 2 greater than $\phi(2)$ for $(\mathcal{V}, \mathcal{W}', \mathcal{E}')$. Since $\eta(\hat{\mathcal{V}}_2^* \cup \{v^1, v^2\})$ is a solution to $\gamma(\mathcal{W})$ that is feasible to (6.6) for $(\mathcal{V}, \mathcal{W}, \mathcal{E})$ that additionally satisfies

$$|\hat{\mathcal{V}}_w^* \cup \{v^{r+1}\}| = |\mathcal{W}'| + (k - 1) + 2 = |\mathcal{W}| + k,$$

we have that $\phi(2)$ for $(\mathcal{V}, \mathcal{W}, \mathcal{E})$ is $|\mathcal{W}| + k$.

□

We mention that this proof does not readily extend to graphs with cycles. The problem is that cycles can share nodes, and hence the removal of a longest cycle can destroy other cycles. Although a proof currently alludes the authors, we suspect the following is true.

Conjecture 6.1. *Let $(\mathcal{V}, \mathcal{W}, \mathcal{E}, \eta, \gamma)$ be a diversity graph and $(\mathcal{V}', \mathcal{W}', \mathcal{E}')$ be the subgraph with all cycles removed. Then, if k is the number of paths identified by Algorithm 6.1, we have that $\phi(2) = |\mathcal{W}| + k$.*

The insight from Theorem 6.8 is that the solutions of a restricted form of the Pure Parsimony problem are representable as a collection of paths. However, this technique has two shortcomings. First, the longest path problem is NP-Complete, making each step of the algorithm difficult. So, the technique describes the nature of a solution but does not provide an efficient solution procedure. The second shortcoming is that the algorithm is not capable of finding every optimal solution. To see this, consider the following collection of genotypes,

$$\left.\begin{array}{l} \mathbf{g}^1 = \gamma(w^1) = (2, 0, -2, -2, -2, -2) \\ \mathbf{g}^2 = \gamma(w^2) = (0, 2, 0, 0, -2, -2) \\ \mathbf{g}^3 = \gamma(w^3) = (-2, 0, 2, 0, -2, 0) \\ \mathbf{g}^4 = \gamma(w^4) = (-2, 0, 0, 2, 0, -2) \\ \mathbf{g}^5 = \gamma(w^5) = (-2, -2, -2, 0, 2, -2) \\ \mathbf{g}^6 = \gamma(w^6) = (-2, -2, 0, -2, -2, 2). \end{array}\right\} \tag{6.7}$$

Assume that $|\mathcal{V}| = 2^6$, that $\eta(\mathcal{V}) = \{-1, 1\}^6$, and that $\mathcal{E}$ is the largest edge set possible. Notice that a path may contain the sequence $\mathbf{g}^i, \mathbf{h}^i, \mathbf{g}^{i+1}$ if and only if there is no SNP where $\mathbf{g}^i$ has a value of 2 or -2 and $\mathbf{g}^{i+1}$ has the other value. So, in the above example there is no $\mathbf{h}$ such that the path $\mathbf{g}^1, \mathbf{h}, \mathbf{g}^3$ exists in the diversity graph because the first SNP of $\mathbf{g}^1$ is a 2 and the first SNP of $\mathbf{g}^3$ is a -2. However, there is an $\mathbf{h}$ so that the path $\mathbf{g}^1, \mathbf{h}, \mathbf{g}^2$, is in the diversity graph because there is no SNP where $\mathbf{g}^1$ and $\mathbf{g}^2$ have different values of 2 and -2. If we compare each pair of genotypes in a similar fashion, we find that the paths pass through the genotypes as indicated in Fig. 6.3. From this figure we see that there is not a path or cycle through every genotype, but that there are several two path solutions. From Theorem 6.8 we know that $\phi(2) = 6+2 = 8$. Up to reversing the order of the genotypes, there are four optimal progressions through the genotypes, see Table 6.1. Our algorithm finds the first solution indicated in Table 6.1, as the first path is as long as possible. None of the other paths have this property, and so the algorithm is not capable of finding these solutions.

Table 6.1. Ways in which the genotypes for the example in (6.7) can be listed in two distinct paths.

First Path's Genotype Progression	**Second Path's Genotype Progression**
$(\mathbf{g}^1, \mathbf{g}^2, \mathbf{g}^3, \mathbf{g}^4, \mathbf{g}^5)$	$(\mathbf{g}^6)$
$(\mathbf{g}^1, \mathbf{g}^2, \mathbf{g}^4, \mathbf{g}^5)$	$(\mathbf{g}^3, \mathbf{g}^6)$
$(\mathbf{g}^1, \mathbf{g}^2, \mathbf{g}^3, \mathbf{g}^6)$	$(\mathbf{g}^4, \mathbf{g}^5)$
$(\mathbf{g}^6, \mathbf{g}^3, \mathbf{g}^4, \mathbf{g}^5)$	$(\mathbf{g}^1, \mathbf{g}^2)$

Our last discussion approaches the pure parsimony problem through lattice theory and requires the more general $\oplus$ as discussed in Section 6.2. Let $\preceq$ be a binary relation such that $2 \preceq 2$, $2 \preceq 0$, $-2 \preceq -2$, $-2 \preceq 0$, and $0 \preceq 0$. Then

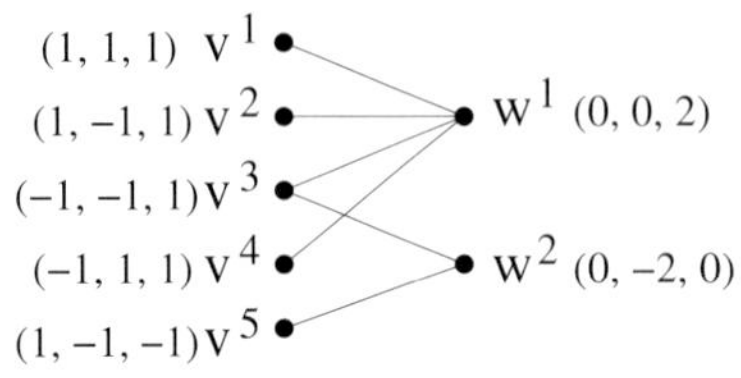

Fig. 6.2. In this example $\eta(\{v^1, v^3, v^5\} = \mathcal{V}_2^*$ and $\phi(2) = 3$.

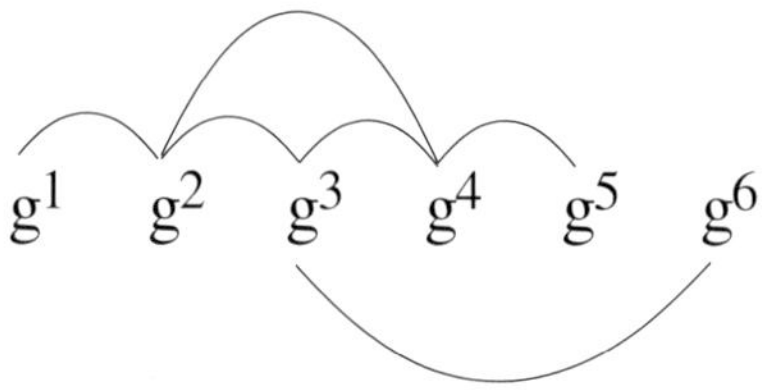

Fig. 6.3. The consecutive genotypes of a path in $(\mathcal{H}, G, \mathcal{E})$ are indicated with an arc. So, there is path that contains the sequence g^2, h', g^4, h'', g^5, but there is no path that contains g^1, h, g^3.

$\{-2, 0, 2\}^n$ is a poset under componentwise comparisons of $\preceq$. A subset of $\mathcal{G}'$ for which all elements are comparable forms a chain of genotypes. For example, the following four genotypes form a chain,

$$(-2, 2, 0, 2, -2) \preceq (0, 2, 0, 0, -2) \preceq (0, 2, 0, 0, 0) \preceq (0, 0, 0, 0, 0).$$

Chains have the property that as we look up the chain from smaller to greater elements that once a 2 or -2 becomes a 0 it remains 0. The following lemma and theorem solve the pure parsimony problem in the special case that $\mathcal{G}'$ is a chain.

Lemma 6.4. *For the diversity graph $(\mathcal{V}, \mathcal{W}, \mathcal{E}, \eta, \gamma)$ assume that $\gamma(\mathcal{W}) = \mathcal{G}'$ forms a chain under $\preceq$. Let $\eta(\mathcal{V}')$ be a minimal solution and assume that $\mathbf{g} \in \mathcal{G}$ has the property that $\gamma(w) \prec \mathbf{g}$, for all $w \in \mathcal{W}$. Then there does not exist v' and v'' in $\mathcal{V}'$ such that $\eta(v') + \eta(v'') = \mathbf{g}$.*

Proof. If $|\mathcal{G}'| = 1$, then $\mathcal{V}' = \{v', v''\}$ and the result follows because $\eta(v') + \eta(v'') \in \mathcal{G}'$ but $\mathbf{g} \notin \mathcal{G}'$. So, assume that $|\mathcal{G}'| \geq 2$. Suppose for the sake of attaining a contradiction that there is a v' and v'' in $\mathcal{V}'$ such that $\eta(v') + \eta(v'') = \mathbf{g}$. Because $\eta(\mathcal{V}')$ is a minimal solution to $\mathcal{G}'$, there are no isolated nodes in $\mathcal{V}'$. Since $\mathbf{g} \notin \mathcal{G}'$, this implies that there exists $\hat{v}'$ and $\hat{v}''$ in $\mathcal{V}'$ such that $\eta(v') + \eta(\hat{v}')$ and $\eta(v'') + \eta(\hat{v}'')$ are distinct elements of $\mathcal{G}'$. Without loss of generality, we assume that $\eta(v') + \eta(\hat{v}') \prec \eta(v'') + \eta(\hat{v}'')$.

Since $\eta(v'') + \eta(\hat{v}'') \prec \mathbf{g}$ and $\mathbf{g} = \eta(v') + \eta(v'')$, we have from the definition of $\oplus$ that

$$\eta(v') \oplus \eta(v'') \oplus \eta(\hat{v}'') = \eta(v') \oplus \eta(v'').$$

Similarly, because $\eta(v') + \eta(\hat{v}') < \eta(v'') + \eta(\hat{v}'')$ and

$$\eta(v') \oplus \eta(\hat{v}') \oplus \eta(v'') \oplus \eta(\hat{v}'') = \eta(v'') \oplus \eta(\hat{v}''),$$

we have that

$$\eta(v') \oplus \eta(v'') \oplus \eta(\hat{v}'') = \eta(v'') \oplus \eta(\hat{v}'').$$

It follows that

$$\mathbf{g} = \eta(v') + \eta(v'') = \eta(v'') + \eta(\hat{v}''),$$

which is a contradiction. □

Theorem 6.9. *Assume that* $(\mathcal{V}, \mathcal{W}, \mathcal{E}, \eta, \gamma)$ *is a diversity graph with* $\eta(\mathcal{V}) = \mathcal{H}$ *and* $\mathcal{E}$ *as large as possible and that* $\gamma(\mathcal{W}) = \mathcal{G}'$ *is a chain under* $\preceq$*. Then a minimum solution has cardinality* $|\mathcal{G}'| + 1$.

Proof. List the elements of $\mathcal{G}'$ as $\mathbf{g}^1, \mathbf{g}^2, \ldots, \mathbf{g}^{|\mathcal{G}'|}$ and let $w^1, w^2, \ldots, w^{|\mathcal{G}'|}$ be such that $\gamma(w^i) = \mathbf{g}^i$, for $i = 1, 2, \ldots, |\mathcal{G}'|$. We first construct a solution to $\mathcal{G}'$ with cardinality $|\mathcal{G}'| + 1$. Choose $v \in \mathcal{V}$ so that $\eta(v) \prec \gamma(w^1) = \mathbf{g}^1$. Then for every $\mathbf{g}^i \in \mathcal{G}'$ there is a unique $v^i \in \mathcal{V}$ such that $\eta(v) + \eta(v^i) = \gamma(w^i) = \mathbf{g}^i$. Then, $\eta(\{v, v^1, v^2, \ldots, v^{|\mathcal{G}'|}\})$ solves $\mathcal{G}'$ and has cardinality $|\mathcal{G}'| + 1$.

We now show by induction on $|\mathcal{G}'|$ that there does not exists a solution with cardinality less than $|\mathcal{G}'| + 1$. This fact is clear if $|\mathcal{G}'| = 1$, and we assume the claim is true when $|\mathcal{G}'| \leq k$. Assume that $\mathcal{G}'$ is a chain of length $k + 1$, and let $\eta(\mathcal{V}')$ be a minimum solution. Let $\mathcal{G}'' = \mathcal{G}' \backslash \{\gamma(w^{k+1})\}$, where we assume that the elements of $\mathcal{G}'$ are ordered so that

$$\gamma(w^1) \prec \gamma(w^2) \prec \ldots \prec \gamma(w^k) \prec \gamma(w^{k+1}).$$

Since $\eta(\mathcal{V}')$ solves $\mathcal{G}''$ and a minimum solution to $\mathcal{G}''$ has cardinality $|\mathcal{G}''| + 1 = k + 1$, we have that $|\eta(\mathcal{V}')| \geq k + 1$. Suppose for sake of attaining a contradiction that $|\eta(\mathcal{V}')| = k + 1$. From the induction hypothesis $\eta(\mathcal{V}')$ is a minimum solution to $\mathcal{G}'$. However, $\gamma(w^i) \prec \gamma(w^{k+1})$ for $i = 1, 2, \ldots, k$, and from Lemma 6.4, this leads to the contradiction that there is no v' and v'' in $\mathcal{V}'$ such that $\eta(v') + \eta(v'') = \gamma(w^{k+1})$. Hence $|\mathcal{V}'| \geq k + 2 = |\mathcal{G}'| + 1$. Since we have already demonstrated that a solution of size $|\mathcal{G}'| + 1$ exists, the proof is complete. □

Corollary 6.1 follows immediately from Theorem 6.9 and provides a bound on the pure parsimony problem.

Corollary 6.1. *Let* $(\mathcal{V}, \mathcal{W}, \mathcal{E}, \eta, \gamma)$ *be a diversity graph such that* $\eta(\mathcal{V}) = \mathcal{H}$ *and* $\mathcal{E}$ *is as large as possible. Partition* $\mathcal{W}$ *into* $\mathcal{W}^1, \mathcal{W}^2, \ldots, \mathcal{W}^q$*, where each* $\gamma(\mathcal{W}^i)$ *is a chain ordered by* $\preceq$*. Then a minimum solution has cardinality no greater than* $|\mathcal{G}'| + q$.

The best bound provided by Corollary 6.1 is the one that minimizes q. An interesting question for future research is whether or not calculating the minimum value of q actually solves the pure parsimony problem in some cases.

Instead of addressing the smallest size of a solution to a chain, the next result and its corollary considers how large a minimal solution can be.

Theorem 6.10. *Let $(\mathcal{V}, \mathcal{W}, \mathcal{E}, \eta, \gamma)$ be a diversity graph such that $\eta(\mathcal{V}) = \mathcal{H}$, $\mathcal{E}$ is as large as possible, and $\gamma(\mathcal{W}) = \mathcal{G}'$ is a chain with respect to $\prec$. Assume that the elements of $\mathcal{W}$ are ordered so that*

$$\gamma(w^1) \prec \gamma(w^2) \prec \ldots \prec \gamma(w^{|\mathcal{G}'|}).$$

Assuming that $\gamma(w^2)$ has 3 or more zero SNPs, we have that there is a minimal solution to $\mathcal{G}'$ with cardinality $2|\mathcal{G}'|$.

Proof. The proof is by induction on $|\mathcal{G}'|$. The result clearly holds if $|\mathcal{G}'| = 1$, and we assume the result is true for $|\mathcal{G}'| \leq k$. Assume that $|\mathcal{G}'| = k + 1$, and let $\eta(\mathcal{V}'')$ be a minimal solution to $\mathcal{G}'' = \mathcal{G}' \backslash \{\gamma(w^{k+1})\}$ whose cardinality is $2|\mathcal{G}''|$. From Lemma 6.4, $\eta(\mathcal{V}'')$ does not solve $\mathcal{G}'$ because there is no v' and v'' in $\mathcal{V}''$ such that $\eta(v') + \eta(v'') = \gamma(w^{k+1})$. We show that there is a minimal solution $\eta(\mathcal{V}')$ such that $\mathcal{V}'' \subseteq \mathcal{V}'$ and $|\mathcal{V}''| + 2 = |\mathcal{V}'|$.

Because $\gamma(w^{k+1})$ is the $k+1$ element in a chain, it has at least $k+1$ ambiguous SNPs, and thus $|N(w^{k+1})| \geq 2^{k+1}$. Since we assumed that $\gamma(w^2)$ has at least 3 zero SNPs, $|N(w^{k+1})| > 2^{k+1}$. For $j \geq 1$, we have $2j \leq 2^j$, and thus $2k < |N(w^{k+1})|/2$. Since $|N(w^{k+1})|/2$ is the number of adjacent pairs to w^{k+1} and $|\mathcal{V}''| = 2k$, there is a v' and v'' in $N(w^{k+1}) \backslash N(\mathcal{V}'')$ such that $\eta(v') + \eta(v'') = \gamma(w^{k+1})$. This means that $\eta(\mathcal{V}'' \cup \{v', v''\})$ is a minimal solution to $\mathcal{G}'$ whose cardinality is $2|\mathcal{G}'|$. □

The condition of $\gamma(w^2)$ having at least 3 zero SNPs is not imposed because this proof requires it, but rather, it is needed by any proof due to the following example. Let $\mathcal{G}' = \{(-2, 0), (0, 0)\}$. Then a four element solution would have the form $\{(-1, 1), (-1, -1), (x, 1), (y, -1)\}$, where x is either 1 or -1 and y is the other. In either case an element is duplicated, and hence there is no solution of size 4.

The following corollary establishes that under the conditions of Theorem 6.10, there is a minimal solution of every cardinality between the minimum value of $|\mathcal{G}'| + 1$ and the maximum value of $2|\mathcal{G}'|$.

Corollary 6.2. *Let $(\mathcal{V}, \mathcal{G}', \mathcal{E}, \eta, \gamma)$ be a diversity graph satisfying the condition of Theorem 6.10. Then, there is a minimal solution of cardinality j for $|\mathcal{G}'| + 1 \leq j \leq 2|\mathcal{G}'|$.*

Proof. For $1 \leq i \leq |\mathcal{G}'|$, let

$$\mathcal{G}'_1 = \{\gamma(w^1), \gamma(w^2), \ldots, \gamma(w^{|\mathcal{G}'|-i+1})\}$$

and

$$\mathcal{G}'_2 = \{\gamma(w^{|\mathcal{G}'|-i+2}), \gamma(w^{|\mathcal{G}'|-i+3}), \ldots, \gamma(w^{|\mathcal{G}'|})\}$$

be subchains of $\mathcal{G}'$. By Theorem 6.9 there is a solution $\eta(\mathcal{V}_1)$ to $\mathcal{G}'_1$ of cardinality $|\mathcal{G}'| - i + 2$ and by Theorem 6.10 there is a solution $\eta(\mathcal{V}_2)$ to $\mathcal{G}'_2$ of cardinality $2i - 2$. If $i = 1$, notice that $\mathcal{G}'_1 = \mathcal{G}'$ and that $\mathcal{G}'_2 = \emptyset$. In this case Theorem 6.9 establishes that we can indeed find a solution of cardinality $|\mathcal{G}'| + 1$. For other values of i we have that if $\mathcal{V}_1$ and $\mathcal{V}_2$ are disjoint, then $\mathcal{V}_1 \cup \mathcal{V}_2$ is a minimal solution whose cardinality is $|\mathcal{G}'| + i$, for $1 \leq i \leq |\mathcal{G}'|$. So, all that is left to show is that $\mathcal{V}_1$ and $\mathcal{V}_2$ may be selected so that they are disjoint. We accomplish this by showing that as i increases to $i + 1$ that there are always enough elements of $\mathcal{V}$ to allow $\mathcal{V}_1$ and $\mathcal{V}_2$ to be disjoint.

For $i = 1, 2, \ldots, |\mathcal{G}'|$ we have that $|\mathcal{G}'| - i + 2 \leq 2^{|\mathcal{G}'|-i+2}$. As in the proof of Theorem 6.10, we have that

$$2^{|\mathcal{G}'|-i+2} < |N(w^{|\mathcal{G}'|-i+2})|/2,$$

which guarantees that there are v' and v'' in $N(w^{|\mathcal{G}'|-i+2}) \backslash \eta(\mathcal{V}_1)$ such that $\eta(v') + \eta(v'') = \gamma(w^{|\mathcal{G}'|-i+2})$. So, as i increases from i to $|\mathcal{G}'|$, we are guaranteed to be able to select disjoint $\mathcal{V}_1$ and $\mathcal{V}_2$. □

6.5. Directions for Future Research

The goal of this paper was to establish an initial investigation into the structure of haplotyping problems by studying the underlying graph theory. We have shown that the structural requirements of the problem are meaningful and that the majority of bipartite graphs are incapable of representing the underlying biology We have further established a solution to the pure parsimony problem in a few cases, and in particular we have shown that ordering the genotypes with $\preceq$ and decomposing $\mathcal{G}'$ into chains bounds the problem. During the writing of this paper the authors had other questions that were left unanswered, many of which promise to be fruitful continued research:

- Although the matrix equation and the logical decomposition stated in Theorem 6.4 characterize the graphs that support diversity, we would have enjoyed a more graph theoretical characterization. A theorem like $(\mathcal{V}, \mathcal{W}, \mathcal{E})$ supports diversity if and only if it does not contain a certain structure would have been particularly appealing.

- We conjecture that m^* is 2 for acyclic diversity graphs, which means that the pure parsimony problem is solve by Algorithm 6.1. This together with a proof of Conjecture 6.1 may highlight the class of diversity graphs for which $m^* = 2$.
- Decomposing the genotypes into chains ordered by $\preceq$ bounds the optimal value of the pure parsimony problem, but this bound can likely be reduced by investigating how the solutions to the individual chains can interact. Moreover, we do not yet know how to decompose the genotypes into the fewest number of chains. This bound could be useful in the integer programming formulation of the problem, and numerical work should be explored.
- Investigating how $\phi(m)$ decreases and estimating m^* are exciting new avenues. If we can accomplish both of these, then we will be able to estimate the solution to the pure parsimony problem.

References

[1] A. S. Asratian, T. M. J. Denley, and R. Häggkvist. *Bipartite Graphs and Their Applications*. Cambridge University, New York, NY, 1998.

[2] V. Bafna, D. Gusfield, G. Lancia, and S. Yooseph. Haplotyping as perfect phylogeny: A direct approach. Technical Report CSE-2002-21, University of California, Davis, Computer Science, 2002. Augmented version to appear in the Journal of Computational Biology.

[3] R. H. Chung and D. Gusfield. Empirical exploration of perfect phylogeny haplotyping and haplotypers. Technical report, University of California, Computer Science, 2003. To appear in the Proceedings of the 2003 Cocoon Conference.

[4] R. H. Chung and D. Gusfield. Perfect phylogeny haplotyper: Haplotype inferral using a tree model. *Bioinformatics*, 19(6):780–781, 2003.

[5] A. G. Clark. Inference of haplotypes from PCR-amplified samples of diploid populations. *Molecular Biology and Evolution*, 7(2):111–122, 1990.

[6] D. Gusfield. A practical algorithm for optimal inference of haplotypes from diploid populations. *Proceedings of the Eight International Conference on Intelligent Systems for Molecular Biology*, 2000.

[7] D. Gusfield. Inference of haplotypes from samples of diploid populations: Complexity and algorithms. *Journal of Computational Biology*, 8(3):305–324, 2001.

[8] D. Gusfield. Haplotyping as perfect phylogeny: Conceptual framework and efficient solutions. *Proceedings of RECOMB 2002: The Sixth Annual International Conference on Computational Biology*, pages 166–175, 2002.

[9] D. Gusfield. Haplotyping by pure parsimony. Technical Report CSE-2003-2, University of California, Davis, 2003. To appear in the Proceedings of the 2003 Combinatorial Pattern Matching Conference.

[10] G. Lancia, V. Bafna, S. Istrail, R. Lippert, and R. Schwartz. SNPs problems, complex-

ity and algorithms. In *European Symposium on Algorithms*, volume 2161 of *Lecture Notes in Computer Science*, pages 182–193. Springer-Verlag, 2001.

[11] G. Lancia and M. Perlin. Genotyping of pooled microsatellite markers by combinatorial optimization techniques. *Discrete Applied Mathematics*, 88(1-3):291–314, 1998.

[12] G. Lancia, M. Pinotti, and R. Rizzi. Haplotyping populations by pure parsimony. complexity, exact and approximation algorithms. *INFORMS Journal on Computing*, 16(4):348–359, 2004.

[13] S. Lin, D. J. Cutler, M. E. Zwick, and A. Chakravarti. Haplotype inference in random population samples. *American Journal of Human Genetics*, 71:1129–1137, 2002.

[14] R. Lippert, R. Schwartz, G. Lancia, and S. Istrail. Algorithmic strategies for the SNPs haplotype assemply problem. *Briefings in Bioinformatics*, 3(1):23–31, 2001.

[15] T. Niu, Z. S. Quin, X. Xu, and J. S. Liu. Bayesian haplotype inference for multiple linked single-nucleotide polymorphisms. *American Journal of Human Genetics*, 70:157–169, 2002.

[16] D. Qian and L. Beckmann. Minimum-recombinant haplotyping in pedigrees. *American Journal of Human Genetics*, 70:1434–1445, 2002.

[17] R. Rizzi, V. Bafna, S. Istrail, and G. Lancia. Practical algorithms and fixed-parameter tractability for the single individual SNP haplotyping problem. In R. Guigo and D. Gusfield, editors, *Algorithms in Bioinformatics: Proceedings of the Second International Workshop on Algorithms on Bioinformatics, WABI 2002, Rome, Italy, September 17-21, 2002*, volume 2452 of *Lecture Notes in Computer Science*, pages 29–43. Springer-Verlag Berlin Heidelberg, 2002.

[18] M. Stephens, N. J. Smith, and P. Donnelly. A new statistical method for haplotype reconstruction from population data. *American Journal of Human Genetics*, 68:978–989, 2001.

[19] C. F. Xu, K. Lewis, K. L. Cantone, P. Khan, C. Donnelly, N. White, N. Crocker, P. R. Boyd, D. V. Zaykin, and I. J. Purvis. Effectiveness of computational methods in haplotype prediction. *Human Genetics*, 110:148–156, 2002.

Chapter 7

Identifying Critical Nodes in Protein-Protein Interaction Networks

Vladimir Boginski

Research & Engineering Education Facility (REEF)
Department of Industrial & Systems Engineering
University of Florida
Shalimar, FL 32579, USA

Clayton W. Commander

Air Force Research Laboratory
Munitions Directorate
Eglin AFB, FL 32542, USA

In recent years, the study of biological networks has increased dramatically. These problems have piqued the interest of researchers in many disciplines from biology to mathematics. In particular, many problems of interest to biological scientists can be modeled as combinatorial optimization problems and studied by operations researchers. In this chapter, we consider the problem of identifying the critical nodes of a network and its potential applications to protein-protein interaction networks. More specifically, we are interested in determining the smallest set of nodes whose removal from the graph maximally disconnects the network. Recent techniques for identifying critical nodes in telecommunication networks are applied to the study of protein-protein interaction graphs and the results are analyzed.

7.1. Introduction

Optimization problems abound in the study of biological networks. This is a timely research topic and has been the focus of great attention in the recent literature [1, 4, 6, 19, 21–23, 27]. In this chapter, we investigate the detection of *critical nodes* in *protein-protein interaction networks*. The CRITICAL NODE DETECTION PROBLEM (CNDP) is a combinatorial optimization problem recently introduced by Arulselvan et al. [2]. Given a graph $G = (V, E)$ and an integer $k \in \mathbb{Z} \setminus |V|$, the objective is to determine a subset $A \subseteq V$, such that $|A| = k$, whose deletion from the graph results in a minimum cohesion and ensures a minimum difference in the sizes of the components. A related problem, the CARDINALITY CONSTRAINED CNDP (CC-CNDP) seeks to determine a minimum cardinality subset of nodes whose deletion ensures that the number of nodes reachable from any other node in the network does not exceed some threshold value.

The CNDP has applications in many fields including social network analysis, quality assurance and risk management in telecommunication networks, transportation science, and control of social contagion [2, 3, 7]. Our proposition is that identifying the critical nodes in protein-protein interaction networks can have applications in computational biology, in particular in drug design. The interpretation of the critical nodes in the context of protein-protein interaction networks is that these nodes represent a minimum cardinality set of proteins whose removal would destroy the primary interactions and thus help neutralize potentially harmful organisms (e.g., bacteria or viruses).

The organization of this chapter is as follows. In the next section, protein-protein interaction networks are discussed. In Section 7.3, we provide mathematical programming formulations for both variants of the CRITICAL NODE DETECTION PROBLEM described above. In Section 7.4, we discuss the implementation of several heuristics for both problems, and provide some preliminary computational results of critical node detection on real-world protein-protein interaction networks in Section 7.5. Conclusions and future directions of research are identified in Section 7.6.

7.2. Protein-Protein Interaction Networks

In recent years, the biological research community's interest in studying proteins from different aspects has steadily increased. As a result, the field of *proteomics*, which investigates proteins' structures and functions, has been developed. In particular, *protein-protein interactions* have been extensively studied using various advanced techniques. [5, 14, 15, 26, 28] Many biological functions involve interactions between proteins at different levels, including signal transduction in cells (i.e., conversion of one kind of a signal to another inside a cell, which may play an important role in biological processes, including disease development), formation of protein complexes (i.e., stable over time structures involving multiple proteins), brief interactions between proteins involving the processes of modification of one protein by another, etc.

Protein-protein interactions can be represented in terms of graph theory as a set of vertices (proteins) and edges (certain types of interactions between proteins). These structures are referred to as *protein-protein interaction networks*.

These networks play an important role in computational biology. In many cases, they can be easily visualized and are convenient for understanding the complex nature of different types of interactions between proteins. As a result of extensive research efforts in studying protein-protein interactions for different biological organisms (e.g., certain types of bacteria), massive amounts of data have been obtained. In particular, detailed and comprehensive data on protein-protein interactions is available for the yeast *Saccharomyces cerevisiae*, which has been considered in a number of works. [16, 17, 25]

Moreover, protein-protein interactions are studied from the perspective of *drug design* applications. [11, 24] In particular, drugs that target specific types of proteins can be developed. This research direction has significantly grown recently in the context of identifying *target proteins* responsible for certain diseases based on the available protein-protein interaction data. Nowadays, experimental studies in this area are extensively conducted by scientists in the pharmaceutical industry and research communities. [20]

On the other hand, protein-protein interaction networks can be investigated from the network optimization perspective. In this chapter, we make the first attempt to put the aforementioned problem of identifying target proteins in protein-protein interaction networks in the framework of combinatorial optimization. Specifically, we propose to apply the recently introduced CRITICAL NODE DETECTION PROBLEM (CNDP) to analyze protein-protein interactions and detect the nodes (proteins) that are the most important for the connectivity of these networks. We believe that identifying these critical nodes can provide information that can be used in drug design and other applications.

Next, we discuss mathematical programming formulations of the considered problems and present computational results obtained for some available protein-protein interaction datasets.

7.3. Optimization Approaches for Critical Node Detection

Denote a graph $G = (V, E)$ as a pair consisting of a set of vertices V, and a set of edges E. All graphs in this chapter are assumed to be undirected and unweighted. For a subset $W \subseteq V$, let $G(W)$ denote the subgraph induced by W on G. A set of vertices $I \subseteq V$ is called an *independent* or *stable set* if for every $i, j \in I$, $(i, j) \notin E$. That is, the graph $G(I)$ induced by I is edgeless. An independent set is *maximal* if it is not a subset of any larger independent set (i.e., it is maximal by inclusion), and *maximum* if there are no larger independent sets in the graph.

7.3.1. *The Critical Node Detection Problem*

Given a graph $G = (V, E)$, let $u : V \times V \mapsto \{0, 1\}$, where $u_{ij} = 1$ if nodes i and j are in the same component of V. Then the objective of the CNDP is to find a subset $A \subseteq V$ of nodes such that $|A| \leq k$, whose deletion results in the minimum value of $\sum u_{ij}$ in the edge induced subgraph $G(V \setminus A)$. This objective function results in a minimum cohesion in the network, while also ensuring a minimum difference in the sizes of the components. An integer programming formulation of the CNDP has been formulated by Arulselvan et al. [2]

Let u be defined as above and define $v : V \mapsto \{0, 1\}$ as

$$v_i := \begin{cases} 1, \text{ if node } i \text{ is deleted in the optimal solution,} \\ 0, \text{ otherwise.} \end{cases} \tag{7.1}$$

Then the CRITICAL NODE DETECTION PROBLEM is given as

$$\textbf{(CNDP)} \quad \text{Minimize} \sum_{i,j \in V} u_{ij} \tag{7.2}$$

s.t.

$$u_{ij} + v_i + v_j \geq 1, \ \forall\, (i, j) \in E, \tag{7.3}$$

$$u_{ij} + u_{jk} - u_{ki} \leq 1, \ \forall\, (i, j, k) \in V, \tag{7.4}$$

$$u_{ij} - u_{jk} + u_{ki} \leq 1, \ \forall\, (i, j, k) \in V, \tag{7.5}$$

$$-u_{ij} + u_{jk} + u_{ki} \leq 1, \ \forall\, (i, j, k) \in V, \tag{7.6}$$

$$\sum_{i \in V} v_i \leq k, \tag{7.7}$$

$$u_{ij} \in \{0, 1\}, \ \forall\, i, j \in V, \tag{7.8}$$

$$v_i \in \{0, 1\}, \ \forall\, i \in V. \tag{7.9}$$

Constraints (7.3) ensure that if $(i, j) \in E$ and nodes i and j are in separate components, then one or both of them is deleted. The set of constraints (7.4-7.6) ensure that if nodes i and j are in the same component and nodes j and k are in the same component, then necessarily i and k belong to the same component. Finally, (7.7) constrains the maximum number of nodes to be deleted. The CNDP was shown to be $\mathcal{NP}$-hard [12] by a reduction of MAXIMUM INDEPENDENT SET to an instance of the CNDP [3].

7.3.2. *Cardinality Constrained Problem*

Given a graph $G = (V, E)$, the *connectivity index* of a node is defined as the number of nodes reachable from that vertex (see Fig. 7.1 for examples). To constrain

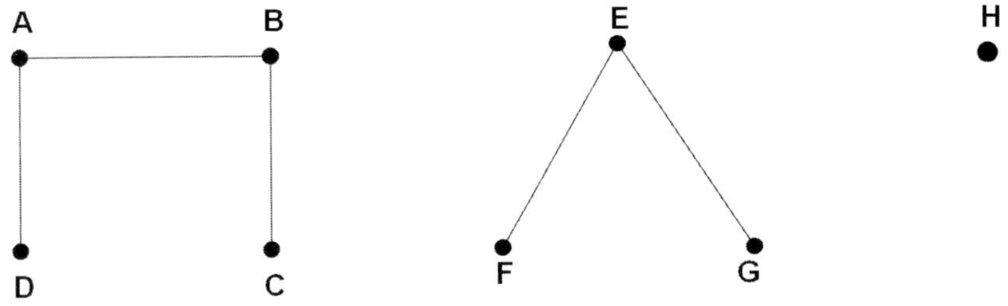

Fig. 7.1. Connectivity Index of nodes A,B,C,D is 3. Connectivity Index of E,F,G is 2. Connectivity Index of H is 0.

the network connectivity in optimization models, we can impose constraints on the connectivity indices of the nodes [8].

The CARDINALITY CONSTRAINED CNDP can be formulated in a similar manner to the the CNDP above. Recall that in this problem, we are given an integer $L \in \mathbb{Z}$, and we are interested in determining a minimum cardinality subset $A \subseteq V$ such that the *connectivity index* of the remaining nodes in the vertex deleted subgraph $G(V \setminus A)$ does not exceed L.

Using the same definition of the variables as in the previous subsection, we can formulate the CC-CNDP as the following integer linear programming problem.

$$\textbf{(CC-CNDP)} \quad \text{Minimize} \sum_{i \in V} v_i \tag{7.10}$$

s.t.

$$u_{ij} + v_i + v_j \geq 1,\ \forall\ (i,j) \in E, \tag{7.11}$$

$$u_{ij} + u_{jk} - u_{ki} \leq 1,\ \forall\ (i,j,k) \in V, \tag{7.12}$$

$$u_{ij} - u_{jk} + u_{ki} \leq 1,\ \forall\ (i,j,k) \in V, \tag{7.13}$$

$$-u_{ij} + u_{jk} + u_{ki} \leq 1, \forall\ (i,j,k) \in V, \tag{7.14}$$

$$\sum_{i,j \in V} u_{ij} \leq L, \tag{7.15}$$

$$u_{ij} \in \{0,1\},\ \forall\ i,j \in V, \tag{7.16}$$

$$v_i \in \{0,1\},\ \forall\ i \in V, \tag{7.17}$$

where L is the maximum allowable connectivity index for any node in the vertex deleted subgraph $G(V \setminus A)$. Notice that the objective is to minimize the number of nodes deleted. Constraints (7.11) follow exactly as in the CNDP model. Also, Constraints (7.12)–(7.14) are equivalent to the constraint $u_{ij} + u_{jk} +$

$u_{ki} \neq 2, \ \forall \ (i, j, k) \in V$, which ensures that if nodes i and j are in the same component and nodes j and k are in the same component, then necessarily i and k belong to the same component (as in the CNDP model). Constraints (7.15) ensure that the connectivity indices of all nodes does not exceed L. The CC-CNDP is also $\mathcal{NP}$-hard as proved by Arulselvan et al [3]. For the application of interest in this chapter, the CC-CNDP is the most applicable critical node optimization model. Therefore as we consider solutions approaches and computational experiments in the following sections, this is the problem on which we will focus.

7.4. Heuristic Approaches for Critical Node Detection

Due to the computational complexity of the mathematical programming formulations presented above, the prior work in this area has concentrated on the development of efficient heuristics for critical node detection problems [2, 3, 7].

7.4.1. *Multi-Start Combinatorial Heuristic*

Arulselvan et al. [3] have proposed an efficient combinatorial heuristic for the CNDP. Pseudo-code for the proposed algorithm is provided in Fig. 7.2. The heuristic starts off by identifying a maximal independent set (MIS). This array holds the set of non-critical nodes. In the main loop from lines 4-16, the procedure iterates through the vertices and determines which of them can be added back to the graph while still maintaining feasibility. If vertex i can be replaced, MIS is augmented to include i in step 7, otherwise NoAdd is incremented. When NoAdd is equal to $|V| - |\text{MIS}|$, then no nodes can be returned to the graph and OPT is set to `TRUE`. The loop is then exited and the algorithm returns the set of critical nodes, i.e., $V \backslash \text{MIS}$.

The solution found by this procedure is then further improved through a local search step and incorporated within a multi-start mechanism (see Fig. 7.3). Computational testing was carried out on a benchmark network of the social interactions of the terrorists involved in the $9 \setminus 11$ hijacking [18] and on some networks generated by the authors. Their tests indicated that the heuristic is very effective in producing optimal solutions in modest computing time as compared to solving the mixed integer formulation of the problem using the branch-and-bound based solver CPLEX [9].

```
procedure CriticalNode(G, L)
    MIS ← MaximalIndepSet(G)
    OPT ← FALSE
    NoAdd ← 0
    while (OPT .NOT.TRUE) do
      for (i = 1 to |V|) do
        if ( |s|(|s|−1)/2 ≤ L ∀ s ∈ S ⊆ G(MIS ∪ {i}) : i ∈ V \ MIS ) then
          MIS ← MIS ∪ {i}
        else
          NoAdd ← NoAdd +1
        end if
        if (NoAdd = |V| − |MIS|) then
          OPT ← TRUE
          BREAK
        end if
      end for
    end while
    return V \ MIS    /* set of nodes to delete */
end procedure CriticalNode
```

Fig. 7.2. Combinatorial heuristic for the CARDINALITY CONSTRAINED CRITICAL NODE PROBLEM.

7.4.2. *Genetic Algorithms*

Genetic algorithms (GAs) represent a broad class of heuristics for global optimization problems. Intuitively, they are designed to mimic the biological process of evolution, and follow Darwin's Theory of Natural Selection [10]. GAs store a set of solutions, or a *population*, and the population *evolves* by replacing these solutions with better ones based on certain fitness criteria represented by the objective function value. In successive iterations, or *generations*, the population evolves by *reproduction*, *crossover*, and *mutation*.

Reproduction is the probabilistic selection of the next generations elements determined by their fitness level (i.e., objective function value). Crossover is the combination of two current solutions, called *parents*, which produces one or more other solutions, referred to as their *offspring*. Finally, mutation is the random perturbation of the offspring and is implemented as an escape mechanism to avoid getting trapped at local optima. [13]. In successive generations, only those solutions having the best *fitness* are carried to the next generation in a process which mimics the fundamental principle of natural selection, *survival of the fittest* [10].

```
procedure CriticalNodeLS(G, k)
    X* ← ∅
    f(X*) ← ∞
    for j = 1 to MaxIter do
        X ← CriticalNode(G, k)
        X ← LocalSearch(X)
        if f(X) < f(X*) then
            X* ← X
        end if
    end
    return (V \ X*)    /* set of k nodes to delete */
end procedure CriticalNodeLS
```

Fig. 7.3. The multi-start framework for the critical node heuristic.

```
procedure GeneticAlgorithm
    Generate population P_k
    Evaluate population P_k
    while terminating condition not met do
        Select individuals from P_k and copy to P_{k+1}
        Crossover individuals from P_k and put in P_{k+1}
        Mutate individuals from P_k and put in P_{k+1}
        Evaluate population P_{k+1}
        P_k ← P_{k+1}
        P_{k+1} ← ∅
    end while
    return best individual in P_k
end procedure GeneticAlgorithm
```

Fig. 7.4. Pseudo-code for a generic genetic algorithm.

Figure 7.4 provides pseudo-code for a standard genetic algorithm. Genetic algorithms were introduced in 1977 by Holland, and were greatly invigorated by the work of Goldberg. [13]

7.5. Computational Experiments

In this section, we present some preliminary computational results on real protein-protein interaction networks obtained from the literature. In particular, three

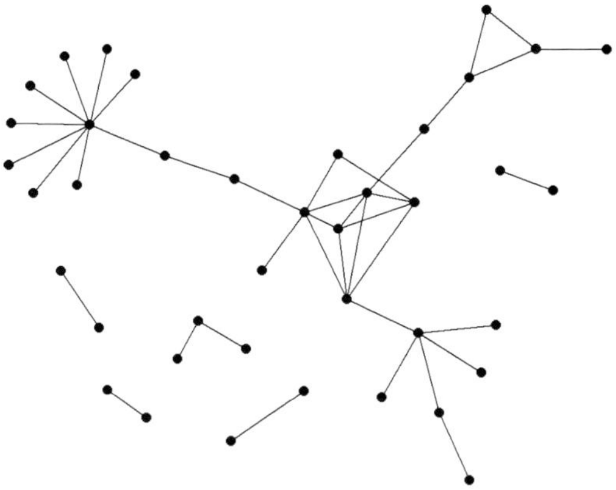

(a) The original network.

(b) The optimal subgraph for $L = 3$.

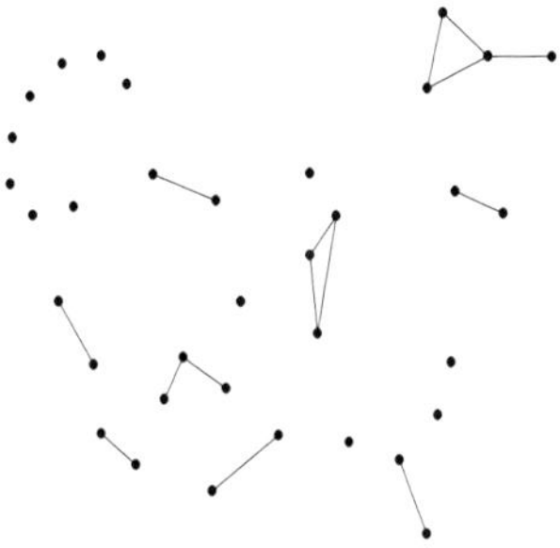

(c) The optimal subgraph for $L = 4$.

Fig. 7.5. The CARDINALITY CONSTRAINED CRITICAL NODE DETECTION PROBLEM is solved for the 46 primary interactions of the *S. cerevisiae* cell cycle.

graphs are tested with various values of the connectivity index bound. Both of the aforementioned heuristics were implemented in the `C++` programming language and complied using GNU `g++` version 4.1.2, using optimization flag `-O4`. They were compiled on a Linux workstation equipped with a 3.0 GHz Intel® Xeon® processor and 1.0 gigabytes of RAM. For more information on specific tuning parameters of the heuristics, the reader is referred to the papers by Arulselvan et al. [2, 3]. The major focus of this chapter is to introduce the concept and related techniques of finding critical nodes in a graph to the computational biology research community.

We begin by examining the 46 primary interactions of the yeast *S. cerevisiae* cell cycle [29]. The graph of the original network can be seen in Fig. 7.5(a). The optimal solutions for the case where $L = 3$ and $L = 4$ are provided in

Fig. 7.5(b) and 7.5(c) respectively. The corresponding numerical results are provided in Table 7.1. The table contains the objective function values and the corresponding computation times required by the algorithms. We provide the optimal solutions as computed by CPLEX and the solutions provided by the heuristics mentioned above. For this small, and relatively unconnected example, we see that all the methods were able to obtain optimal solutions in a negligible amount of time.

Table 7.1. Results of the IP model and the genetic algorithm for the 46 interactions of *S. cerevisiae*.

Instance	IP Model		Genetic Alg		Comb. Alg	
Max Conn. Index (L)	Obj Val	Comp Time (s)	Obj Val	Comp Time (s)	Obj Val	Comp Time (s)
2	8	0.18	8	0.05	8	0.04
3	6	1.19	6	0.01	6	0.00
4	5	2.57	5	0.05	5	0.00
5	4	1.05	4	0.01	4	0.00
6	4	2.63	4	0.04	4	0.00

The next network considered is from the 78 protein-protein interactions from the development of *D. melanogaster* [29]. The graph of the network can be seen in Fig. 7.6(a). Similar to the previous example, we provide the graphs corresponding to the optimal solution for the cases of $L = 5$ and $L = 4$. Table 7.2 reflects the computational results for this instance. As above, we see that the heuristics were able to provide the optimal solutions for each value of L tested. However, we see that even for this relatively small instance, the required computation time to compute the optimal solution has increased by two orders of magnitude from the previous example.

As a final test case we examine the network comprising 186 yeast two-hybrid system interactions of *S. cerevisiae* proteins [17]. The original network is shown in Fig. 7.7. For this case, CPLEX was unable to compute optimal solutions for any values of L. Therefore, we only provide solutions for the two heuristics in Table 7.3. Notice that both heuristics computed the same objective function value in each case. However, the combinatorial algorithm required over 30 seconds for the case where $L = 2$.

Though promising, these preliminary results indicate the need for advanced heuristics and exact solution methods for computing critical nodes in protein-protein interaction networks. The primary challenge to computing optimal solutions in real-world networks is that the sizes of the networks prohibit optimal solutions from being calculated using standard branch-and-bound techniques. The test cases presented represent relatively small instances of protein-protein interaction

(a) The original network.

(b) The optimal subgraph for $L = 5$.

(c) The optimal subgraph for $L = 4$.

Fig. 7.6. The CARDINALITY CONSTRAINED CRITICAL NODE DETECTION PROBLEM is solved for the 77 primary interactions during the development of *D. melanogaster*.

Table 7.2. Results of the IP model and the genetic algorithm for the 77 primary interactions of *D. melanogaster* development.

Instance	IP Model		Genetic Alg		Comb. Alg	
Max Conn. Index (L)	Obj Val	Comp Time (s)	Obj Val	Comp Time (s)	Obj Val	Comp Time (s)
2	17	0.87	17	0.17	17	0.04
3	14	36.5	14	0.27	17	0.03
4	12	276.29	12	0.19	17	0.01
5	10	382.88	10	0.28	17	0.02

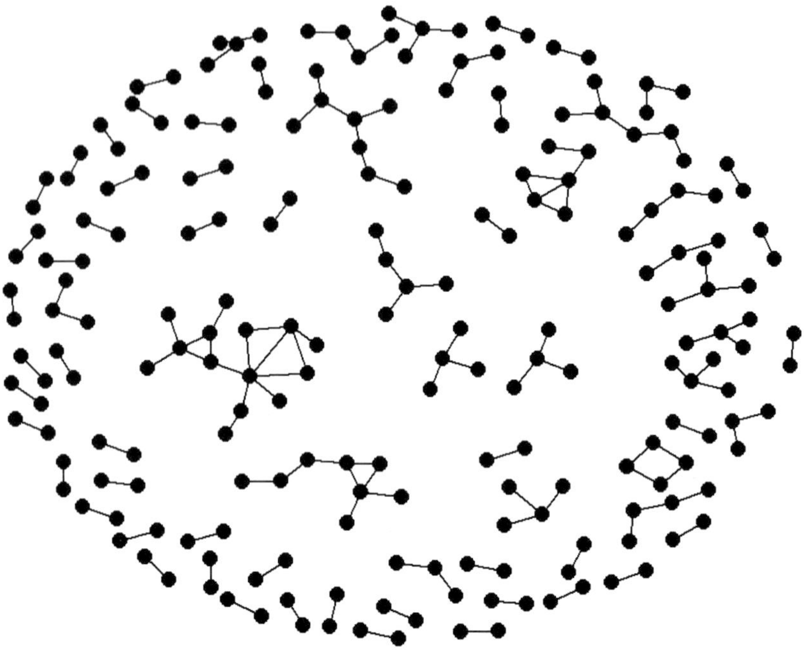

Fig. 7.7. The graph shows 186 yeast two-hybrid system interactions of *S. cerevisiae* proteins.

Table 7.3. Results of the genetic algorithm and the combinatorial heuristic for the 186 yeast two-hybrid system interactions of *S. cerevisiae*.

Instance	Genetic Alg		Comb. Alg	
Max Conn. Index (L)	Obj Val	Comp Time (s)	Obj Val	Comp Time (s)
2	31	7.64	31	23.63
3	20	2.88	20	0.08
4	7	1.92	7	0.03
5	6	3.35	6	0.03
6	3	3.49	3	0.01

networks found in the literature. It is not uncommon for these graphs to contain tens of thousands of nodes and arcs [16]. Clearly, more sophisticated algorithms are required for graphs of this size.

7.6. Conclusions

In this chapter, we have identified an important practical application of the recently introduced problem of detecting critical nodes to protein-protein interaction networks. As indicated above, in many cases, potentially harmful biological organisms, such as bacteria and viruses that cause diseases, can be studied in terms of their protein-protein interaction patterns. Therefore, finding the critical nodes corresponding to the proteins that are the most important for the integrity of the network would be very helpful in terms of identifying the proteins that need to be targeted in the efficient process of destroying this network and neutralizing the corresponding organisms. This approach can potentially be used in drug design applications, e.g., in developing drugs that target specific proteins that are the most "important" in the considered networks. It would be interesting to consider protein-protein interaction networks corresponding to different dangerous viruses, such as HIV (although obtaining detailed information on these interactions certainly represents another challenge) and identify key proteins that need to be targeted to ensure that these networks are sufficiently disconnected.

In addition to their potential important practical applications, the considered problems need to be studied from the computational perspective as well. As indicated above, these problems are $\mathcal{NP}$-hard, and the available exact and heuristic methods do not always perform sufficiently well, especially on large problem instances. Clearly, large-scale protein-protein interaction networks can provide valuable information about the structure of complex molecules and organisms; therefore, efficient techniques for solving the considered problems on massive networks need to be developed.

Overall, we believe that the area of research proposed in this chapter is promising and challenging due to multiple reasons; therefore, this research clearly needs to be conducted further, including both biological and mathematical aspects.

References

[1] C. Alves, P. M. Pardalos, and L. N. Vicente, editors. *Optimization in Medicine*. Springer, 2008.

[2] A. Arulselvan, C. W. Commander, L. Elefteriadou, and P. M. Pardalos. Detecting critical nodes in sparse graphs. *Computers and Operations Research*, under revision, 2008.

[3] A. Arulselvan, C. W. Commander, P. M. Pardalos, and O. Shylo. Managing network risk via critical node identification. In N. Gulpinar and B. Rustem, editors, *Risk Management in Telecommunication Networks*. Springer, 2008.

[4] B. Balasundaram, S. Butenko, and S. Trukhanov. Novel approaches for analyzing biological networks. *Journal of Combinatorial Optimization*, 10:23–29, 2005.
[5] A. M. Bonvin. Flexible protein-protein docking. *Current Opinion in Structural Biology*, 16(2):194–200, April 2006.
[6] S. Butenko and W. Wilhelm. Clique-detection models in computational biochemistry and genomics. *European Journal of Operational Research*, to appear, 2008.
[7] C. W. Commander. *Optimization Problems in Telecommunications with Military Applications*. PhD thesis, University of Florida, 2007.
[8] C. W. Commander, P. M. Pardalos, V. Ryabchenko, S. Uryasev, and G. Zrazhevsky. The wireless network jamming problem. *Journal of Combinatorial Optimization*, 14(4):481–498, 2007.
[9] ILOG CPLEX. http://www.ilog.com/products/cplex, accessed March 2008.
[10] C. Darwin. *The Origin of Species*. Murray, sixth edition, 1872.
[11] J. Drews. Drug discovery: A historical perspective. *Science*, 287:1960–1964, March 2000.
[12] M. R. Garey and D. S. Johnson. *Computers and Intractability: A Guide to the Theory of NP-Completeness*. W.H. Freeman and Company, 1979.
[13] D. E. Goldberg. *Genetic Algorithms in Search, Optimization, and Machine Learning*. Addison-Wesley, 1989.
[14] J. J. Gray. High-resolution protein-protein docking. *Current Opinion in Structural Biology*, 16(2):183–93, April 2006.
[15] C. Herzberg, L. A. Weidinger, B. Dorrbecker, S. Hubner, J. Stulke, and F. M. Commichau. Spine: A method for the rapid detection and analysis of protein-protein interactions *in vivo*. *Proteomics*, 7(22):4032–5, November 2007.
[16] T. Ito, T. Chiba, R. Ozawa, M. Yoshida, M. Hattori, and Y. Sakaki. A comprehensive two-hybrid analysis to explore the yeast protein interactome. *Proceedings of the National Academy of Sciences*, 98(8):4569–74, 2001.
[17] T. Ito, K. Tashiro, S. Muta, R. Ozawa, T. Chiba, M. Nishizawa, K. Yamamoto, S. Kuhara, and Y. Sakaki. Toward a protein-protein interaction map of the budding yeast: A comprehensive system to examine two-hybrid interactions in all possible combinations between the yeast proteins. *Proceedings of the National Academy of Sciences*, 97(3):1143–1147, 2000.
[18] V. Krebs. Uncloaking terrorist networks. *First Monday*, **7**, 2002.
[19] R. P. Mondaini and P. M. Pardalos, editors. *Mathematical Modeling of Biosystems*. Springer, 2008.
[20] Conference on Drug Discovery Chemistry. La Jolla, CA, April 28-30 2008.
[21] P. M. Pardalos, V. Boginski, and A. Vazacopoulos, editors. *Data Mining in Biomedicine*. Springer, 2007.
[22] P. M. Pardalos and J. Principe, editors. *Biocomputing*. Kluwer Academic Publishers, 2002.
[23] P. M. Pardalos, J. C. Sackellares, P. Carney, and L. Iasemidis, editors. *Quantitative Neuroscience*. Kluwer Academic Publishers, 2004.
[24] S. L. Schreiber. Target-oriented and diversity-oriented organic synthesis in drug discovery. *Science*, 275(5460):1964–1969, 2000.
[25] B. Schwikowski, P. Uetz, and S. Fields. A network of protein-protein interactions in yeast. *Nature Biotechnology*, 18(12):1257–61, December 2000.

[26] M. Selbach and M. Mann. Protein interaction screening by quantitative immunoprecipitation combined with knockdown (quick). *Nature Methods*, 3(12):981–3, December 2006.

[27] O. Seref, E. Kundakcioglu, and P. M. Pardalos, editors. *Data Mining, Systems Analysis and Optimization in Biomedicine*. Springer, 2008.

[28] M. Suchanek, A. Radzikowska, and C. Thiele. Photo-leucine and photo-methionine allow identification of protein-protein interactions in living cells. *Nature Methods*, 2(4):261–7, April 2005.

[29] `http:\\www.genome.jp`, accessed March 2008.

Chapter 8

Faster Algorithms for Constructing a Concept (Galois) Lattice

Vicky Choi
Department of Computer Science, Virginia Tech, USA
vchoi@cs.vt.edu

In this paper, we present a fast algorithm for constructing a concept (Galois) lattice of a binary relation, including computing all concepts and their lattice order. We also present two efficient variants of the algorithm, one for computing all concepts only, and one for constructing a frequent closed itemset lattice. The running time of our algorithms depends on the lattice structure and is faster than all other existing algorithms for these problems.

8.1. Introduction

Formal Concept Analysis (FCA) [13] has found many applications since its introduction. As the size of datasets grows, such as data generated from high-throughput technologies in bioinformatics, there is a need for efficient algorithms for constructing concept lattices. The input of FCA consists of a triple $(\mathcal{O}, \mathcal{M}, \mathcal{I})$, called *context*, where $\mathcal{O}$ is a set of objects, $\mathcal{M}$ is a set of attributes, and $\mathcal{I}$ is a binary relation between $\mathcal{O}$ and $\mathcal{M}$. In FCA, the context is structured into a set of *concepts*. The set of all concepts, when ordered by set-inclusion, satisfies the properties of a *complete lattice*. The lattice of all concepts is called *concept* [33] or *Galois* [4] lattice. When the binary relation is represented as a bipartite graph, each concept corresponds to a maximal bipartite clique (or maximal biclique). There is also a one-one correspondence of a closed itemset [37] studied in data mining and a concept in FCA. The one-one correspondence of all these terminologies – concepts in FCA, maximal bipartite cliques in theoretical computer science (TCS), and closed itemsets in data mining (DM) – was known [6, 37]. There is extensive work of the related problems in these three communities, in TCS [1, 6, 10, 14, 15, 20, 28] in FCA [3, 5, 11–13, 16–19, 21–23, 30, 31], and in DM [2, 7, 9, 24–27, 32, 34–37]. In general, in TCS, the research focuses on efficiently enumerating all maximal bipartite cliques (of a bipartite graph); in FCA, one is interested in the lattice structure of all concepts; in DM, one is often interested in computing frequent closed itemsets only.

Time complexity. Given a bipartite graph, it is not difficult to see that there can be exponentially many maximal bipartite cliques. For problems with potentially exponential (in the size of the input) size output, in their seminal paper [14], Johnson et al introduced several notions of *polynomial time* for algorithms for these problems: *polynomial total time*, *incremental polynomial time*, *polynomial delay time*. An algorithm runs in polynomial total time if the time is bounded by a polynomial in the size of the input and the size of the output. An algorithm runs in incremental polynomial time if the time required to generate a successive output is bounded by the size of input and the size of output generated thus far. An algorithm runs in polynomial delay time if the generation of each output is only polynomial in the size of input. It is not difficult to see that polynomial delay is stronger than incremental polynomial (namely an algorithm with polynomial delay time is also running in incremental polynomial), which is stronger than polynomial total time. polynomial delay algorithm, we can further distinguish if the space used is polynomial or exponential in the input size.

Previous work. Observe that the maximal bipartite clique (MBC) problem is a special case of the maximal clique problem in a general graph. Namely, given a bipartite graph $G = (V_1, V_2, E)$, a maximal bipartite clique corresponds to a maximal clique in $\breve{G} = (V_1 \cup V_2, \breve{E})$ where $\breve{E} = E \cup (V_1 \times V_1) \cup (V_2 \times V_2)$. Consequently, any algorithm for enumerating all maximal cliques in a general graph [14, 28], also solves the MBC problem. In fact, the best known algorithm in enumerating all maximal bipartite cliques, which was proposed by Makino and Uno [20] that takes $O(\Delta^2)$ polynomial delay time where Δ is the maximum degree of G, was based on this approach. The fact that the set of maximal bipartite cliques constitutes a lattice was not observed in the paper and thus the property was not utilized for the enumeration algorithm.

In FCA, much of research has been devoted to study the properties of the lattice structure. There are several algorithms [19, 22, 31], that construct the lattice, i.e. computing all concepts together with its lattice order. There are also some algorithms that compute only concepts [13, 21]. (We remark that the idea of using a total *lectical order* on concepts Ganter's algorithm [13] is also used for enumerating maximal (bi)cliques.) [14, 20] See [18] for a comparison studies of these algorithms. The best polynomial total time algorithm was by Nourine and Raynaud [22] with $O(nm|\mathcal{B}|)$ time and $O(n|\mathcal{B}|)$ space, where $n = |\mathcal{O}|$ and $m = |\mathcal{M}|$

and $\mathcal{B}$ denote the set of all concepts. This algorithm can be easily modified to run in $O(mn)$ incremental time [23]. Observe that the space of total size of all concepts is needed if one is to keep the entire structure explicitly. There were several other algorithms [13, 19], all run in $O(n^2m)$ polynomial delay. There is another algorithm [31] that is based on divide-and-conquer approach, but the analytical running time of the algorithm is unknown as it is difficult to analyze.

There are several algorithms in data mining for computing frequent closed itemsets, such as CHARM(-L) [35, 36], CLOSET(+) [25, 32] and LCM [29]. Boros et al. [6] gave an algorithm with $O(m^2n)$ incremental polynomial running time, where $n = |\mathcal{O}|$ and $m = |\mathcal{M}|$.

Our Results. In this paper, by making use of the lattice structure of concepts, we present a simple and fast algorithm for computing all concepts together with its lattice order. The main idea of the algorithm is that given a concept, when all of its successors are considered together (i.e. in a batch manner), they can be efficiently computed. We compute concepts in the Breadth First Search (BFS) order – the ordering given by BFS traversal of the lattice. When computing the concepts in this way, not only do we compute all concepts but also we identify all successors of each concept. Another idea of the algorithm is that we make use of the concepts generated to dynamically update the adjacency relations. The running time of our algorithm is $O(\sum_{a\in \mathsf{ext}(C)} |\mathsf{cnbr}(a)|)$ polynomial delay for each concept C (see Section 8.2 for related background and terminology), where $\mathsf{cnbr}(a)$ is the reduced adjacency list of a. Our algorithm is faster than the best known algorithms for constructing a lattice because the algorithm is faster than a basic algorithm that runs in $O(\sum_{a\in \mathsf{ext}(C)} |\mathsf{nbr}(a)|)$, where $|\mathsf{nbr}(a)|$ is number of attributes adjacent to the object a, and this basic algorithm is already as fast as the current best algorithms for the problem.

We also present two variants of the algorithm: one is computing all concepts only and another is constructing the frequent closed itemset lattice. Both algorithms are faster than the current start-of-the-art program for these problems.

Outline. The paper is organized as follows. In Section 8.2, we review some background and notation on FCA. In Section 8.3, we describe some basic properties of concepts that we use in our lattice-construction algorithm. In Section 8.4, we first describe the high level idea of our algorithm. Then we describe how to efficiently implement the algorithm. In Section 8.5, we describe two variants of the algorithm. One is for computing all concepts only and another is for constructing a frequent closed itemset lattice. We conclude with discussion in Section 8.6.

8.2. Background and Terminology on FCA

In FCA, a triple $(\mathcal{O}, \mathcal{M}, \mathcal{I})$ is called a *context*, where $\mathcal{O} = \{g_1, g_2, \ldots, g_n\}$ is a set of n elements, called *objects*; $\mathcal{M} = \{1, 2, \ldots, m\}$ is a set of m elements, called *attributes*; and $\mathcal{I} \subseteq \mathcal{O} \times \mathcal{M}$ is a binary relation. The context is often represented by a *cross-table* as shown in Fig. 8.1. A set $X \subseteq \mathcal{O}$ is called an *object set*, and a set $J \subseteq \mathcal{M}$ is called an *attribute set*. Following the convention, we write an object set $\{a, c, e\}$ as ace, and an attribute set $\{1, 3, 4\}$ as 134.

For $i \in \mathcal{M}$, denote the adjacency list of i by $\mathsf{nbr}(i) = \{g \in \mathcal{O} : (g, i) \in \mathcal{I}\}$. Similarly, for $g \in \mathcal{O}$, denote the adjacency list of g by $\mathsf{nbr}(g) = \{i \in \mathcal{M} : (g, i) \in \mathcal{I}\}$.

Definition 8.1. The function $\mathsf{attr} : 2^{\mathcal{O}} \longrightarrow 2^{\mathcal{M}}$ maps a set of objects to their common attributes: $\mathsf{attr}(X) = \cap_{g \in X} \mathsf{nbr}(g)$, for $X \subseteq \mathcal{O}$. The function $\mathsf{obj} : 2^{\mathcal{M}} \longrightarrow 2^{\mathcal{O}}$ maps a set of attributes to their common objects: $\mathsf{obj}(J) = \cap_{j \in J} \mathsf{nbr}(j)$, for $J \subseteq \mathcal{M}$.

It is easy to check that for $X \subseteq \mathcal{O}$, $X \subseteq \mathsf{obj}(\mathsf{attr}(X))$, and for $J \subseteq \mathcal{M}$, $J \subseteq \mathsf{attr}(\mathsf{obj}(J))$. Note $\mathsf{obj}(\emptyset) = \mathcal{O}$ and $\mathsf{attr}(\emptyset) = \mathcal{M}$.

Definition 8.2. An object set $X \subseteq \mathcal{O}$ is *closed* if $X = \mathsf{obj}(\mathsf{attr}(X))$. An attribute set $J \subseteq \mathcal{M}$ is closed if $J = \mathsf{attr}(\mathsf{obj}(J))$.

The composition of obj and attr induces a *Galois connection* between $2^{\mathcal{O}}$ and $2^{\mathcal{M}}$. Readers are referred to [13] for properties of the Galois connection.

Definition 8.3. A pair $C = (A, B)$, with $A \subseteq \mathcal{O}$ and $B \subseteq \mathcal{M}$, is called a *concept* if $A = \mathsf{obj}(B)$ and $B = \mathsf{attr}(A)$.

For a concept $C = (A, B)$, by definition, both A and B are closed. The object set A is called the *extent* of C, written as $A = \mathsf{ext}(C)$, and the attribute set B is called the *intent* of C, and written as $B = \mathsf{int}(C)$. The set of all concepts of the context $(\mathcal{O}, \mathcal{M}, \mathcal{I})$ is denoted by $\mathcal{B}(\mathcal{O}, \mathcal{M}, \mathcal{I})$ or simply $\mathcal{B}$ when the context is understood.

Let (A_1, B_1) and (A_2, B_2) be two concepts in $\mathcal{B}$. Observe that if $A_1 \subseteq A_2$, then $B_2 \subseteq B_1$. We order the concepts in $\mathcal{B}$ by the following relation $\prec$:

$$(A_1, B_1) \prec (A_2, B_2) \Longleftrightarrow A_1 \subseteq A_2 (B_2 \subseteq B_1).$$

It is not difficult to see that the relation $\prec$ is a partial order on $\mathcal{B}$. In fact, $\mathcal{L} =< \mathcal{B}, \prec>$ is a complete lattice and it is known as the *concept* or *Galois* lattice of the context $(\mathcal{O}, \mathcal{M}, \mathcal{I})$. For $C, D \in \mathcal{B}$ with $C \prec D$, if for all $E \in \mathcal{B}$ such that

$C \prec E \prec D$ implies that $E = C$ or $E = D$, then C is called the *successor* [a](or *lower neighbor*) of D, and D is called the *predecessor* (or *upper neighbor*) of C. The diagram representing an ordered set (where only successors/predecessors are connected by edges) is called a *Hasse diagram* (or a line diagram). See Fig. 8.1 for an example of the line diagram of a Galois lattice.

For a concept $C = (\mathsf{ext}(C), \mathsf{int}(C))$, $\mathsf{ext}(C) = \mathsf{obj}(\mathsf{int}(C))$ and $\mathsf{int}(C) = \mathsf{attr}(\mathsf{ext}(C))$. Thus, C is uniquely determined by either its extent, $\mathsf{ext}(C)$, or by its intent, $\mathsf{int}(C)$. We denote the concepts restricted to the objects $\mathcal{O}$ by $\mathcal{B}_{\mathcal{O}} = \{\mathsf{ext}(C) : C \in \mathcal{B}\}$, and the attributes $\mathcal{M}$ by $\mathcal{B}_{\mathcal{M}} = \{\mathsf{int}(C) : C \in \mathcal{B}\}$. For $A \in \mathcal{B}_{\mathcal{O}}$, the corresponding concept is $(A, \mathsf{attr}(A))$. For $J \in \mathcal{B}_{\mathcal{M}}$, the corresponding concept is $(\mathsf{obj}(J), J)$. The order $\prec$ is completely determined by the inclusion order on $2^{\mathcal{O}}$ or equivalently by the reverse inclusion order on $2^{\mathcal{M}}$. That is, $\mathcal{L} =< \mathcal{B}, \prec>$ and $\mathcal{L}_{\mathcal{M}} =< \mathcal{B}_{\mathcal{M}}, \supseteq>$ are order-isomorphic. We have the property that $(\mathsf{obj}(Z), Z)$ is a successor of $(\mathsf{obj}(X), X)$ in $\mathcal{L}$ if and only if Z is a successor of X in $\mathcal{L}_{\mathcal{M}}$. Since the set of all concepts is finite, the lattice order relation is completely determined by the covering (successor/predecessor) relation. Thus, to construct the lattice, it is sufficient to compute all concepts and identify all successors of each concept.

8.3. Basic Properties

In this section, we describe some basic properties of the concepts on which our lattice construction algorithms are based.

Proposition 8.1. *Let C be a concept in $\mathcal{B}(\mathcal{O}, \mathcal{M}, \mathcal{I})$. For $i \in \mathcal{M} \setminus \mathsf{int}(C)$, if $E_i = \mathsf{ext}(C) \cap \mathsf{nbr}(i)$ is not empty, E_i is closed. Consequently, $(E_i, \mathsf{attr}(E_i))$ is a concept.*

Proof. For $i \in \mathcal{M} \setminus \mathsf{int}(C)$, suppose that $E_i = \mathsf{ext}(C) \cap \mathsf{nbr}(i)$ is not empty. We will show that $\mathsf{obj}(\mathsf{attr}(E_i)) = E_i$. Since $E_i \subseteq \mathsf{obj}(\mathsf{attr}(E_i))$, it remains to show that $\mathsf{obj}(\mathsf{attr}(E_i)) \subseteq E_i$. By definition, $\mathsf{obj}(\mathsf{int}(C) \cup \{i\}) = (\cap_{j \in \mathsf{int}(C)} \mathsf{nbr}(j)) \cap \mathsf{nbr}(i) = \mathsf{ext}(C) \cap \mathsf{nbr}(i) = E_i$. Thus, $(\mathsf{int}(C) \cup \{i\}) \subseteq \mathsf{attr}(\mathsf{obj}(\mathsf{int}(C) \cup \{i\})) = \mathsf{attr}(E_i)$. Consequently, $\mathsf{obj}(\mathsf{attr}(E_i)) \subseteq \mathsf{obj}(\mathsf{int}(C) \cup \{i\}) = E_i$. □

Example. Consider the concept $C = (abcd, \emptyset)$ of context in Fig. 8.1, we have $E_1 = abc, E_2 = bd, E_3 = ac, E_4 = bd$.

[a]Some authors called this as immediate successor.

	1	2	3	4
a	×		×	
b	×	×		×
c	×		×	
d		×		×

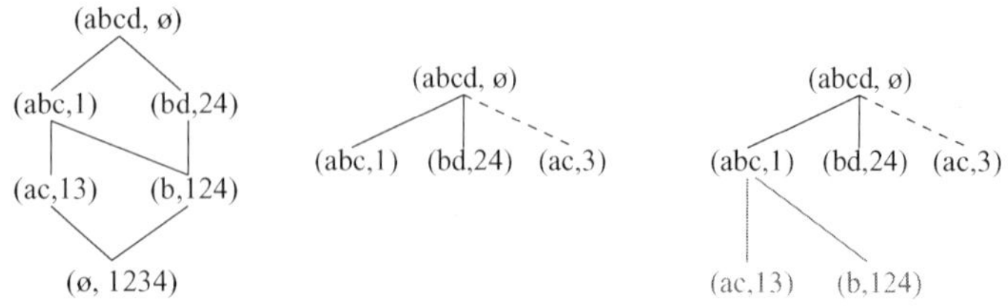

(a) (b) (c)

Fig. 8.1. (a) A context $(\mathcal{O}, \mathcal{M}, \mathcal{I})$ with $\mathcal{O} = \{a, b, c, d\}$ and $\mathcal{M} = \{1, 2, 3, 4\}$. The cross $\times$ indicates a pair in the relation $\mathcal{I}$. (b) The corresponding Galois/concept lattice. (c) $\mathsf{Child}(abcd, \emptyset) = \{(abc, 1), (bd, 24), (ac, 3)\}$; $\mathsf{Child}(abc, 1) = \{(ac, 13), (b, 124)\}$.

8.3.1. *Defining the Equivalence Classes*

For a closed attribute set $X \subset \mathcal{M}$, denote the set of remaining attributes $\{i \in \mathcal{M} \setminus X : \mathsf{obj}(X) \cap \mathsf{nbr}(i) \neq \emptyset\}$ by $\mathsf{res}(X)$. Consider the following equivalence relation $\sim$ on $\mathsf{res}(X)$: $i \sim j \iff \mathsf{obj}(X) \cap \mathsf{nbr}(i) = \mathsf{obj}(X) \cap \mathsf{nbr}(j)$, for $i \neq j \in \mathsf{res}(X)$.

Let $S_1, \ldots, S_t$ be the equivalence classes induced by $\sim$, i.e. $\mathsf{res}(X) = S_1 \cup \ldots \cup S_t$, and $\mathsf{obj}(X) \cap \mathsf{nbr}(i) = \mathsf{obj}(X) \cap \mathsf{nbr}(j)$ for any $i \neq j \in S_k$, $1 \leq k \leq t$. We denote the set $\{S_1, \ldots, S_t\}$ by $\mathsf{AttrChild}(X)$. We call S_j the sibling of S_i for $j \neq i$. For convenience, we will write $X \cup S_i$ by XS_i. When there is no confusion, we abuse the notation by writing $X \cup \mathsf{AttrChild}(X) = \{XS : S \in \mathsf{AttrChild}(X)\}$. Note that by definition, $\mathsf{obj}(XS_k) = \mathsf{obj}(X) \cap \mathsf{obj}(S_k) = \mathsf{obj}(X) \cap \mathsf{nbr}(i)$ for some $i \in S_k$. We denote the pairs $\{(\mathsf{obj}(XS_1), XS_1), \ldots, (\mathsf{obj}(XS_1), XS_1)\}$ by $\mathsf{Child}(\mathsf{obj}(X), X)$.

Recall that $\mathcal{L} =< \mathcal{B}, \prec>$ and $\mathcal{L}_{\mathcal{M}} =< \mathcal{B}_{\mathcal{M}}, \supseteq>$ are order-isomorphic. We have the property that $(\mathsf{obj}(Y), Y)$ is a successor of $(\mathsf{obj}(X), X)$ in $\mathcal{L}$ if and only if Y is a successor of X in $\mathcal{L}_{\mathcal{M}}$. For each $S \in \mathsf{AttrChild}(X)$, we call XS a child of X and X a parent of XS. By the definition of the equivalence class, for each Z that is a successor of X, there exists a $S \in \mathsf{AttrChild}(X)$ such that $Z = XS$. That is, if Z is a successor of X, Z is a child of X.

Let $\mathsf{Succ}(X)$ denote all the successors of X, then we have $\mathsf{Succ}(X) \subseteq X \cup \mathsf{AttrChild}(X)$. However, not every child of X is a successor of X. For the example in Fig. 8.1, $\mathsf{AttrChild}(\emptyset) = \{1, 24, 3\}$, where 1 and 24 are successors of $\emptyset$ but 3 is not. $\mathsf{Succ}(\emptyset) = \{1, 24\} \subset \mathsf{AttrChild}(\emptyset)$; while $\mathsf{AttrChild}(1) = \{24, 3\}$, $\mathsf{Succ}(a) = \{124, 13\} = 1 \cup \mathsf{AttrChild}(1)$. Similarly, if P is a predecessor of X,

	1	2	3	4	5	6	7
a	×					×	
b	×		×	×	×	×	
c	×			×		×	
d		×	×		×		
e		×					×

(a)

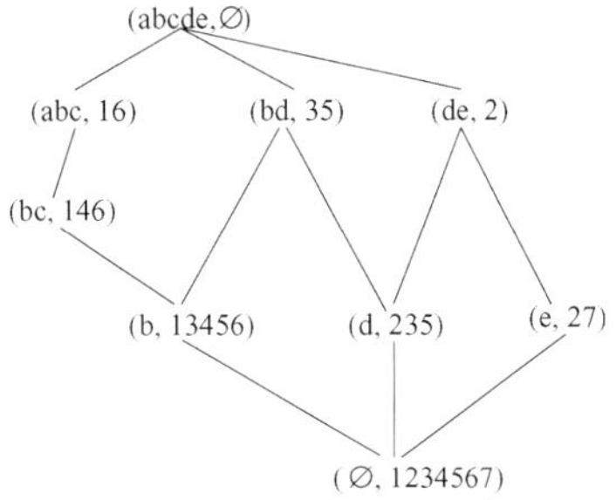

(b)

	16	2	**35**	4	7
a	×				
b	×		×	×	
c	×			×	
d		×	×		
e		×			×

(c)

Fig. 8.2. (a) A context. (b) The corresponding concept lattice. (c) Reduced cross-table of $\mathsf{Child}(abcde, \emptyset)$ of the context.

level manner. More specifically, for a concept $C = (\mathsf{obj}(X), X)$, we first compute all its children $\mathsf{Child}(C)$. Then we dynamically update the adjacency lists by representing the attributes in each child of C with one single element. We then use these condensed adjacency lists to process each child of C. That is, instead of using the global adjacency lists, when processing $(\mathsf{obj}(XS), XS)$, we use the condensed adjacency lists of its parent. It takes $O(\sum_{S \in \mathsf{AttrChild}(X)} |\mathsf{obj}(XS)|)$ for C to generate its condensed adjacency lists $\mathsf{cnbr}()$ (see Algorithm 8.3 in the Appendix for the pseudo-code). And the time for the procedure SPROUT is $O(\sum_{a \in \mathsf{obj}(X)} |\mathsf{cnbr}(a)|)$ (see Algorithm 8.2 in the Appendix for the pseudo-code). Notice that $\sum_{a \in \mathsf{obj}(X)} |\mathsf{cnbr}(a)| > \sum_{S \in \mathsf{AttrChild}(X)} |\mathsf{obj}(XS)|$, the time for updating the adjacency lists is subsumed by the time required for procedure SPROUT. Therefore, our new running time is $O(\sum_{a \in \mathsf{obj}(X)} |\mathsf{cnbr}(a)|)$ for each concept $(\mathsf{obj}(X), X)$. See Algorithm 8.4 in the Appendix for the pseudo-code and Fig. 8.3 for a step-by-step illustration of the algorithm.

8.5. Variants of the Algorithm

For some applications, one is not interested in the entire concept lattice. In the following, we will describe how to modify our algorithm to solve two special cases: enumerating all concepts only and constructing a frequent closed itemset lattice.

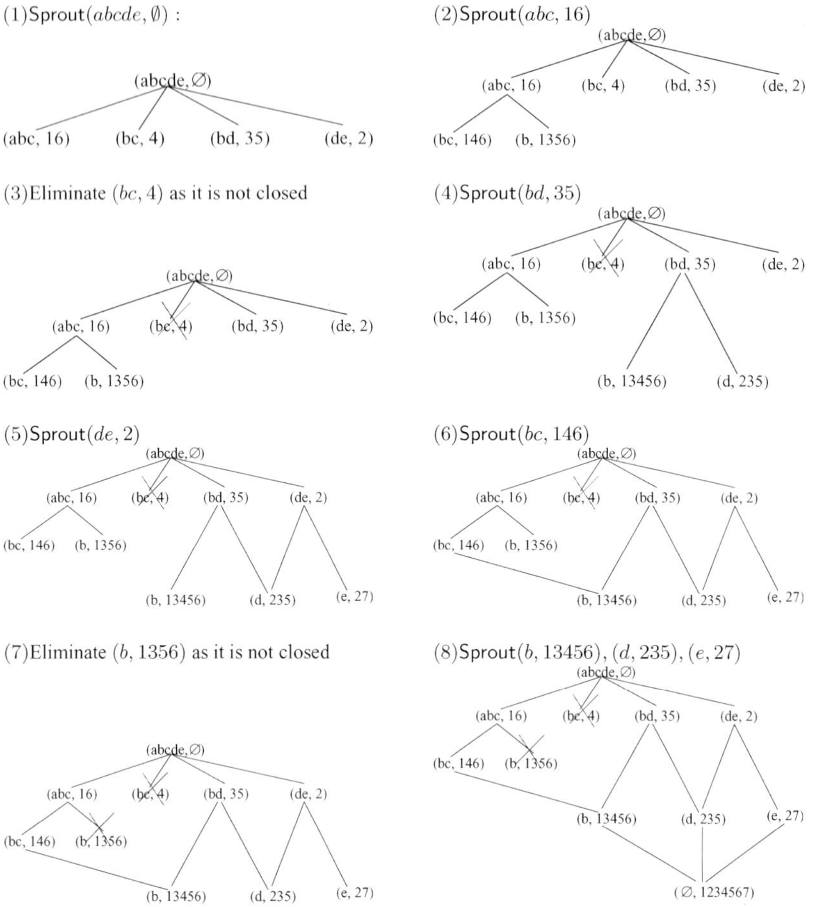

Fig. 8.3. Step by step illustration of the 2-level BFS lattice construction algorithm. The context and the corresponding lattice are shown in Fig. 8.2.

8.5.1. *Algorithm 2: Computing All Concepts or Maximal Bipartite Cliques*

If one is interested in computing all the concepts and not in their lattice order, as in enumerating all maximal bicliques studied in [20]. We can easily modify our algorithm to give an even faster algorithm for this purpose. This is because in our algorithm, each concept is generated many times, more precisely, at least number of its predecessors times. For example in Fig. 8.3, $(d, 235)$ is generated twice, one by each of its predecessor. However, when we need all concepts only, we do not need regenerate the concepts again and again. This can be easily accomplished by considering the right siblings only in the procedure SPROUT, i.e. changing the line 3 to for $i \in \mathsf{nbr}(a)$ AND $i > s$ do, while the other parts of the algorithm remain the same. Depending on the lattice structure, this can significantly speed up the algorithm as the number of siblings is decreasing in a cascading fashion. A more careful analysis is needed for the running time of this algorithm.

8.5.2. *Algorithm 3: Constructing a Closed Itemset Lattice*

In data mining, one is interested in large concepts, i.e. $(\mathsf{obj}(X), X)$ where $|\mathsf{obj}(X)|$ is larger than a threshold. Although our algorithm can naturally be modified to construct such a closed itemset lattice: we stop processing a concept when the size of its object set is less than the given threshold, where objects correspond to transactions and attributes correspond to items. Theoretically, when the memory requirement is not a concern, our algorithm is faster than all other existing algorithms (including the state-of-art program CHARM-L) for constructing such a frequent closed itemset lattice. However, in practice, for large data sets (as those studied in data mining), the data structure – a trie on objects (transactions) – requires huge memory and this may threaten the algorithm's practical efficiency. However, it is not difficult to modify our algorithm so that a trie on attributes (items) instead is used. Recall that a trie on objects are required in two steps of our algorithm: testing the closure of an attribute set and testing the existence of a concept. As noted above, the existence of a concept can also be tested on its intent (i.e. attributes), thus we can use a trie on attributes for testing the existence of a concept. To avoid using a trie on objects for testing the closure of an attribute set, we can use the first characterization in Proposition 8.3 instead, that is, we test the closure of an attribute set by using *subset testing* of its object set against its siblings' object set, as described in Section 8.3. Further, we can employ the practically efficient technique *diffset* as in CHARM(-L) for both our SPROUT procedure and subset testing operations. We are testing the performance of the diffset based implementation on the available benchmarks and the results will be reported elsewhere.

8.6. Discussion

Our interest in FCA stems from our research in microarray data analysis [8]. We have implemented an not yet optimized version of our algorithm (with less than 500 effective lines in C++). The program is very efficient for our applications, in which our data consists of about 10000 objects and 29 attributes. It took less than 1 second for the program to produce the concept lattice (about 530 vertices/concepts and 1500 edges) in a Pentium IV 3.0GHz computer with 2G memory running under Fedora 2 linux OS.

As FCA finds more and more applications, especially in bioinformatics, efficient algorithms for constructing concept/Galois lattices are much needed. Our algorithm is faster than the existing algorithms for this problem, nevertheless, it seems to have much room to improve. Furthermore, our algorithm can be easily modified to compute the minimal generators for redescription mining [38] directly.

Acknowledgment

We would like to thank Reinhard Laubenbacher for introducing us FCA. We thank Naren Ramakrishnan for the discussion about redescription mining.

References

[1] J. Abello, A. Pogel, L. Miller. Breadth first search graph partitions and concept lattices. *J. of Universal Computer Science*, 10(8), p934–954, 2004.

[2] F. Afrati, A. Gionis, H. Mannila. Approximating a collection of frequent sets. *Proc. 10th ACM SIGKDD International Conference on Knowledge Discovery and Data mining*, p12–19, 2004.

[3] F. Baklouti, R.E. Grarvy. A fast and general algorithm for Galois lattices building. *J. of Symbolic Data Analysis*, 3(1), p19-31, 2005. www.icons.rodan.pl/publications/

[4] M. Barbut and B. Montjardet. Ordre et Classifications: Algebre et combinatoire. Hachette, 1970.

[5] J. Besson, C. Robardet, J-F. Boulicaut. Constraint-based mining of formal concepts in transactional data. *Proceedings of the 8th Pacific-Asia Conference on Knowledge Discovery and Data Mining PaKDD04*, 2004. Springer-Verlag LNCS 3056, pp. 615-624.

[6] E. Boros, V. Gurvich, L. Khachiyan, K. Makino. On maximal frequent and minimal infrequent sets in binary matrices. *Annals of Mathematics and Artificial Intelligence*, 39, p211–221, 2003. (STACS 2002)

[7] T. Calders, C. Rigotti, J.F. Boulicaut. A survey on condensed representations for frequent sets. *Constrained-based Mining*, Springer, 3848, 2005.

[8] V. Choi, Y. Huang, V. Lam, D. Potter, R. Laubenbacher, K. Duca. Using Formal Concept Analysis for Microarray Data Comparison. *DIMACS Workshop on Clustering Problems in Biological Networks.* Piscataway, New Jersey, May 9–11, 2006. Manuscript in preparation.

[9] G. Cong, K.L. Tan, A.K.H. Tung, F. Pan. Mining frequent closed patterns in microarray data. *Proc. 4th IEEE International Conference on Data Mining*, p363–366, 2004.

factors for the 20 dimensions are random numbers uniformly distributed between 1 and 20.

Table 9.1. Dimensions of the Input Clusters

Input	Dimensions	Points
A	0, 2, 3, 10, 16, 17, 18	16768
B	9, 10, 12, 16, 17, 18, 19	23859
C	1, 3, 9, 12, 13, 14, 16	25678
D	1, 5, 9, 10, 12, 13, 14	23093
E	3, 6, 9, 10, 11, 12, 14	5602
Outlier		5000

Table 9.2 shows a typical result of the dimensions of the output clusters found by PROCLUS and IPROCLUS. We can find a good correspondence between the sets of dimensions of the output clusters found by IPROCLUS and their corresponding input clusters.

Table 9.2. Dimensions of the Output Clusters for IPROCLUS and PROCLUS

IPROCLUS	Dimensions	Points
1	1, 5, 9, 10, 12, 13, 14	23304
2	0, 2, 3, 10, 14, 16, 18	2802
3	9, 10, 12, 16, 17, 18, 19	25034
4	1, 3, 9, 12, 13, 14, 16	25982
5	0, 2, 3, 10, 16, 17, 18	15991
Outliers		6887

PROCLUS	Dimensions	Points
1	0, 2, 3, 4, 9, 12, 16, 17, 18	17660
2	3, 9, 12, 16, 17, 18	24971
3	3, 4, 9, 12, 16, 189	11844
4	1, 3, 9, 12, 13, 14, 16	19130
5	0, 3, 4, 8, 9, 12, 18	8236
Outliers		18159

Table 9.3 gives the confusion matrix for the output clusters in Table 9.2. Confusion matrix is defined in the same way as in PROCLUS paper. Entry (i, j) is equal to the number of data points assigned to output cluster i, which were generated as part of input cluster j. IPROCLUS discovers output clusters in which the majority of points come from one input cluster. In other words, it recognizes the natural clustering of the points. More specifically, we calculate the accuracy from the confusion matrix. In order to define the accuracy, for each output cluster i, we identify the input cluster j with which it shares the largest number of points. We say that output cluster i corresponds to input cluster j. All points in their common intersection are clustered correctly. All the other points in output cluster i are clus-

tered incorrectly. Therefore, we define the accuracy as the percentage of points that are correctly clustered by the algorithm. For the typical result, the accuracy of IPROCLUS is 91.45% and that of PROCLUS is 70.46%. Both algorithms are tested multiple times. On average, IPROCLUS and PROCLUS can achieve the accuracy of 91.90% and 70.18% respectively. In a word, IPROCLUS can achieve much better accuracy than PROCLUS for the scaled dataset in the random case.

Table 9.3. Confusion Matrix for IPROCLUS and PROCLUS

IPROCLUS

Input / Output	A	B	C	D	E	Outliers
1	3	0	97	22893	39	272
2	1525	1	60	120	428	668
3	0	23854	162	0	463	555
4	1	4	25031	15	348	583
5	14800	0	311	65	422	393
Outliers	439	0	17	0	3902	2529

PROCLUS

Input / Output	A	B	C	D	E	Outliers
1	16626	0	2	406	366	260
2	2	22794	12	1346	461	356
3	6	1	2565	6953	1221	1098
4	3	0	19116	0	0	11
5	131	361	310	4803	1450	1181
Outliers	0	703	3673	9585	2104	2094

Then we consider the accuracy for the unscaled dataset in the random case. The data used are generated in exactly the same way as in PROCLUS paper. The two algorithms have compatible results. The average accuracy of IPROCLUS and PROCLUS are 93.02% and 93.94% respectively.

Table 9.4 gives the average accuracy result for the two extreme cases. Similar to the random case, IPROCLUS exhibits much better accuracy than PROCLUS for the scaled datasets and compatible accuracy for the unscaled datasets.

Table 9.4. The Accuracy Result for the Two Extreme Cases

Data	PROCLUS	IPROCLUS
All-same-dimensions_scaled	75.27%	96%
All-same-dimensions_unscaled	95.91%	94.57%
No-common-dimensions_scaled	51.08%	87.75%
No-common-dimensions_unscaled	87.67%	86.87%

In summary, IPROCLUS can achieve much better accuracy than PROCLUS for the scaled data in all the three cases. When the unscaled data is considered, the two algorithms have compatible accuracy. This can be explained by the use of modified Manhattan segmental distance in IPROCLUS. Modified Manhattan segmental distance can effectively deal with scaled data while for unscaled data, it is quite close to Manhattan segmental distance which is used in PROCLUS. When comparing the three cases together, it can be seen that the no-common-dimensions case has the lowest accuracy rate and the all-same-dimensions case has the highest accuracy rate. It can be explained by the difference in the average number of dimensions in a cluster. When the average number of dimensions is low (in no-common-dimensions, l=4), there is a higher probability that data points are assigned to the wrong cluster since points are just correlated on 4 dimensions. However, in the all-same-dimensions case, where l=10, it is easier to correctly cluster points since the correlation between data points is stronger.

Second, we present the result of testing the dependence on l. The dependence is evaluated by the least square error of the number of dimensions. The same datasets in the three cases as in the accuracy test are used.

Figure 9.4 shows the results we get for the unscaled dataset in the random case. We get similar results for the extreme cases. We can see that IPROCLUS has less dependence on l than PROCLUS in terms of the number of dimensions. For the scaled data in the three cases, we have got similar trend and we don't give the figures here since PROCLUS has much higher error rate for the scaled datasets.

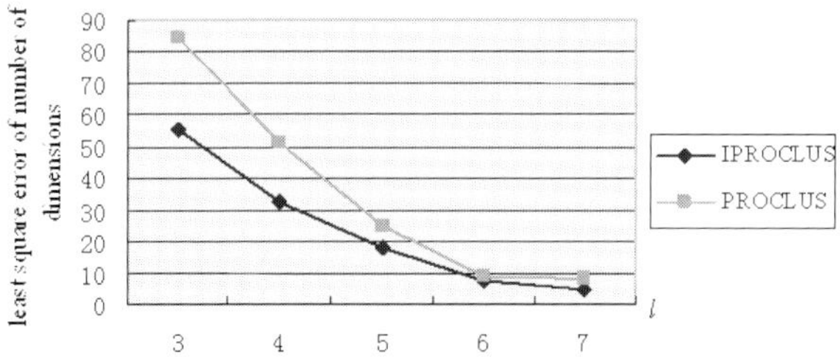

Fig. 9.4. Dependence of the Number of Dimensions on l

In summary, IPROCLUS greatly reduces the dependence on parameter l, since the dimension tuning process checks for additional dimensions for each cluster in order to add any dimension that can enable better clustering, while PROCLUS decide the average number of dimensions solely based on l.

For the running time test, we apply IPROCLUS and PROCLUS on different number of points. The datasets are generated in the same way as the datasets used in the previous two tests. The result we get is that the execution time of these two algorithms is comparable for all the three different cases. Since the simplified replacing logic offsets the running time increase from the additional steps for the dimension tuning, IPROCLUS as a whole doesn't cause any increase in the execution time.

9.4.3. *Results on the Colon Tumor Dataset*

One challenge of gene expression data to clustering algorithms is the huge number of genes (dimensions) involved. Projected clustering algorithms are designed to deal with high dimensionalities. We compared the performance of IPROCLUS and PROCLUS on the colon tumor dataset [4]. This dataset consists of the expression values on 2000 genes of 40 tumor and 22 normal colon tissue samples. Since each cell is either a tumor or a normal cell, we removed the outlier logic in both PROCLUS and IPROCLUS algorithms.

Table 9.5 gives a typical result for the two algorithms. The result is obtained when the k value is set to 2 since there are only two clusters and the l value is set to 124 which is based on experimental analysis. For the colon tumor dataset, IPROCLUS can correctly classify 52 out of the 62 tissues, achieving the accuracy of 83.9%, while PROCLUS can only achieve the accuracy of 53.2% (33 correctly classified). We can see that IPROCLUS can achieve much better accuracy on the colon tumor set than PROCLUS.

Table 9.5. Confusion Matrix for IPROCLUS and PROCLUS on the Colon Tumor Dataset

(a)IPROCLUS

Input / Output	Normal	Tumor
1	18	6
2	4	34

(b)PROCLUS

Input / Output	Normal	Tumor
1	5	17
2	12	28

9.5. Conclusion

Projected clustering in high dimensional space is an interesting research topic and several algorithms have been proposed. We have introduced two existing methods in this topic and mentioned their strengths and weaknesses. We have proposed an effective and efficient algorithm, IPROCLUS, which is based on PROCLUS. We have significantly improved the accuracy by proposing modified Manhattan segmental distance. We have reduced the dependence on user input l by adding the dimension tuning process at the end of the refinement phase and we have proposed a simplified replacing logic in the iterative phase to offset the running time increase caused by the dimension tuning process.

Empirical results have shown that IPROCLUS is able to accurately discover clusters embedded in lower dimensional subspaces. For the synthetic datasets, it can achieve much higher accuracy than PROCLUS for the scaled datasets while keeping compatible performance with PROCLUS for the unscaled datasets in all the three cases. Moreover, IPROCLUS has lower dependence on l than PROCLUS. We also apply our algorithm on the colon tumor dataset, IPROCLUS still achieves much higher accuracy than PROCLUS.

References

[1] C. C. Aggarwal, C. Procopiuc, J. L. Wolf, P. S. Yu, and J. S. Park. Fast algorithms for projected clustering. In *ACM SIGMOD International Conference on Management of Data*, 1999.

[2] C. C. Aggarwal and P. S. Yu. Finding generalized projected clusters in high dimensional spaces. In *ACM SIGMOD International Conference on Management of Data*, 2000.

[3] R. Agrawal, J. Gehrke, D. Gunopulos, and P. Raghavan. Automatic subspace clustering of high dimensional data for data mining applications. In *ACM SIGMOD International Conference on Management of Data*, 1998.

[4] U. Alon, N. Barkai, D. A. Notterman, K. Gish, S. Ybarra, D. Mack and A. J. Levine. Broad patterns of gene expression revealed by clustering analysis of tumor and normal colon tissues probed by oligonucleotide arrays. *Proc. Natl. Acad. Sci. U.S.A.*, 96, 6745-6750, 1999, `http://microarray.princeton.edu/oncology/affydata/index.html`

[5] K. Beyer, J. Goldstein, R. Ramakrishnan, and U. Shaft. When is "nearest neighbor" meaningful? *ICDT Conference*, 1999.

[6] R. O. Duda and P.E. Hart. *Pattern classification and scene analysis*. John Wiley and Sons, 1973.

[7] M. Easter, H. P. Kriegel, J. Sander, and X. Xu. A density-based algorithm for discovering clusters in large spatial databases with noise. In *Proceedings of the 2nd International Conference on Knowledge Discovery and Data Mining*, Portland, Oregon, August 1996.

[8] T. Gonzalez. Clustering to minimize the maximum intercluster distance. *Theoretical Computer Science*, 38: 293-306, 1985.

[9] S. Guha, R. Rastogi, and K. Shim. CURE: An efficient clustering algorithm for large databases. In *Proceedings of ACM SIGMOD International Conference Management of Data*, 1998.

[10] T. Ibaraki and N. Katoh. *Resource Allocation Problems: Algorithmic Approaches*. MIT Press, Cambridge, Massachusetts, 1988.

[11] R. Kohavi and D. Sommerfield. Feature subset selection using the wrapper method: Overfitting and dynamic search space topology. In *Proceedings of the 1st International Conference on Knowledge Discovery and Data Mining*, 1995.

[12] R. T. Ng and J. Han. Efficient and effective clustering methods for spatial data mining. In *Proceedings of the 20th International Conference Very Large Data Bases*, 1994.

[13] C. M. Procopiuc, M. Jones, P. K. Agarwal, and T. M. Murali. A Monte Carlo algorithm for fast projective clustering. In *ACM SIGMOD International Conference on Management of Data*, 2002.

[14] K. Y. Yip, D. W. Cheung, M. K. Ng. A highly-usable projected clustering algorithm for gene expression profiles. *BIOKDD*, 2003.

Chapter 10

Graph Algorithms for Integrated Biological Analysis, with Applications to Type 1 Diabetes Data

John D. Eblen

Department of Electrical Engineering and Computer Science, University of Tennessee, Knoxville, TN 37996–3450, USA

Ivan C. Gerling

University of Tennessee Health Science Center, Memphis, TN, 38163, USA

Arnold M. Saxton

Department of Animal Science, University of Tennessee, Knoxville, TN 37996–4574, USA

Jian Wu

University of Tennessee Health Science Center, Memphis, TN, 38163, USA

Jay R. Snoddy

Biomedical Informatics Department, Vanderbilt University, Nashville, TN 37232, USA

Michael A. Langston*

Department of Electrical Engineering and Computer Science, University of Tennessee, Knoxville, TN 37996–3450, USA

Graph algorithms can be effective tools for analyzing the immense data sets that frequently arise from high-throughput biological experiments. A major computational goal is to identify dense subgraphs, from which one can often infer some form of biological meaning. In this paper, new techniques are devised and analyzed in an effort to improve the quality and relevance of these subgraphs, and

*Corresponding author: langston@cs.utk.edu

to extend the utility of clique-centric methods that may produce them. Using non-obese diabetic mice as a target organism, the paraclique algorithm is tested on transcriptomic data under various parameters in order to determine how it can best be tuned to applications. The use of proteomic anchors is also discussed in an effort to help guide subgraph selection in the presence of inhomogeneous data, which is an important but notoriously difficult problem in its own right.

10.1. Overview

Many inbred strains of *Mus musculus*, the common house mouse, are employed in biomedical research. The non-obese diabetic (NOD) mouse is particularly useful as a model of type 1 diabetes mellitus (also called juvenile onset, or insulin dependent, diabetes). In both mice and humans, this disease is characterized by persistent hyperglycemia (elevated blood sugar level) that is induced in genetically susceptible individuals and modified by a variety of environmental triggers including food and infections. It is caused by an abnormal and self-destructive immune response (autoimmunity), which allows mononuclear leukocytes to target the insulin producing beta cells in the pancreas [26, 27, 32]. Eventually this process destroys so many of the beta cells that the body is unable to produce sufficient insulin to retain normal blood glucose levels and diabetes is observed. Our studies in the NOD mouse focus on the very early leukocyte abnormalities that may be associated with initiation of the autoimmune process [14]. If we can gain a better understanding of the initiation of autoimmunity, then we may be able to develop rational intervention strategies that can stop the disease process in its preclinical phase effectively and with minimal side effects.

The importance of melding experimental research with continuing advances in computational analysis is well understood [17, 18, 21]. In the work reported here, we begin with high-throughput NOD mouse data and apply novel clique-centric methods to analyze it. Fixed parameter tractability [1, 8] and various realizations of the paraclique algorithm [6] form the basis of techniques we use to extract dense putative networks from the vast sea of correlations that arise in the analysis of comprehensive transcriptomic data [4, 19]. Proteomics data is added to the mix, thereby introducing challenging new problems in inhomogeneous data interpretation [16]. The results we obtain are evaluated in terms of both statistical quality and biological relevance.

10.2. Description of Data

To define abnormalities in the early phases of autoimmunity, we have conducted comprehensive studies of gene expression in young NOD mice and mice from control strains (NON and C57BL/6) that do not develop diabetes or autoimmunity to beta cells [13, 14]. Genes encoded in DNA are transcribed into mRNA, which is then translated into proteins that are the major determinants of a cell's activation and function. To gain a comprehensive picture of how the genetic differences between our strains can affect the development of autoimmunity, we evaluated gene expression at the mRNA level using Affymetrix MOE430A/B arrays, and at the protein level using 2D-gel electrophoresis. We collected mononuclear spleen leukocytes from each of the three strains at both two and four weeks of age. This is a critical window for our analysis, because it represents the prepathology stage before leukocytes begin to infiltrate the islets of Langerhans, which typically occurs in NOD pups when they are about five weeks old. See Fig. 10.1. From each of the six strain/age groups, we collected five independent samples for a total of 30 samples in the complete dataset. Experimental details regarding the analysis of mRNA and protein expression levels have been published previously [14].

Fig. 10.1. At birth, NOD mice have normal blood glucose levels, with no indication of destruction of insulin-producing beta cells (located in the islets of Langerhans). At five weeks of age, the first signs of pathology occur with leukocytes invading the space around the islets. This infiltration progresses to involve additional leukocytes with more invasive and destructive character. By twelve weeks of age or later, so many beta-cells have been destroyed that insulin production capacity has been severely diminished. Because blood glucose can no longer be regulated normally, the mice become diabetic. We sampled leukocytes in the early and late prepathology stage to evaluate defects at the molecular level associated with initiation of the pathology.

Because the data is biological, it has a fairly high level of noise. At the time of sample collection, individual mice may or may not have just eaten, been fighting, been scared, been sleeping, etc. These biological parameters can be difficult to control, and can have an influence on expression levels of some genes. In addition to this biologically derived noise, there are also technical sources of noise to be considered. The mRNA gene expression arrays have a very effective normalization and scaling process and very good technical reproducibility on identical samples with percent coefficients of variance usually in the low single digits [24, 29]. In contrast, protein expression data involves technologies that are more complicated and difficult to standardize. Technical reproducibility of protein expression data collected from identical samples often has percent coefficients of variance in the low double digit range [22, 31].

10.3. Correlation Computations

We employ the aforementioned 30 samples to compute a correlation matrix. The matrix entry at location (i, j) denotes the correlation coefficient between the ith and j^{th} items (genes or proteins), normalized to the range [-1.0,1.0]. Because mRNA arrays alone can measure over 45,000 different values, we may be faced with making sense of over a trillion correlate pairs. Close examination of the data reveals a paucity of outliers, so that we are able to use the well-known Pearson's method for the computation of correlation coefficients. Because we are searching for putative pathways and networks, both positive and negative correlations are of equal interest. We therefore take absolute correlation values. Recall that this is biological and hence noisy data. Not every probe set is reliably measured in every sample. Thus we move away from simple correlation and compute a p-value for each pair of correlates, which is the probability that they have a correlation different from zero [33]. See Fig. 10.2.

From this we can build a simple, unweighted graph as needed with the use of a cut-off value (we favor the use of p=0.01) and a high-pass filter. An edge whose weight is less then the cut-off is discarded. Other edges are retained, but their weights are now ignored.

10.4. Clique and Its Variants

We assume the reader is familiar with standard concepts in graph and complexity theory [25, 30]. We begin with the well-known clique problem. A clique is a densest possible subgraph. Each pair of its vertices is connected by an edge. A clique is maximum if it is a largest clique in a graph. A clique is maximal if it is not contained wholly within a larger clique. A clique on five vertices is illustrated in Fig. 10.3. Protein correlations are too weak to find relevant relationships at this level, and so for them we turn to other methods as will be described in Section 10.6. The correlation matrix is transformed into a complete, weighted correlation graph by using a vertex for each transcript and protein, and by weighting the edge between each pair of items with the corresponding correlation matrix entry. Clique is widely acknowledged for its many applications in computational molec-

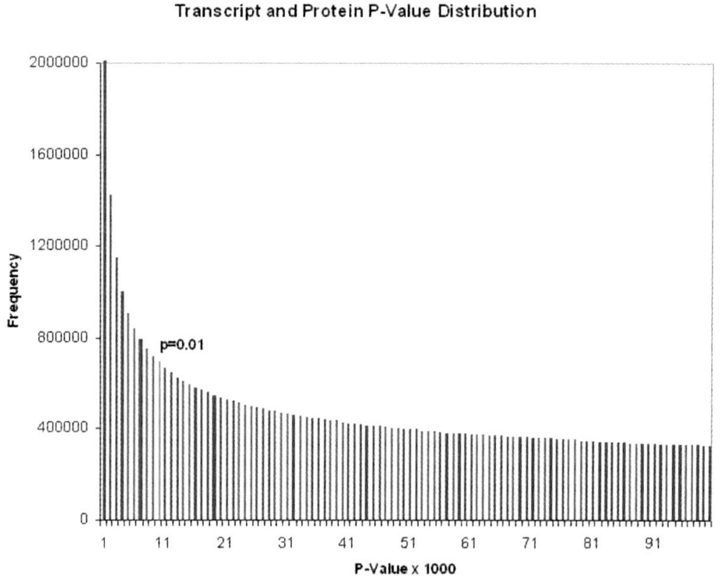

Fig. 10.2. The transcriptomic data used in this study provides a broad spectrum of p-values. A threshold p-value of 0.01, for example, creates an unweighted graph with 22750 vertices and roughly 11 million edges.

ular biology [23]. In the present setting, its advantages include cluster purity (all edges are present), cluster overlap (genes and gene products are pleiotropic), and resistance to false positives (the bane of many clustering methods). Contrasts with other techniques can be found in [28]. The classic *decision* version of clique is $\mathcal{NP}$-complete. Finding approximate solutions appears no easier, because ensuring solutions within n^{ϵ} in polynomial time implies $\mathcal{P} = \mathcal{NP}$ for any $\epsilon > 0$ [9].

We are of course more interested in *search* and *optimization*. By transforming clique to vertex cover, we can apply notions from fixed-parameter tractability [1, 8] and many years of basic research [11, 12] to solve the *maximum* clique problem effectively in practice. With novel implementations and high performance platforms, we are currently able to find maximum cliques with hundreds of vertices in graphs with tens of thousands of nodes. We must often also solve the *maximal* clique problem [5]. Even when the maximum clique size is modest, we frequently find that the number of maximal cliques is staggering. Thus it is that space as well as time is a critical resource for solving maximal clique, even when supercomputing technologies are used. Our work on this general subject, as well as its application to transcriptomic data analysis, is chronicled in [1, 4, 6, 7, 19, 20, 28, 34].

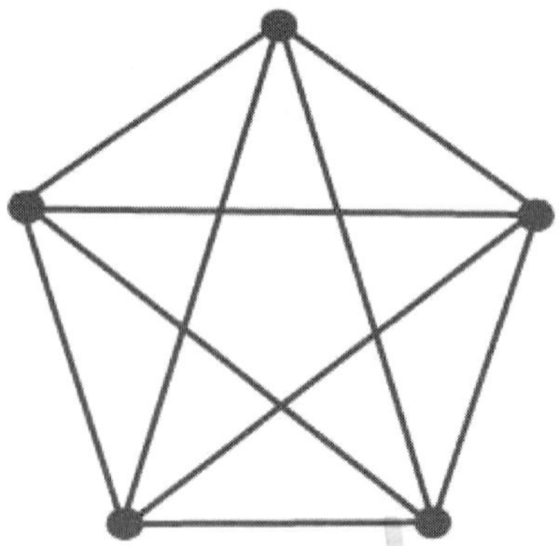

Fig. 10.3. A clique of size five.

The *paraclique* algorithm was recently introduced in [6], where it was shown to have advantages in the amelioration of noise inherent in high throughput biological data. Clique by itself is highly resistant to false positives. Under certain experimental conditions, however, it can be subject to false negatives. This is because, if even a single edge is missing, the clique is lost. Moreover, we frequently encounter enormous numbers of overlapping cliques [19]. To coalesce these into fewer but larger clusters, and to reduce the significance of noise, paraclique solves something similar to the dense-k-subgraph problem [10], which is $\mathcal{NP}$-complete even on graphs of maximum degree three. Roughly speaking, a paraclique is a clique augmented with non-clique vertices in a highly controlled manner. A user-defined *glom factor*, g, is provided to increase cluster size while limiting the number of missing edges permitted. We glom onto a non-clique vertex only if it is adjacent to at least g clique/paraclique members. This notion is depicted in Fig. 10.4. Correlations between non-adjacent vertices may be taken into consideration as well. We refer the reader to [6] for details. Thus, when the application permits, we employ the paraclique algorithm and sacrifice overlap in order to build robust clusters.

Paraclique is also useful from a computational standpoint because it can, depending on the application, obviate the need for maximal clique enumeration. To illustrate, the processing of an NOD file whose maximum clique size was only 20 produced a list containing over four million maximal cliques and requiring over two gigabytes of memory before the enumeration was terminated by the operating system. In contrast, only 25 paracliques were generated. We therefore identify a maximum clique, use paraclique to decompose the graph, and then iterate the process on the remaining subgraph. We halt the process when maximum clique size

[10] D. Eppstein. Arboricity and bipartite subgraph listing algorithm. *Information Processing Letters*, 51(4), p207–211, 1994.
[11] Formal Concept Analysis Homepage. http://www.upriss.org.uk/fca/fca.html
[12] B. Ganter, G. Stumme, R. Wille (eds.). Formal Concept Analysis: Foundations and Applications. *Lecture Notes in Computer Science*, vol 3626, Springer, 2005.
[13] B. Ganter, R. Wille. Formal Concept Analysis: Mathematical Foundations. Springer Verlag, 1996 (Germany version), 1999 (English version).
[14] D.S. Johnson, M. Yannakakis, C.H. Papadimitriou. On generating all maximal independent sets. *Information Processing Letters*, 27, p119–123, 1988.
[15] T. Kashiwabara, S. Masuda, K. Nakajima, T. Fujisawa. Generation of maximal independent sets of a bipartite graph and maximum cliques of a circular-arc graph. *J. Algorithms*, 13, p161–174, 1992.
[16] S.O. Kuznetsov. Complexity of learning in context lattices from positive and negative examples. *Discrete Applied Mathematics*, 142, p111–125, 2004.
[17] S.O. Kuznetsov. On computing the size of a lattice and related decision problems. *Order*, 18, p313–321, 2001.
[18] S.O. Kuznetsov and S.A. Obedkov. Comparing performance of algorithms for generating concept lattices. *Journal of Experimental and Theoretical Artificial Intelligence*, 14(23), p189–216, 2002.
[19] C. Lindig. Fast concept analysis. Gerhard Stumme ed. *Working with Conceptual Structures - Contributions to ICCS 2000*, Shaker Verlag, Aachen, Germany, 2000.
[20] K. Makino, T. Uno. New algorithms for eumerating all maximal cliques. *Proc. 9th Scand. Workshop on Algorithm Theory (SWAT 2004)*, p260–272. Springer Verlag, Lecture Notes in Computer Science 3111, 2004.
[21] E.M. Norris. An algorithm for computing the maximal rectangles in a binary relation. *Revue Roumaine de mathematiques Pures et Appliquees*, 23(2), p243–250, 1978.
[22] L. Nourine, O. Raynaud. A fast algorithm for building lattices. *Information Processing Letters*, 71, p199–204, 1999.
[23] L. Nourine, O. Raynaud. A fast incremental algorithm for building lattices. *J. of Experimental and Theoretical Artificial Intelligence*, 14, p217–227, 2002.
[24] N. Pasquier, Y. Bastide, R. Taouil, L. Lakhal. Efficient mining of association rules using closed itemset lattices. *Information Systems*, 24(1), p25–46, 1999.
[25] J. Pei, J. Han, R. Mao. CLOSET: An efficient algorithm for mining frequent closed itemsets. *Proc. ACM SIGMOD Workshop on Research Issues in Data Mining and Knowledge Discovery*, p21–30, 2000.
[26] F. Rioult, J.F. Boulicaut, B. Cremilleux, J. Besson. Using transposition for pattern discovery from microarray data. *Proc. 8th ACM SIGMOD workshop on Research issues in Data Mining and Knowledge Discovery*, p73–79, 2003.
[27] G. Stumme, R. Taouil, Y. Bastide, N. Pasquier, L. Lakhal. Fast Computation of Concept Lattices using data mining techniques. *Proc. 7th International Workshop on Knowledge Representation meets Databases*, p129–139, 2000.
[28] S. Tsukiyama, M. Ide, H. Ariyoshi, and I. Shirakawa. A new algorithm for generating all the maximal independent sets *SIAM Journal on Computing*, 6(3), p505–517, 1977.
[29] T. Uno and T. Asai, H. Arimura and Y. Uchida. LCM: An Efficient Algorithm for Enumerating Frequent Closed Item Sets. *IEEE ICDM'04 Workshop FIMI'03 (International Conference on Data Mining, Frequent Itemset Mining Implementations)*.

[30] P. Valtchev, R. Missaoui, R. Godin, M. Meridji. Generating frequent itemsets incrementally: two novel approaches based on Galois lattice theory. *Journal of Experimental and Theoretical Artificial Intelligence*, 14(2-3), p115–142, 2002.

[31] P. Valtchev, R. Missaoui, P. Lebrun. A partition-based approach towards constructing Galois (concept) lattices. *Discrete Mathematics*, 256(3), p801–829, 2002

[32] J. Wang, J. Han, J. Pei. CLOSET+: Searching for the best straegies for mining frequent closed itemsets. *Proc. 9th ACM SIGKDD international conference on Knowledge Discovery and Data mining*, p236–245, 2003.

[33] R. Wille. Restructuring the lattice theory: An approach based on hierarchies of concepts. In I. Rival, editor, *Ordered sets*, pages 445-470, Dordrecht-Boston, 1982, Reidel.

[34] G. Yang. The complexity of mining maximal frequent itemsets and maximal frequent patterns. *Proc. 10th ACM SIGKDD International Conference on Knowledge Discovery and Data mining*, p344–353, 2004.

[35] M.J. Zaki, C.-J. Hsiao. CHARM: An efficient algorithm for closed association rule mining. *Proc. 2nd SIAM International Conference on Data Mining*, 2002.

[36] M.J. Zaki, C.-J. Hsiao. Efficient Algorithms for mining closed itemsets and their lattice structure. *IEEE Trans. Knowledge and Data Engineering*, 17(4), p462–478, 2005.

[37] M.J. Zaki, M. Ogihara. Theoretical foundations of association rules. *Proc. 3rd ACM SIGMOD Workshop on Research Issues in Data Mining and Knowledge Discovery*, p1–7, 1998.

[38] M.J. Zaki, N. Ramakrishnan. Reasoning about setes using redescription mining. *Proc. 11th ACM SIGKDD International Conference on Knowledge Discovery and Data mining*, 2005.

Appendix

Algorithm 8.2. Sprout-Insert

Input: s, content and nbr
$(\mathsf{obj}(X), X)$ is the sth child of G. Let $K = \{1, \dots, k\}$ be all the children of G.
Output: $\mathsf{Child}(\mathsf{obj}(X), X) = \{(\mathsf{obj}(XS_i), XS_i) : 1 \leq i \leq t\}$ and update the global trie T: update the attribute set of $\mathsf{obj}(XS_i)$ (insert if it does not exist)
For each $i \in K$, set $C_i = \emptyset$.
for $a \in \mathsf{obj}(X)$ **do**
    for $i \in \mathsf{nbr}(a) \setminus \{s\}$ **do**
        Append a to C_i;
    end for
end for
The following takes $O(\sum_{i \in K} |C_i|) = O(\sum_{a \in C} |\mathsf{nbr}(a)|)$ time.
Initialize a local trie T_C over objects;
for $i \in K$ **do**
    if C_i does not exist in T_C **then**
        Insert C_i into T_C;
        $S_i = \mathsf{content}(i)$;
    else
        Merge S_i with $\mathsf{content}(i)$;
    end if
end for
Let $\mathsf{Child}(\mathsf{obj}(X), X)$ be all the pairs in T_C: $\{(\mathsf{obj}(XS_j), XS_j) : 1 \leq j \leq t\}$
for $i = 1$ to t **do**
    Search $\mathsf{obj}(XS_i)$ in T;
    if $\mathsf{obj}(XS_i)$ does not exist **then**
        Insert $\mathsf{obj}(XS_i)$ into T, and associate it with the attribute set XS_i;
    else
        if the size of attribute set associate with $\mathsf{obj}(XS_i)$ is smaller than XS_i **then**
            Associate XS_i with $\mathsf{obj}(XS_i)$;
        end if
    end if
end for

Algorithm 8.3. CondenseAdjacentlists

Input: $\mathsf{Child}(\mathsf{obj}(X), X) = \{(\mathsf{obj}(XS_i), XS_i) : 1 \leq i \leq t\}$
Output: $\mathsf{content}(i)$ for $i = 1 \ldots t$, and new adjacency lists, $\mathsf{nbr}(a)$, $a \in \mathsf{obj}(X)$
For each $a \in \mathsf{obj}(X)$, $\mathsf{nbr}(a) = \emptyset$;
for $i = 1$ to t **do**
  $\mathsf{content}(i) = S_i$;
  for each $a \in \mathsf{obj}(XS_i)$, append i to $\mathsf{nbr}(a)$;
end for

Algorithm 8.4. Concept-Lattice Construction – 2-level BFS

Compute the top concept $C = (\mathcal{O}, \mathsf{attr}(\mathcal{O}))$;
Initialize a queue $Q = \{C\}$;
Initialize a trie T for the object set $\mathcal{O}$;
$\mathsf{content}(i) = \{i\}$ for $i \in \mathcal{M}$;
$\mathsf{Child(C)} = \text{SPROUT-INSERT}(0, \mathsf{content}, \mathsf{nbr})$;
while Q is not empty **do**
  $C =$ dequeue(Q);
  Sort the pairs in $\mathsf{Child}(C)$ according to its extent size in decreasing order: $(\mathsf{obj}(XS_i), XS_i), 1 \leq i \leq k$.
  $(\mathsf{content}, \mathsf{nbr}) = \text{CONDENSEADJACENTLISTS}(\mathsf{Child}(C))$;
  for $i = 1$ to k **do**
    Search $\mathsf{obj}(XS_i)$ in T;
      Denote $(\mathsf{obj}(XS_i), XS_i)$ by K; $\{\}K$ is not necessary a concept.
    if the size of attribute set associate with $\mathsf{obj}(XS_i)$ is not greater than XS_i **then**
      Identify K as the successor of C;
      if K is not sprouted **then**
        $\mathsf{Child}(K) = \text{SPROUT-INSERT}(i, \mathsf{content}, \mathsf{nbr})$;
        Enqueue K into Q;
      end if
    end if
  end for
end while

Chapter 9

A Projected Clustering Algorithm and Its Biomedical Application

Ping Deng

Computer Science
University of Illinois at Springfield, USA
pdeng2@uis.edu

Qingkai Ma

Economic Crime and Justice Studies
Utica College, USA
qma@utica.edu

Weili Wu *

Computer Science
The University of Texas at Dallas, USA
weiliwu@utdallas.edu

Projected clustering is concerned with clustering data in high dimensional space where data is more likely correlated in subspaces of full dimensions. Recently, several projected clustering algorithms that focus on finding specific projection for each cluster have been proposed. We find that, besides distance, the closeness of points in different dimensions also depends on the distributions of data along those dimensions. Based on this, we propose a projected clustering algorithm, IPROCLUS (Improved PROCLUS), which is efficient and accurate in handling data in high dimensional space. According to the experimental results on randomly generated synthetic data, our algorithm shows much higher accuracy for the scaled datasets and lower dependence on one of user inputs than PROCLUS. We also apply IPROCLUS on real biomedical data and show that it can achieve much better accuracy than PROCLUS.

*This work was partially supported by NSF grant # ACI-0305567. The content of this work does not necessarily reflect the position or policy of the government and no official endorsement should be inferred.

9.1. Introduction

Clustering is described as finding groups of data points so that the points in each group are similar to each other according to their attributes (dimensions). Clustering has been discussed thoroughly by both statistics and database communities due to its numerous applications in problems such as classification, machine learning, data mining, indexing, trend analysis, customer segmentation, and pattern recognition. Most known clustering algorithms proposed in the last few years, such as CLARANS [12], DBSCAN [7], and CURE [9] cluster data points based on the distance function that uses all dimensions of the data. They are proved to work well when datasets are low dimensional. However, when dimensional space grows higher, the above algorithms lose their efficiency and accuracy because of the inherent sparsity of data [3]. It is shown in [5] that the distance of a point to the nearest neighbor approaches the distance to its farthest neighbor as dimensionality increases, which means computing the distance based on full dimensions is not meaningful in high dimensional space. Actually, natural clusters might exist in subspaces. In other words, data points may be close to each other in a few dimensions, but not all dimensions. Furthermore, data points in different clusters may be correlated with respect to different subsets of dimensions. Feature selection [11] is one technique that has been proposed to deal with data in high dimensions. It reduces the dimensionality of the space before clustering. On the other hand, it may lead to a loss of information by choosing a subset of dimensions ahead of time and it cannot select different subsets of dimensions for different clusters. Another solution is called dimension reduction. One widely used approach is principal component analysis [6] which is concerned with projecting all data points on the same subspace to minimize the loss of information. One drawback is that the reduced new dimensions might be difficult to interpret, while most applications require simple descriptions of clusters. Moreover, it doesn't handle well the case where clusters may exist in different subspaces of full dimensions. In Fig. 9.1, we illustrate two projections on different dimensions for a set of points in 3-dimensional space. It is obvious that there is no cluster in Fig. 9.1(a). However, when the data points are projected in $x - z$ plane, cluster P and Q in Fig. 9.1(b) are discovered and cluster R and S in Fig. 9.1(c) are found when they are projected in $y - z$ plane.

Projected clustering has been proposed recently to effectively deal with high dimensionalities. The projected clustering problem is known as finding clusters and their relevant attributes from a dataset. Instead of projecting entire dataset on the same subspace in dimension reduction techniques, projected clustering focuses on finding specific projection for each cluster such that the similarity is

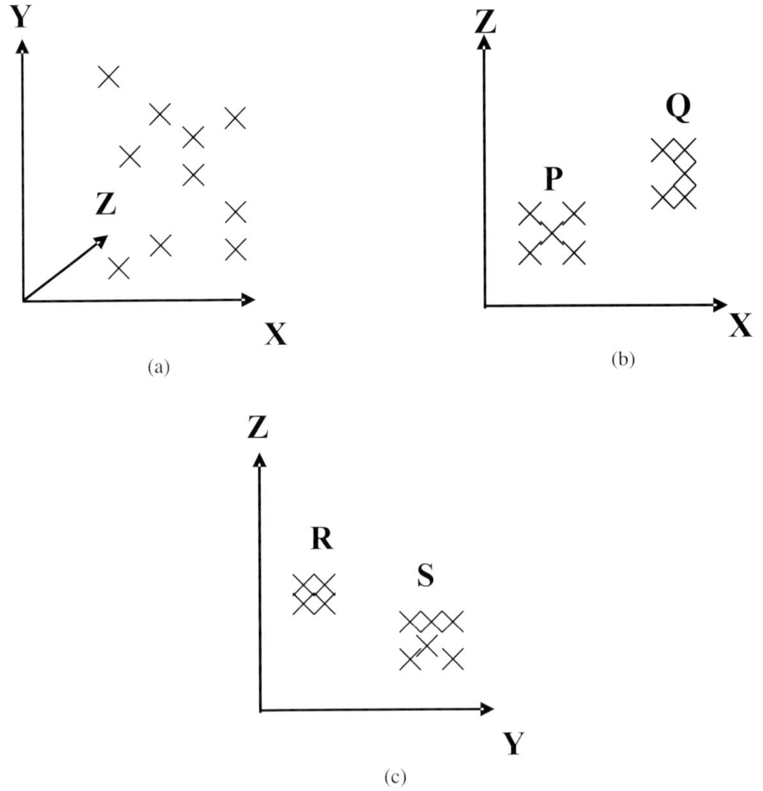

Fig. 9.1. Illustrations of Different Clusters in Different Subspaces (a) Data in 3-Dimensional Space. (b) Projection in X-Z Plane. (c) Projection in Y-Z Plane.

reserved as much as possible. Three criteria have been proposed to evaluate clusters [14]. A good cluster should have as many points as possible. Its dimensionalities should be as large as possible and the distance between points in the cluster should be as small as possible. Actually, there is a tradeoff among these criteria. If one criterion is fixed, the other two criteria are at odds.

In this paper, we propose our algorithm, IPROCLUS, which is based on PROCLUS [1]. We find that the closeness of points in different dimensions not only depends on the distance between them, but also relates to the distributions of points along those dimensions. PROCLUS uses the Manhattan segmental distance which loses its effectiveness when points in different dimensions have very different variance. We propose the modified Manhattan segmental distance which is more accurate and meaningful in projected clustering. PROCLUS strongly depends on two user inputs. In order to reduce the dependence on one of the user parameters, we propose a dimension tuning process. Compared to PROCLUS, we propose a simplified replacing logic as well. We compared the performance of PROCLUS and IPROCLUS on randomly generated synthetic data and real biomedical data.

The rest of this chapter is organized as follows. Related works on projected clustering are discussed in Section 9.2. We then present our algorithm, IPROCLUS (Improved PROCLUS) in Section 9.3. The experimental evaluation of our algorithm is given in Section 9.4. Section 9.5 concludes this chapter.

9.2. Related Works

We classify projected clustering algorithms into two categories. One is density-based algorithms [3, 13]. The other is distance-based algorithms [1, 2]. In the following sections, we introduce two typical algorithms for each category.

9.2.1. *Density-based Algorithms*

Density-based algorithms define a cluster as a region that has a higher density of data points than its surrounding regions. From the point of view of projected clustering, we want to find dense regions only in their corresponding subspaces. Two algorithms are discussed in this category.

CLIQUE (CLustering In QUEst) [3] is based on the following property. Dense regions in a particular subspace still create dense regions when they are projected onto lower dimensional subspaces. It depends on two parameters, the partition threshold ξ and the density threshold τ. The algorithm consists of three steps: identification of dense units, identification of clusters,and generation of minimal descriptions. While this algorithm is insensitive to the order of records and scales linearly with the size of inputs, like many density-based algorithms, it has exponential dependence on the number of dimensions. Since there are large overlaps among dense regions, it doesn't return disjoint clusters that are required in many applications. In the density-units prune procedure, some potential clusters are likely to be lost. Furthermore, the user parameters, ξ and τ, are hard to pre-select and partitioning each dimension into ξ intervals is not flexible.

DOC (Density-based Optimal projective Clustering) [13] gives a mathematical definition for the concept of optimal projective cluster and uses a Monte Carlo algorithm to search for a good approximation of the optimal projective cluster. It has one user parameter, w, which is the width of the bounding hyper-cubes. This algorithm doesn't miss small clusters. However, it still suffers from a critical user parameter that is the fixed global interval width w for each dimension and the accuracy of the algorithm depends on whether the definition for the optimal projective cluster is good.

9.2.2. *Distance-based Algorithms*

Distance-based algorithms define a cluster as a partition such that the distance between objects within the same cluster is minimized and the distance between objects from different clusters is maximized. Compared to density-based methods in which each data point is assigned to all clusters with a different probability, distance-based methods assign data to a cluster with probability 0 or 1. A distance measure is defined between data points. The most commonly used measure is the Euclidean distance that is not effective for projected clustering since data points in two different clusters may have different number of dimensions. We can find the solution in the following two distance-based algorithms.

PROCLUS (PROjected CLUStering) [1] allows the selection of different subsets of dimensions for different clusters. It requires two user parameters: the number of clusters denoted by k and the average cardinality of the subsets of full dimensions denoted by l. The Manhattan segmental distance is used in this method instead of the Euclidean distance since the number of dimensions has been normalized away in the Manhattan segmental distance when comparing data points in two different clusters that have different number of dimensions. The Manhattan segmental distance between p_1 and p_2 relative to dimension set D is defined as: $(\sum_{i \in D} |p_{1,i} - p_{2,i}|)/|D|$. In Fig. 9.2, we show the Manhattan segmental distance by an example. The distance between x_1 and x_2 for dimension (x, y) is $(a+b)/2$. PROCLUS is based on k-medoid techniques and has three phases: an initialization phase, an iterative phase, and a refinement phase. In the initialization phase, a random sample of data points with size $A \times k$ is chosen and then a greedy algorithm is applied to find a superset of medoids denoted by M with size $B \times k$ (A, B are constants, and $A > B$). In the iterative phase, the best k medoids are found from M by the following steps. It first randomly finds a set of k medoids from M and then finds an appropriate set of dimensions for each medoid in the set. It then forms a cluster corresponding to each medoid. Bad medoids are chosen by the following two rules: the medoid of the cluster with the least number of points is bad; the medoid of any cluster with less than $(N/k) \times minDeviation$ points is bad, where *minDeviation* is a constant smaller than 1. The replacing logic is that the current clustering is compared with the case that the bad medoids are replaced with random points in M. The result is replaced if the new clustering is better. Otherwise, the bad medoids are replaced with other random points in M. If the result doesn't change after a certain number of attempts have been tried, the iterative phase terminates and the best k medoids are reported. In the refinement phase, the process in the iterative phase is redone once by using the data points distributed by the result clusters. Once the new dimensions have been computed, the points are reassigned to the medoids with respect to these new sets of dimensions. Outliers are also handled in this phase. This algorithm returns a partition of data points, together with sets of dimensions on which data points in each cluster are correlated. Nevertheless, the problem of pre-selecting user parameters still hasn't been solved. It relies on random sampling in the initialization phase. Hence, small clusters are likely to be missed.

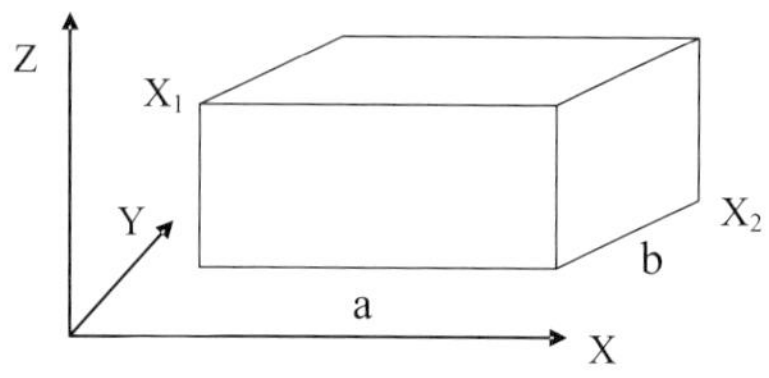

Fig. 9.2. Manhattan Segmental Distance

ORCLUS (arbitrarily ORiented projected CLUSter generation) [2] uses arbitrarily projected subspaces for finding clusters due to the fact that real data often contains inter-attribute correlations, which leads to projections that are not parallel to the original axis system. It also asks for two user parameters, the number of clusters k and the cardinality of the dimensions for each cluster l. ORCLUS modifies the PROCLUS algorithm by adding a merging process of clusters and asks each cluster to select principal components instead of attributes. It improves PROCLUS in that it can construct clusters in arbitrarily aligned subspaces of lower dimensionality. However, ORCLUS requires all projected clusters to exist in the same number of dimensions and it also relies on random sampling in the initialization phase. Moreover, like CLIQUE and PROCLUS, it still needs some user parameters though the guidance in finding a good value of l has been proposed in this method.

9.3. The IPROCLUS Algorithm

Our algorithm, IPROCLUS, is based on PROCLUS. It takes the number of clusters k and the average number of dimensions l in a cluster as inputs. It has three phases: an initialization phase, an iterative phase, and a cluster refinement phase. Compared to PROCLUS, we propose the modified Manhattan segmental distance that is more accurate and meaningful in projected clustering. We add one more step in the last phase to reduce the dependence on the user parameter l which is the average number of dimensions in a cluster. We also propose a new logic of replacing bad medoids in the iterative phase, which is more time efficient. The overall pseudo code of our algorithm is given in Algorithm 9.1. The steps that are new in IPROCLUS are underlined and will be discussed extensively in this section. The detailed information about methods used in both PROCLUS and IPROCLUS can be found in [1].

Algorithm 9.1. IPROCLUS(No. of Clusters: k, Avg. Dimensions: l)

$\{C_i$ is the ith cluster$\}$
$\{D_i$ is the set of dimensions associated with cluster $C_i\}$
$\{M_{current}$ is the set of medoids in current iteration$\}$
$\{M_{best}$ is the best set of medoids found so far$\}$
$\{N$ is the final set of medoids with associated dimensions$\}$
$\{A;B$ are constant integers$\}$

begin

$\{$1. Initialization Phase$\}$
S = random sample of size $A \times k$
<u>Calculate the normalization factors for each dimension</u>
M = Greedy(S, $B \times k$)

$\{$First Iteration$\}$
$M_{current}$ = Random set of medoids $\{m_1, m_2, \ldots, m_k\} \subset M$
$\{$Approximate the optimal set of dimensions$\}$
for each medoid m_i in $M_{current}$ **do**
  <u>Let δ_i be the modified Manhattan segmental distance to the nearest medoid from m_i</u>
  L_i = Points in sphere centered at m_i with radius δ_i
end for
$L = \{L_1, \ldots, L_k\}$
$(D_1, D_2, \ldots, D_k)$ = FindDimensions(k, l, L)
$\{$Form the clusters$\}$
$(C_1, \ldots, C_k)$ = AssignPoints($D_1, D_2, \ldots, D_k$)
bestObjective = EvaluateClusters($C_1, \ldots, C_k, D_1, D_2, \ldots, D_k$)
$M_{best} = M_{current}$
compute the bad medoids in M_{best}

{2. Iterative Phase}
repeat
  Compute $M_{current}$ by replacing the bad medoids in M_{best} with random points from M
  {Approximate the optimal set of dimensions}
  for each medoid m_i in $M_{current}$ **do**
    compute δ_i, which is the distance to nearest medoid from m_i
    compute B as the set of clusters whose δ_i or medoids changed
    only compute L_i = Points in sphere centered at m_i with radius δ_i for all the clusters in B
  end for
  $L = \{L_1, \ldots, L_k\}$
  $(D_1, D_2, \ldots, D_k)$ = FindDimensions(k, l, L)

  {Form the clusters}
  $(C_1, \ldots, C_k)$ = IterativeAssignPoints($C_1, \ldots, C_k, D_1, D_2, \ldots, D_k, B$)
  ObjectiveFunction = EvaluateClusters($C_1, \ldots, C_k, D_1, D_2, \ldots, D_k$)
  if ObjectiveFunction < BestObjective **then**
    BestObjective = ObjectiveFunction
    $M_{best} = M_{current}$
    Compute the bad medoids in M_{best}
  end if
until (termination criterion)

{3. Refinement Phase}
$L = \{C_1, \ldots, C_k\}$
$(D_1, D_2, \ldots, D_k)$ = FindDimensions(k, l, L)
$(C_1, \ldots, C_k)$ = AssignPoints($D_1, D_2, \ldots, D_k$)
DimensionTuning($M_{best}, C_1, \ldots, C_k, D_1\ D_2 \ldots, D_k$)
$N = (M_{best}, C_1, \ldots, C_k, D_1, D_2, \ldots, D_k)$
return (N)
end

9.3.1. *Modified Manhattan Segmental Distance*

In our algorithm, we propose the modified Manhattan segmental distance as the distance measure to improve accuracy. We find that the closeness of points in different dimensions not only depends on the distance between them. When it comes to clustering, the distributions of points along different dimensions also matter. For example, the distance of points x_1 and x_2 in dimension i and j is 20 and 200 respectively, but that doesn't necessarily mean that x_1 and x_2 are closer in dimension i than they are in dimension j since data points in dimension i and j may have different distributions. Therefore, there is a need to normalize the distance along each dimension. We want to find a normalization factor n_i for each dimension, and the modified Manhattan segmental distance between x_1 and x_2 relative to dimension set D can be defined as: $(\sum_{i \in D} |p_{1,i} - p_{2,i}|/n_i)/|D|$. In our algorithm, we use the standard deviation of all points in a dataset along dimension i as the normalization factor n_i.

9.3.2. *Initialization Phase*

In the initialization phase, all data points are first chosen by random to form a random data sample set S with size $A \times k$, where A is a constant. Then S is chosen by a greedy algorithm to obtain an even smaller set of points M with size $B \times k$, where B is a small constant. The greedy algorithm is based on the concept that we avoid choosing the medoids from the same cluster; therefore, we choose the set of points which are most far apart. The greedy algorithm has been proposed in [8] and is illustrated in Algorithm 9.2.

Algorithm 9.2. Greedy(Set of points: S, Number of medoids: k)

{ d (.,.) is the distance function}

begin
$M = \{m_1\}$ { m_1 is a random point of S}
{compute distance between each point and medoid m_1}
for each $x \in S \backslash M$ **do**
  $dist(x) = d(x, m_1)$
end for
for i = 2 to k **do**
  {choose medoid m_i to be far from previous medoids}
  let $m_i \in S \backslash M$ be s. t. $dist(m_i) = max\{dist(x) | x \in S \backslash M\}$
  $M = M \cup \{m_i\}$
  {compute distance of each point with closest medoid}
  for each $x \in S \backslash M$ **do**
    $dist(x) = min\{dist(x), d(x, m_i)\}$
  end for

end for
return M
end

9.3.3. *Iterative Phase*

We begin by choosing a random set of k points from M. Then we iteratively replace the bad medoids in the current best medoids set with random points from M until the current best medoids set does not change after a certain number of replacements have been tried.

In each iteration, we first find dimensions for each medoid in the set, and form the cluster corresponding to each medoid. Then we evaluate the clustering and replace the bad medoids in the current best medoids set if the new clustering is better.

In order to find dimensions, we first define several notations. For each medoid m_i, δ_i is the minimum distance from any other medoids to m_i based on full dimensions. We define the locality L_i to be the set of points within the distance of δ_i from m_i and define $X_{i,j}$ to be the average distance to m_i along dimension j. We then calculate $X_{i,j}$ as follows. We divide the average distance from the points in L_i to m_i along dimension j by the normalization factor n_j. There are two constraints when associating dimensions to medoids. The total number of dimensions associated to medoids must be equal to $k \times l$ since k is the number of clusters and l is the average number of dimensions in a cluster. The number of dimensions associated with each medoid must be at least 2, which makes each cluster meaningful. For each medoid i, we compute the mean $Y_i = \left(\sum_{j=1}^{d} X_{i,j}\right) / d$, and the standard deviation $\sigma_i = \sqrt{\sum_j (X_{i,j} - Y_i)^2/(d-1)}$ of the values $X_{i,j}$. Y_i represents the average modified Manhattan segmental distance of the points in L_i relative to the entire space. Thus $Z_{i,j} = (X_{i,j} - Y_i)/\sigma_i$ indicates how the average distance along dimension j associated with the medoid m_i is related to the average modified Manhattan segmental distance associated with the same medoid. We decide dimensions for all clusters by picking the smallest $Z_{i,j}$ values by a greedy algorithm [10] used in PROCLUS such that a total of $k \times l$ values are chosen and at least 2 values are chosen for each medoid. More specifically, all the $Z_{i,j}$ values are sorted in increasing order. We assign the two smallest for each i, which gives a total of $2k$ values, and then greedily pick the remaining lowest $k \times (l-2)$ values.

We do a single pass over the database to assign the points to the medoids. For each i, we compute the modified Manhattan segmental distance relative to D_i between a point and the medoid m_i, and assign the point to the closest medoid.

Then the quality of a set of medoids is evaluated by the average modified Manhattan segmental distance from the points to the centroids of the clusters to which they belong.

9.3.3.1. *Simplified Replacing Logic*

We propose a simplified replacing logic compared to PROCLUS to decide whether it's good to replace the bad medoids in the current best medoids set with new medoids. When replacing the bad medoids, we first calculate δ_i for each medoid m_i. We only recalculate the $X_{i,j}$ values for those medoids whose δ_i values changed (store the $X_{i,j}$ value for current best objective case so that for those medoids whose δ_i values don't change, their $X_{i,j}$ values can be recovered from the stored values). Then we calculate Y_i, σ_i and $Z_{i,j}$. We decide dimensions for all clusters by the $Z_{i,j}$ values. When we assign points to clusters, there are two cases. For the points previously in the clusters whose δ values don't change, we only compare their modified Manhattan segmental distance from the current medoids with their modified Manhattan segmental distance from the medoids whose δ values changed. For the points previously in the clusters whose δ values changed or in the bad medoid's cluster, we compare its distance to all the current medoids to decide which cluster it belongs to. Then the new clusters are evaluated to decide whether the objective value is better. The simplified logic for assigning points in the iterative phase is given in Algorithm 9.3.

Algorithm 9.3. IterativeAssignPoints(C_1, ..., C_k, D_1, D_2, ..., D_k, B)

{B is the set of medoids whose d_i values changed and newly added medoids}
begin
for all the points i **do**
  assume point $i \in C_j$
  if $C_j \in B$ **then**
    compare point i's modified Manhattan segmental distance with all the medoids to decide which cluster it belongs to
  else
    compare point i's modified Manhattan segmental distance with all the medoids in B to decide which cluster it belongs to
  end if
end for
end

9.3.4. *Refinement Phase*

In the refinement phase, we redo the process in the iterative phase once by using the data points distributed by the result clusters at the end of the iterative phase, as opposed to the localities of the medoids. Once the new dimensions have been computed, we reassign the points to the medoids relative to these new sets of dimensions.

9.3.4.1. *Dimension Tuning Process*

We notice that users need to specify the average number of dimensions denoted as l in PROCLUS. Although it has achieved that different clusters have different subsets of dimensions, the number of dimensions for each cluster is still under the control of l, which is not flexible enough. According to criterion 2 discussed in Section 1, we want the number of attributes in a cluster to be as large as possible. Therefore, we propose one more step at the end of the refinement phase to reduce the dependence on l. In this step, for each cluster i, we choose the dimension with the smallest $Z_{i,j}$ value from the dimensions that are not chosen in previous steps and add it to the dimensional space to see if the new cluster is better. If the new cluster is better, we keep the newly added dimension and repeat this process to try to add more dimensions; otherwise, it will be discarded and we stop trying for this cluster. This process is achieved by the DimensionTuning algorithm for which the pseudo-code is given in Algorithm 9.4. The quality of a cluster is evaluated by the combination of criteria 1 and 3. A user-defined threshold is set for criterion 3. Clusters that pass the threshold will be evaluated by criterion 1. By doing this, we introduce criteria 2 and 3 into the algorithm, which gives a more balanced result.

Algorithm 9.4. DimensionTuning(M_{best}, C_1, ..., C_k, D_1, D_2,..., D_k)

begin
for each cluster C_i **do**
  bestEvaluateValue= the average distance to centroid in C_i
  for each cluster C_i **do**
    isGood=false;
    repeat
      add the dimension j ($j \notin D_i$) with the smallest $Z_{i,j}$value to D_i
      reassign the points to clusters $C'_1, \ldots, C'_k$ according to the new D_i
      newEvaluateValue= the average distance to centroid in C'_i
      if newEvaluateValue<bestEvaluateValue and the number of points in C'_i is more than a threshold value **then**

```
          (C_1, ..., C_k )= (C'_1, ..., C'_k )
          isGood=true
          for each cluster C_i do
             update bestEvaluateValue
          end for
       else
          remove dimension j from D_i
          isGood=false
       end if
    until isGood=false
  end for
end for
end
```

Outliers are also handled during the last pass over the data. For each medoid m_i with the dimensions D_i, we find the smallest Manhattan segmental distance Δ_i to any of the other $(k-1)$ medoids with respect to the set of dimensions D_i:

$$\Delta_i = \min_{j \neq i} d_{D_i}(m_i, m_j)$$

Δ_i is the sphere of influence of the medoid m_i. A data point is an outlier if its Manhattan segmental distance to each medoid m_i, relative to the set of dimensions D_i exceeds Δ_i.

9.4. Empirical Results

The experimental evaluation was performed on a Dell Dimension 4600 Intel Pentium IV processor 2.4GHz with 1.00GB of memory, running Windows XP professional with service pack 2. The data was stored on a 7200RPM, 8MB cache, 80G hard drive. The flow chart of the experimental evaluation for a dataset is illustrated in Fig. 9.3.

We test the performance of IPROCLUS and PROCLUS for synthetic data and real biomedical data. Unless otherwise specified, all the results are obtained by running the algorithms on the datasets multiple times and taking the average. Each time a random seed is chosen and the two algorithms are both fed with this seed for the random generator to guarantee fair comparison. We discuss the generation of the synthetic datasets in Subsection 9.4.1. Then we compare our empirical results of running IPROCLUS and PROCLUS on synthetic datasets in Subsection 9.4.2. The performance of IPROCLUS and PROCLUS on a real biomedical dataset, the colon tumor dataset, is analyzed in Subsection 9.4.3.

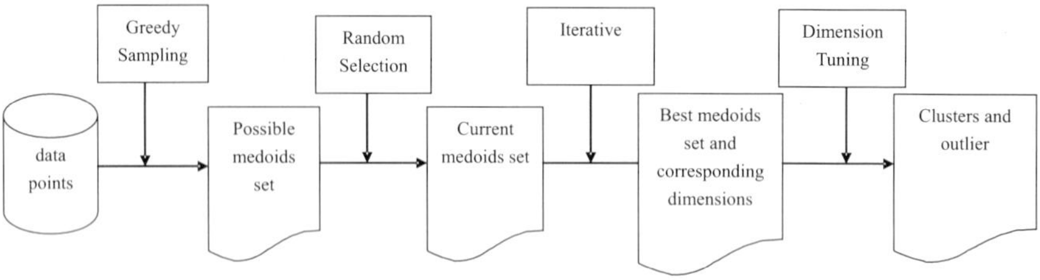

Fig. 9.3. The Flow Chart of the Experiment

9.4.1. *Synthetic Data Generation*

We generate one random case and two extreme cases. In each case, we consider two datasets: unscaled datasets and scaled datasets. The unscaled datasets are generated in exactly the same way as described in [1]. In order to generate the scaled datasets, we assign a dimensional scale factor s_j to each dimension j. The variance of the normal distribution on dimension j is s_j times the variance used in [1].

The only difference between the random datasets and the two extreme datasets is the way of generating dimensions. In the extreme case 1, all the clusters have exactly the same dimensions. In our experimental study, we choose to let all clusters have 10 exactly the same dimensions that are randomly chosen from all the dimensions. We also call it all-same-dimensions case. In the extreme case 2, we consider clusters with no common dimensions. The dimensions are chosen randomly and we just make sure that there are no shared dimensions between clusters. We choose to let each cluster have 4 dimensions. We also call it no-common-dimensions case.

9.4.2. *Results on Synthetic Datasets*

We compare the performance of IPROCLUS and PROCLUS on synthetic datasets in three aspects: the accuracy, the dependence on l, and the running time.

We first discuss the accuracy performance. In PROCLUS paper, the scale factors in the generated synthetic data are random numbers in the range [1, 2] for all dimensions. That's not necessarily the case for real data. In order to simulate real data, we use the dimensional scale factor introduced in the previous section. First, we consider a scaled dataset for the random case. We conduct our experiment when different dimensions have different dimensional scale factors. Table 9.1 shows the generated dimensions of the input clusters for the random case. Each cluster has 7 dimensions which are generated randomly. The dimensional scale

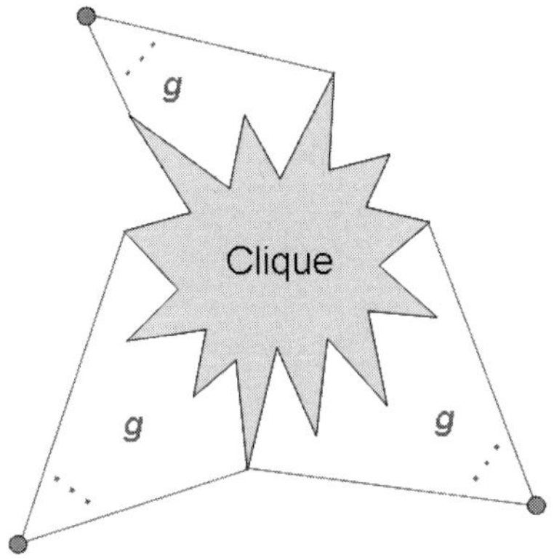

Fig. 10.4. Paraclique augments a clique with non-clique vertices in a controlled manner to increase size, decrease overlap and maintain density.

falls below some reasonable cutoff value (we set this value at 50). In this way, paraclique eliminates the need to compute and store enormous lists of maximal cliques.

10.5. Statistical Evaluation and Biological Relevance

Edge density is arguably the most telling statistical clustering metric. Clique, of course, maximizes density at 100% by definition. With the paraclique algorithm, density will tend to decrease as new nodes are glommed onto a starting clique. How precipitously density falls depends heavily on g. Table 10.1 summarizes the results for paraclique when it was run over the NOD data of this study. Clique size, paraclique size, and edge density are averaged over the paracliques generated. Note the manner in which paraclique increases cluster size with only a gradual reduction in density. (In contrast, we find that enlarging cliques using simple 1- and 2-neighborhoods quickly drops density into the single digits.) As a practical matter, we must balance the desire to handle noise and expand paracliques with the real need to maintain suitably high edge densities. As a rule of thumb, therefore, we seek to maintain a minimum density of at least 90% and henceforth set g at $|C| - 5$. We emphasize that this choice is highly data-dependent, and tunable to each application by design.

Density alone, however, tells only part of the story. To test for biological relevance, we used the Ingenuity Pathways Analysis (IPA) package from Ingenuity® Systems, `www.ingenuity.com`. IPA allows subscribers to upload and test lists

Table 10.1. Paraclique Parameter Variation

Glom Factor	Number of Paracliques	Clique Size	Paraclique Size	Edge Density	Lowest Edge Density
$\|C\| - 1$	32	99.4	104.8	99.8%	99.5%
$\|C\| - 2$	30	99.9	118.8	99.0%	97.9%
$\|C\| - 3$	28	101.6	137.4	97.8%	96.0%
$\|C\| - 4$	27	101.4	151.4	96.4%	92.3%
$\|C\| - 5$	24	106.1	173.8	94.9%	90.3%
$\|C\| - 6$	24	104.7	186.8	92.9%	86.7%
$\|C\| - 7$	22	108.5	205.7	91.4%	83.1%
$\|C\| - 8$	21	110.2	221.1	90.0%	80.0%
$\|C\| - 9$	21	109.3	231.1	88.6%	77.9%
$\|C\| - 10$	19	114.7	250.5	87.7%	76.6%

of genes (in our case Affymetrix probesets) against a manually curated biological interaction database. Probe sets known by the database are mapped to genes, which are then termed *focus genes*. Other probe sets are ignored. Focus genes are analyzed to determine how they are connected to one another based on evidence from the biomedical literature. Based on this analysis, one or more molecular networks are produced. Each typically consists of a mixture of focus genes, sprinkled with additional database genes and gene products that are needed to connect the focus genes and complete the network. We term a focus gene that is placed in such a network a *focus gene utilized*. In general, one cannot expect that all focus genes will become members of a network. The database may have very little information about a focus gene's connectivity. Alternately, a focus gene may be only distantly related to other focus genes. Due to technical constraints, IPA imposes a limit on network size, which is currently set to 35 nodes. As a result, lists with large numbers of focus genes often create multiple networks. Fortunately, these can often be fused together into a single common network using commands that are available on the Ingenuity website and that are designed for this purpose. The more closely connected a group of focus genes are biologically, the more likely it is that the database can connect them all into a network. Thus, an important metric is the *percent focus genes utilized*. This number alone can be misleading, however, because we must bear in mind that IPA may spread the genes across more than one network. A group of 40 focus genes, for example, would be considered more closely related if they could be connected in two networks than if four networks are needed to connect them all. We will therefore also calculate and examine *focus genes utilized per network*, a metric that normalizes for this effect.

As a control, we also tested K-means clustering, a traditional and highly popular algorithm. We invoked it via the R programming language, with the "kmeans" function from the "amap" package [15]. Input values were log transformed. Pearson correlations were employed. We sought to generate 500 clusters, because that should yield clusters of roughly the same size as those produced by the paraclique algorithm. Iteration was performed until convergence. IPA requires that each network be analyzed separately (no batch mode is available), a process that can be quite time consuming. Thus, only a small number, say ten, of clusters could be selected for further analysis. For paraclique, we simply selected the first ten outputs. Deciding on a representative set of K-means outputs was not as straightforward. We therefore chose to select K-means clusters under three different criteria. One criterion was to choose the ten largest clusters. Another was to favor those ten with the highest edge density in the p=0.01 graph. In case this produced unfairly small genesets, we also required that for a cluster to be selected it had to have size at least 50, the same lower bound we use for paraclique. The third criterion was based on paraclique overlap. For this we chose the ten K-means clusters with the highest percentage overlap with some paraclique, again insisting that a cluster had to have at least 50 vertices. Overlap ranged from roughly 45% to 64%, with an average of about 55%. Table 10.2 summarizes these results. All values are averaged over the relevant ten clusters.

Table 10.2. Paraclique versus K-Means

Method	Probe Sets	Edge Density	Focus Genes	Genes Utilized	Percent Utilized	Focus Genes per Network
Paraclique	254.3	97.1%	146.9	140.7	95.5%	14.4
Large K-means	244.0	31.5%	143.1	133.5	93.0%	12.8
Dense K-means	80.9	84.6%	52.3	46.7	89.4%	12.8
Overlap K-means	89.0	79.6%	55.7	49.9	89.8%	12.3

By inspection, paraclique is superior to K-means clustering in terms of density. The case for superior biological relevance is perhaps less obvious. We therefore performed ANOVA tests for statistical significance. The number of focus genes per network was higher ($p < .001$) for paraclique than for any of the K-means methods. And while paraclique did not differ markedly from Large K-means in terms of cluster size, it was more successful than other K-means methods in both size and percent focus genes utilized ($p < .05$).

10.6. Proteomic Data Integration

We now consider the problem of combining quantitative transcript and protein data for analysis. Only a few studies have been reported (see, for example, [?]). The related problem of combining gene expression with measures of function was recently considered in [3]. There gene ontology, phenotypes and protein-

protein interaction were used to devise distance measures and permutation tests for strength of commonality in graphs from these different data sources. Although no quantitative protein values were employed, data derived from *Saccharomyces cerevisiae*, commonly known as baker's or budding yeast, suggested that similarity in expression is related to similarity in function.

Our main goal is to identify biological pathways, each of which is anchored by a protein of interest. We are fortunate that both gene expression array data and protein gel data were collected from the exact same samples. If it were not for the expense involved, we would wonder why this is not done more often. Nevertheless, data integration remains a formidable task. The biggest difficulty we must overcome is probably that transcriptomic and proteomic data are generated by two completely different and unrelated processes. Thus we will not be able to use parametric statistical procedures, including the highly favored Pearson's correlation technique. Another problem is that current technologies for protein sensing are generally inferior to those for transcript detection. Modern expression array platforms can often detect transcripts for more than 50% of the known genes in the relevant organism, and generate highly reproducible quantitative measurements. In contrast, protein identification platforms can seldom cover more than 10% of an organism's estimated number of proteins, and with only moderate quantization and reproducibility. Of course function is a direct consequence of proteins, not mRNA, and so the importance of protein expression cannot be underestimated. Finally, it is well known that gene expression at the mRNA level will not always correlate well with gene expression at the protein level. After all, gene products are subject to post-transcriptional and post-translational modifications, degradation and other factors. Put together, these difficulties make any serious attempt at transcript-protein co-expression analysis a huge challenge. In the sequel, we shall address this challenge with non-parametric methods, graph algorithms and a clique-centric combinatorial approach.

We begin with the establishment of two correlation structures. For transcript-transcript relationships, we retain the Pearson's coefficients already computed. Transcript-protein relationships are typically much weaker and, for reasons already stated, require a non-parametric approach. For these we employ the rank metric provided by Spearman's correlation technique. This naturally leads to the loss of some information; a simple ranked list "flattens" raw data values. Our aim is now two-fold. We still wish to find dense, well-connected subgraphs. Yet these subgraphs must also be anchored as much as possible about some given protein, p, under scrutiny. Of course we could simply choose a putative pathway to be p and those transcripts ranked most highly with it. As we shall show, however, we can do better with the use of graph structure. To accomplish this, we take the transcript graph and add to it a new vertex for protein p. We then use the rank order provided by the Spearman's coefficient list to add edges connecting p with transcript vertices. We add these edges until the subgraph induced by p and its neighbors contains at least 100 maximal cliques each of size at least 40. We then output p along with the 60 or fewer vertices that most highly populate the resultant set of cliques. The values 40, 60 and 100 were chosen based on trial and error combined with our previous experience working with the idiosyncrasies of IPA. Other values may be superior in other applications.

To test this approach, we chose six proteins on which IPA contained information, which were well-expressed in the experimental samples, and which appear to be orthogonal to each other in terms of their biological function. Two of the six, HNRPK and EIF4A1, are of special interest because they are generally known to have increased expression in NOD mice relative to the NON and C57BL/6 strains [14]. The other four are ACTB, GDI2, GNB2L1 and ZBTB1. We also chose three different transcript graphs constructed from respective Pearson correlation thresholds 0.60, 0.70, and 0.80. For each of these 18 tests, maximal cliques were highly overlapping, as expected. As a measure of a cluster's biological relevance, we examine a metric we call *protein links*. Protein links is a count of the number of connections between an anchored protein and the network created by IPA. For each protein, we chose the threshold setting that maximizes protein links, with ties broken in favor of the higher threshold. The lowest threshold, 0.60, had none of the best results. It is probably the case that, in a graph this dense, the transcript-transcript relationships drown out protein-transcript correlations.

As a control, we compared the quality of the transcript sets we produced against the 60 transcripts that simply correlate most highly with the protein. GDI2 and ZBTB1 had fewer than three protein links for all four results (the three threshold values plus the straight correlation list), and so were dropped from further analysis. Results for each of the four remaining proteins are shown in Table 10.3.

Table 10.3. Clique vs Correlates

Protein	Algorithm	Probe Sets	Focus Genes	Protein Links
ACTB	Clique at 0.70	60	42	6
	Correlates List	60	27	6
EIF4A1	Clique at 0.70	59	50	7
	Correlates List	60	41	2
GNB2L1	Clique at 0.70	60	39	6
	Correlates List	60	37	3
HNRPK	Clique at 0.80	55	38	5
	Correlates List	60	42	3

From this table, we see that our clique-centric approach builds subgraphs that are no worse and in fact generally better than those simply defined by ranking

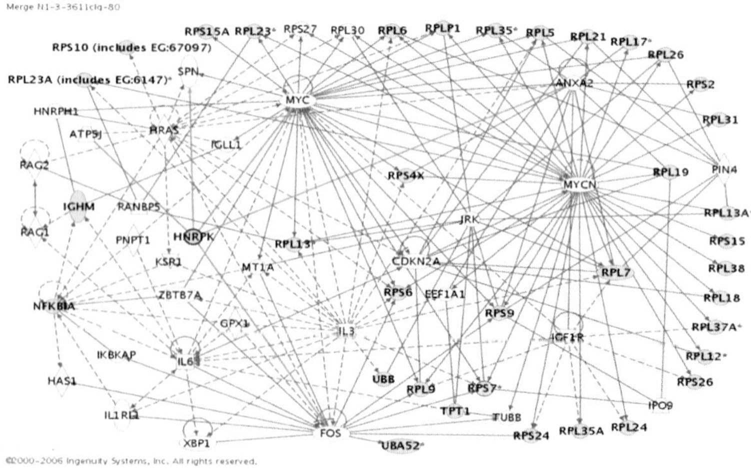

Fig. 10.5. The IPA merged network for HNRPK using a clique-centric algorithm.

and selecting correlates. Although protein links are our primary focus, other metrics are equally revealing. In the case of ACTB, for example, we find that both methods produce six protein links, but the algorithm based on clique is superior in terms of percent focus genes utilized (100% versus 85.2%) and focus genes per network (14 versus 11.5).

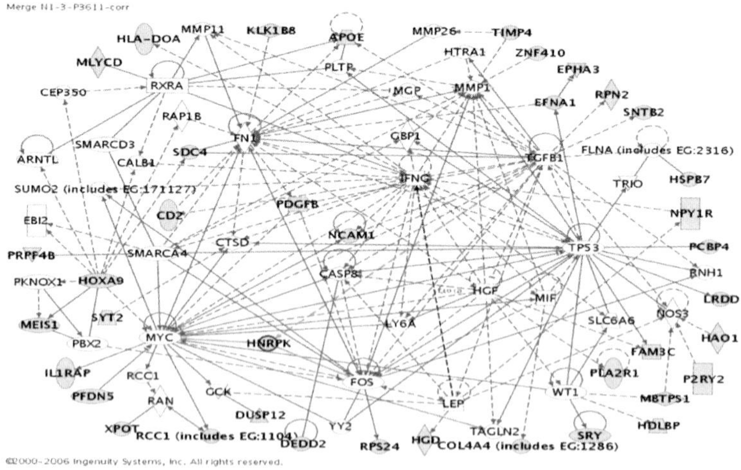

Fig. 10.6. The IPA merged network for HNRPK using simple correlation.

It may also be instructive to compare IPA's outputs visually. Fig. 10.5 and 10.6 contain screenshots of merged network diagrams created by IPA for HNRPK. Fig. 10.5 was generated from the list of transcripts produced by our clique-centric method; Fig. 10.6 was generated from the list produced by mere correlate ranking. Focus genes are depicted in grey. Connections to the anchor protein are rendered in blue. Glyph shapes vary depending on IPA classifiers.

The IPA screenshots shown in Fig. 10.5 and 10.6 demonstrate how the two methods we consider create quite different networks, and how the protein is connected to more genes in the network created by the clique-centric algorithm.

10.7. Remarks

We have studied clique-centric algorithms in the context of effective biological data clustering. Statistical quality based primarily on edge density and biological significance based on curated pathway matching have demonstrated the utility of paraclique and related methods. We have also considered the problem of inhomogeneous data integration. Transcriptomic data from gene expression arrays and proteomics data from 2d gels have been reconciled to identify biological networks for further scrutiny.

We emphasize that this work has been limited in scope to the analysis of inhomogeneous data of relevance to type 1 diabetes. It is not meant to provide a comprehensive guide to the literature. Nor is it intended to serve as an exhaustive comparison of clustering methods. Such a task would be an enormous challenge, requiring the implementation of a huge number of algorithms, and necessitating tests across a great many diverse datasets.

There are a variety of ways to modify and enhance paraclique and the other algorithms we describe. In [6], for example, an optional user-defined *threshold parameter* is provided to help guide the search for edges affected by noise. For simplicity, we have ignored this parameter here and considered only the effect of the glom factor. Another enhancement is to glom vertices in stages, invoking paraclique iteratively until a certain threshold is reached. Initial results suggest that this procedure can further increase paraclique size while maintaining both edge density and biological fitness as measured by IPA.

Finally, we observe that pleiotropism is common in gene and gene products. It is thus a major reason for the popularity of soft clustering methods such as clique: a vertex can lie in more than one clique, just as an oligonucleotide or a protein can lie in more than one pathway. Noise and the need for simpler structures motivate the paraclique algorithm. The clusters produced are robust with respect to a few missing edges. Unfortunately, they no longer overlap with the basic paraclique method. It is possible to modify the algorithm so that overlap is permitted. This is a topic of current research within our group. Optimal ways to accomplish this, however, probably depend on the application. The same may be said for the highly challenging task of inhomogeneous data integration. We are currently working on techniques to integrate multiple proteins in a single step, rather than handling them one at a time. This is not as easy as it might sound, and may require the use of three rather than just two forms of correlate pairs.

Acknowledgments

We wish to thank an anonymous referee, whose careful review of our original submission helped us to improve the final presentation. This research has been supported in part by the National Science Foundation under grant CCR-0311500, by the National Institutes of Health under grants 1-P01-DA-015027-01, 5-U01-AA-013512, 1-R01-DK-062103-01 and 1-R01-MH-074460-01, and by the UT-ORNL Science Alliance. A preliminary version of a portion of this paper was presented at the DIMACS Workshop on Clustering Problems in Biological Networks, held at the DIMACS Center at Rutgers University in May, 2006.

References

[1] F. N. Abu-Khzam, M. A. Langston, P. Shanbhag, and C. T. Symons. Scalable parallel algorithms for FPT problems. *Algorithmica*, 45:269–284, 2006.

[2] J. S. Bader, A. Chaudhuri, J. M. Rothberg, and J. Chant. Gaining confidence in high-throughput protein interaction networks. *Nature Biotechnology*, 22:78–85, 2004.

[3] R. Balasubramanian, T. LaFramboise, D. Scholtens, and R. Gentleman. A graph-theoretic approach to testing associations between disparate sources of functional genomics data. *Bioinformatics*, 20(18):3353–3362, 2004.

[4] N. E. Baldwin, E. J. Chesler, S. Kirov, M. A. Langston, J. R. Snoddy, R. W. Williams, and B. Zhang. Computational, integrative, and comparative methods for the elucidation of genetic coexpression networks. *J Biomed Biotechnol*, 2(2):172–180, 2005.

[5] C. Bron and J. Kerbosch. Algorithm 457: finding all cliques of an undirected graph. In *Proceedings of the ACM*, pages 575–577, 1973.

[6] E. J. Chesler and M. A. Langston. Combinatorial genetic regulatory network analysis tools for high throughput transcriptomic data. In *RECOMB Satellite Workshop on Systems Biology and Regulatory Genomics*, 2005.

[7] E. J. Chesler, L. Lu, S. Shou, Y. Qu, J. Gu, J. Wang, H. C. Hsu, J. D. Mountz, N. E. Baldwin, M. A. Langston, J. B. Hogenesch, D. W. Threadgill, K. F. Manly, and R. W. Williams. Complex trait analysis of gene expression uncovers polygenic and pleiotropic networks that modulate nervous system function. *Nature Genetics*, 37:233–242, 2005.

[8] R. G. Downey and M. R. Fellows. *Parameterized Complexity*. Springer, 1999.

[9] U. Feige, S. Goldwasser, L. Lovasz, S. Safra, and M. Szegedy. Approximating the maximum clique is almost $\mathcal{NP}$-complete. In *IEEE Symposium on the Foundations of Computer Science*, pages 2–12, 1991.

[10] U. Feige, D. Peleg, and G. Kortsarz. The dense k-subgraph problem. *Algorithmica*, 29:410–421, 2001.

[11] M. R. Fellows and M. A. Langston. Nonconstructive tools for proving polynomial-time decidability. *Journal of the ACM*, 35:727–739, 1988.

[12] M. R. Fellows and M. A. Langston. On search, decision, and the efficiency of polynomial-time algorithms. *Journal of Computer and Systems Sciences*, 49:769–779, 1994.

[13] I. C. Gerling, C. Ali, and N. Lenchik. Characterization of early developments in the splenic leukocyte transcriptome of NOD mice. *Ann N Y Acad Sci*, 1005:157–160, 2003.

[14] I. C. Gerling, S. Singh, N. I. Lenchik, D. R. Marshall, and J. Wu. New data analysis and mining approaches identify unique proteome and transcriptome markers of susceptibility to autoimmune diabetes. *Mol Cell Proteomics*, 5(2):293–305, 2006.

[15] R. Ihaka and R. Gentleman. R: A language for data analysis and graphics. *Journal of Computational and Graphical Statistics*, 5:299–314, 1996.

[16] R. Kirova, M. A. Langston, X. Peng, A. D. Perkins, and E. J. Chesler. A systems genetic analysis of chronic fatigue syndrome: combinatorial data integration from snps to differential diagnosis of disease. In *International Conference for the Critical Assessment of Microarray Data Analysis (CAMDA)*, 2006.

[17] H. Kitano. Computational systems biology. *Nature*, 420(6912):206–210, 2002.

[18] H. Kitano. Systems biology: a brief overview. *Science*, 295(5560):1662–1664, 2002.

[19] M. A. Langston, L. Lan, X. Peng, N. E. Baldwin, C. T. Symons, B. Zhang, and J. R. Snoddy. A combinatorial approach to the analysis of differential gene expression data: the use of graph algorithms for disease prediction and screening. In K. F. Johnson and S. M. Lin, editors, *Methods of Microarray Data Analysis IV, Papers from CAMDA '03*, pages 223–238. Kluwer Academic Publishers, 2005.

[20] M. A. Langston, A. D. Perkins, A. M. Saxton, J. A. Scharff, and B. H. Voy. Innovative computational methods for transcriptomic data analysis. In *ACM Symposium on Applied Computing*, 2006.

[21] E. Pennisi. Systems biology. tracing life's circuitry. *Science*, 302(5651):1646–1649, 2003.

[22] I. Seefeldt, G. Nebrich, I. Römer, L. Mao, and J. Klose. Evaluation of 2-de protein patterns from pre- and postnatal stages of the mouse brain. *Proteomics*, 6(18):4932–4939, 2006.

[23] J. C. Setubal and J. Meidanis. *Introduction to Computational Molecular BIology*. PWS Publishing Company, 1997.

[24] L. Shi, L. H. Reid, W. D. Jones, R. Shippy, J. A. Warrington, S. C. Baker, P. J. Collins, F. de Longueville, E. S. Kawasaki, K. Y. Lee, Y. Luo, Y. A. Sun, J. M. Willey, R. A. Setterquist, G. M. Fischer, W. Tong, Y. P. Dragan, D. J. Dix, F. W. Frueh, F. M Goodsaid, D. Herman, R. V. Jensen, C. D. Johnson, E. K. Lobenhofer, R. K. Puri, U. Schrf, J. Thierry-Mieg, C. Wang, M. Wilson, P. K. Wolber, L. Zhang, W. Slikker, L. Shi, L. H. Reid, and M. A. Q. C. Consortium. The microarray quality control (maqc) project shows inter- and intraplatform reproducibility of gene expres-

sion measurements. *Nat Biotechnol*, 24(9):1151–1161, 2006.

[25] M. Sipser. *Introduction to the Theory of Computation*. Course Technology, 1996.

[26] H. E. Thomas and T. W. Kay. Beta cell destruction in the development of autoimmune diabetes in the non-obese diabetic (NOD) mouse. *Diabetes Metab Res Rev*, 16(4):251–261, 2000.

[27] J. Tian, A. P. Olcott, L. R. Hanssen, D. Zekzer, B. Middleton, and D. L. Kaufman. Infectious th1 and th2 autoimmunity in diabetes-prone mice. *Immunol Rev*, 164:119–127, 1998.

[28] B. H. Voy, J. A. Scharff, A. D. Perkins, A. M. Saxton, B. Borate, E. J. Chesler, L. K. Branstetter, and M. A. Langston. Extracting gene networks for low-dose radiation using graph theoretical algorithms. *PLoS Comput Biol*, 2(7):e89, 2006.

[29] M. T. Wayland and S. Bahn. *Chapter 5, Reproducibility of microarray studies: concordance of current analysis methods.*, volume 158, pages 109–125. ScienceDirect, 2006.

[30] D. B. West. *Introduction to Graph Theory*. Prentice Hall, 1996.

[31] A. M. Wheelock and A. R. Buckpitt. Software-induced variance in two-dimensional gel electrophoresis image analysis. *Electrophoresis*, 26(23):4508–4520, 2005.

[32] J. W. Yoon, H. S. Jun, and P. Santamaria. Cellular and molecular mechanisms for the initiation and progression of beta cell destruction resulting from the collaboration between macrophages and t cells. *Autoimmunity*, 27(2):109–122, 1998.

[33] J.H. Zar. *Biostatistical Analysis*. Prentice Hall, 4th edition, 1998.

[34] Y. Zhang, F. N. Abu-Khzam, N. E. Baldwin, E. J. Chesler, M. A. Langston, and N. F. Samatova. Genome-scale computational approaches to memory-intensive applications in systems biology. In *Supercomputing*, 2005.

Chapter 11

A Novel Similarity-based Modularity Function for Graph Partitioning

Zhidan Feng

Acxiom Corporation, USA, dafeng@acxiom.com

Xiaowei Xu

Department of Information Science, UALR, USA, xwxu@ualr.edu

Nurcan Yuruk

Department of Information Science, UALR, USA, nxyuruk@ualr.edu

Thomas Schweiger

Acxiom Corporation, USA, Tom.Schweiger@acxiom.com

Graph partitioning, or network clustering, is an essential research problem in many areas. Current approaches, however, have difficulty splitting two clusters that are densely connected by one or more "hub" vertices. Further, traditional methods are less able to deal with very confused structures. In this paper we propose a novel similarity-based definition of the quality of a partitioning of a graph. Through theoretical analysis and experimental results we demonstrate that the proposed definition largely overcomes the "hub" problem and outperforms existing approaches on complicated graphs. In addition, we show that this definition can be used with fast agglomerative algorithms to find communities in very large networks.

11.1. Introduction

Many problems can be modeled as networks or graphs, and identifying cluster in the network or partitioning the graph is a fundamental problem for computer networks analysis, VLSI design, biological network analysis, social networks analysis [18], business networks analysis, and community detection [6]. In the literature, graph partitioning has many names, and is sometimes called network analysis, network clustering, detecting communities in networks, etc. This area of research has seen a lot of efforts in this problem over decades, and many algorithms have been proposed, studied, and used.

The problem is defined as: Given a graph $G = \{V, E\}$, where V is a set of vertices and E is a set of weighted edges between vertices, optimally divide G into k disjoint sub-graphs $G_i = \{V_i, E_i\}$, in which $V_i \cap V_j = \Phi$ for any $i \neq j$ and $V = \sum_{i=1}^{k} V_i$. The number of sub-graphs, k, may or may not be prior known.

In this paper, we focus on partitioning problem of un-weighted graphs, that is, graphs for which the weight of all edges is 1.

Two question need to be answered, one that is mathematical and one that is algorithmic. First, what is the definition of *optimal* when partitioning a graph, and second, how does one find the optimal partition efficiently.

Regarding the first question, there is no consensus in the literature. Different approaches use different criteria for different applications. Examples are the min-max cut [3, 8, 16], modularity [1, 13, 14] and betweenness [6, 12]. To the second question, since finding the optimal partition in a graph is normally a NP-complete problem, most current approaches use heuristics to reduce the running time and find only near-optimal partitions.

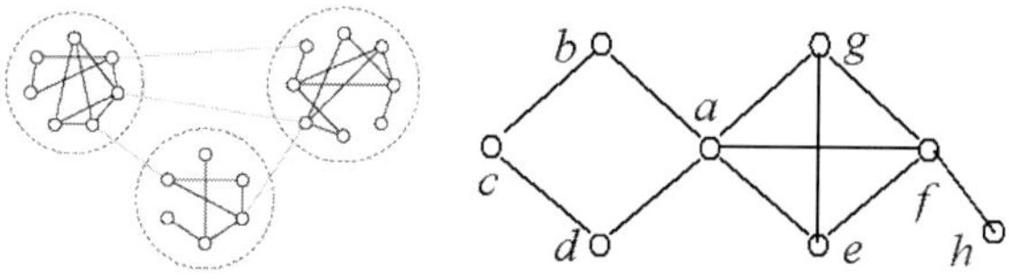

(a). Type I **(b).** Type II

Fig. 11.1. Two types of graphs

These approaches, though work well for some applications, their performance deteriorates as graphs become more confused. Graphs may present three kinds of complications. Figure 11.1 illustrates two types of un-weighted graphs. The first and main source of confusion is "random" interconnections between clusters, illustrated by the Type I graph. These have dense clusters that are sparsely interconnected. As the number of interconnections increase, discerning the underlying structure becomes more challenging. Such structures have been the focus of past study.

The second and third source of confusion is what we call hubs and outliers, and these have been little discussed in the literature. The Type II graph in Fig. 11.1 has clusters that are connected by "hub" vertex 'a' that is difficult to place in one cluster or another. It also has an outlier vertex 'h' that may be best placed in a cluster to itself. Past approached do not deal with the "hub" and "outlier" problems very well. That is, they cannot split two clusters very well that are densely connected by one or more "hub" vertices, and they often fail to detect and isolate outliers.

In this paper, we propose a similarity-based graph partitioning definition that globally measures whether one partitioning is better than another. Through theoretical analysis and experimental results, we demonstrate that the proposed definition outperforms existing approaches on complicated graphs and handles with agility hubs and outliers. Further, we show that this novel definition can be used with the fast agglomerative algorithm [1, 13] to find communities in very large networks.

The rest of this paper is organized as follows: in section 2 we briefly review related works; in section 3 we propose our novel similarity-based modularity measurement; in section 4 and 5 we discuss two techniques for finding optimal graph partitions by maximizing the modularity; in section 6 we apply them to network with known structures and present experimental results; and finally in section 7 we summarize our conclusions.

11.2. Related Work

The most traditional definition of graph partitioning is probably the min-max cut [3, 8, 16], which seeks to partition a graph $G = \{V, E\}$ into two sub graphs A and B. The principle of min-max clustering is minimizing the number of edges between clusters A and B and maximizing the number of edges within each. Instead of number of edges, sum of weights is used for a weighted graph. Thus the connection between A and B the cut:

$$cut(A, B) = W(A, B) = \sum_{u \in A, v \in B} w(u, v)\,. \tag{11.1}$$

It is obvious that $W(A) \equiv W(A, A)$.

A bi-partition of the graph is defined as minimizing the following objective function:

$$M_{cut} = \frac{cut(A, B)}{W(A)} + \frac{cut(A, B)}{W(B)}\,. \tag{11.2}$$

The above function is called the min-max cut function. It minimizes the *cut* between two clusters while maximizing the connections within a cluster. However, a pitfall of this definition is that if we only cut out one node from a graph, M_{cut} will probably get the minimum value. So, in practice, M_{cut} must be accompanied with some constraints, such as A and B should of equal or similar size, or

$|A| \approx |B|$. These constraints are not always applicable for all applications, e.g., in clustering problems where some communities are much larger than the others.

To amend the above issue, a $normalized\ cut$ (N_{cut}) was proposed [16]:

$$N_{cut} = \frac{cut(A,B)}{assoc(A,V)} + \frac{cut(A,B)}{assoc(B,V)}, \tag{11.3}$$

where $assoc(A,V) = \sum_{u \in A, v \in V} w(u,v)$ is the total connection from vertices in A to all vertices in the graph. Using the N_{cut}, cutting out one vertex or some small part of the graph will no longer always yield a small N_{cut} value.

Using the M_{cut} or N_{cut} one can partition a graph into two sub-graphs. As the natural consequence, to divide a graph into k sub-graphs, one has to adopt a top-down approach, i.e., splitting the graph into two sub-graphs, then further splitting these sub-graphs into next level sub-graphs; repeat this process until k sub-graphs have been detected. The disadvantage [5] of this method is that, like all other bisection algorithms, it only partitions a graph into two sub-graphs. Though one can repeat this procedure on the sub-graphs, there is no guarantee of the optimality of partitioning, and one has no clue on when to stop the repeated the bisection procedure or how many sub-graphs should be produced in a graph.

To answer the question "what is the best graph partition", other researchers proposed global measurements, such as M. J. Newman's modularity [1, 7, 13]:

$$Q_N = \sum_{i=1}^{NC} \left[\frac{l_s}{L} - \left(\frac{d_s}{L} \right)^2 \right], \tag{11.4}$$

where NC is the number of clusters, L is the number of edges in the network, l_s is the number of edges between vertices within cluster s, and d_s is the sum of the degrees of vertices in cluster s. The optimal clustering is achieved by maximizing the modularity Q_N. Modularity is defined such that Q_N is 0 at the extremes of all vertices are clustered into a single cluster or of vertices are randomly clustered. As this modularity definition requires no constraints, it is better than the min-max cut definition. Modularity is a quality measure of graph partitioning, while normalized cut is not. Note that with this definition both top-down (divide and conquer) and bottom-up (agglomerative) algorithms can be used for graph partitioning.

Finding the best Q_N is NP-complete. Instead of performing an exhaustive search, Newman used a bottom-up approach [1, 13] which begins by merging two vertices into clusters that increase Q_N, then likewise merging pairs of clusters, until all have merged to form a single cluster. At each stage the value of Q_N is recorded. The partition with the highest Q_N is the solution of this algorithm. This is a typical hill-climbing greedy search algorithm, which generally has high speed but easily falls into a local optimum. Guimera and Amaral [7] use a simulated annealing algorithm to find the optimal partition with the highest Q_N. This is a random search based algorithm, which usually can achieve higher quality than the greedy search (like Guimera and Amaral claimed [7]) but has much lower speed.

According to our study [4], while Newman's modularity works well for type I graphs (Fig. 11.1 (a)); it fails to deal well with type II graphs due to the "hub" and "outliner" vertices. To reiterate, a "hub" vertex is a vertex that densely connects two or more groups in a graph. For example, in Fig. 11.1 (b), the central vertex 'a' connects two sub-groups. Note also that this same vertex has the largest degree. Using Newman's modularity definition, this graph in Fig. 11.1 (b) will be clustered into two groups, $\{a, b, c, d\}$ and $\{e, f, g, h\}$. However, the more reasonable clustering schedule could be $\{b, c, d\}$, $\{a, e, f, g\}$, and $\{h\}$, where 'h' is in a cluster by itself. Newman's modularity definition will not identify outliers; rather it will assign them membership to some larger cluster.

Generally, when there are clear sub-graph structures existed in a graph, i.e., the connections between sub-graphs are sparse while the connections within a sub-graph are much dense, both min-max cut or its variations and Newman's modularity works very well. But when the sub-graph structure becomes more confused, i.e., the between connections become denser, the performance of these methods deteriorates rapidly.

11.3. A Novel Similarity-based Modularity

Based on our observation, it is not very accurate to use connections within clusters and between clusters as the criteria of partitioning for all types of graphs, especially for graphs with complicated cluster structures. Instead, we propose a more general concept, similarity, to measure the graph partitioning quality. A good partition **is** one for which the similarity of vertices within a cluster is higher than the similarity of vertices between clusters.

If two connected vertices also share a lot of neighbors, we say they are similar. However, merely counting the number of shared neighbors doesn't tell us what proportion of the vertices' neighbors is shared, and therefore is insufficient for distinguishing a hub from a normal vertex. To handle this problem, we adopt the normalized similarity [11]. Let Γ_i be the neighborhood of vertex i in a network, the cosine normalization similarity is defined as:

$$S(u, v) = \frac{|\Gamma_u \cap \Gamma_v|}{\sqrt{|\Gamma_u| \cdot |\Gamma_v|}} . \tag{11.5}$$

Then we define the Similarity-based Modularity (Q_S) function as the following:

$$Q_s = \sum_{i=1}^{NC}\left[\frac{LS_i}{TS} - \left(\frac{DS_i}{TS}\right)^2\right] \quad (11.6)$$

with $IS_i = \sum_{u,v \in V_i} S(u,v), DS_i = \sum_{u \in V_i, v \in V} S(u,v)$ and $TS = \sum_{u,v \in V} S(u,v)$, where NC is the number of clusters, IS_i is the total similarity of vertices within cluster i; DS_i is the total similarity between vertices in cluster i and any vertices in the graph; TS is the total similarity between any two vertices in the graph; $S(u,v)$ is defined in (11.5); V is the vertex set of the graph and V_i is the vertex set of cluster i.

If we define

$$\delta(u,i) = \begin{cases} 0 & \text{if } u \notin \text{ cluster } i, \\ 1 & \text{if } u \in \text{ cluster } i, \end{cases}$$

then (11.6) is written as

$$Q_s = \sum_{i=1}^{NC}\left[\frac{\sum_u \sum_v S(u,v)\delta(u,i)\delta(v,i)}{\sum_u \sum_v S(u,v)} - \left(\frac{\sum_u \sum_v S(u,v)\delta(u,i)}{\sum_u \sum_v S(u,v)}\right)^2\right]. \quad (11.7)$$

In other words, Q_S is a measure of the similarity of a partition of the network versus the similarity of the same partition if the network's vertices were randomly connected. As with Newman's modularity, if we put all vertices either into one cluster or place them in clusters randomly, Q_S will be 0.

Referring again to the type II graph in Fig. 11.1 (b), the similarity between the "hub" vertex 'a' and its neighbors is quite low, despite the hub's high degree. Likewise, the similarity between the "outlier" vertex 'h' and its lone neighbor is low. Table 11.1 summarizes Q_N and Q_S respectively for several possible partitioning of the network in Fig. 11.2 (b).

Table 11.1. Q_N and Q_S for different clustering structures

Clustering Schedule	Q_N	Q_S
All vertices in one cluster	0.0	0.0
$\{a\}, \{b,c,d\}, \{e,f,g\}, \{h\}$	0.119835	0.343714
$\{a,b,c,d\}, \{e,f,g\}, \{h\}$	0.177686	0.312469
$\{a,b,c,d\}, \{e,f,g,h\}$	**0.227273**	0.289307
$\{b,c,d\}, \{a,e,f,g\}, \{h\}$	0.185950	**0.368656**
$\{b,c,d\}, \{a,e,f,g,h\}$	0.214876	0.321978

We acknowledge that preferring one graph partitioning over another may be subjective; however, we think the original modularity definition classifies the hub node 'a' into the wrong cluster $\{a,b,c,d\}$ and fails to segregate the outliner node 'h'. The proposed Similarity-based Modularity successfully detects the outlier node 'h' and classifies the hub node 'a' into the more reasonable cluster $\{a, e, f, g\}$, which was our aim.

11.4. A Genetic Graph Partitioning Algorithm

Recall that finding the optimum similarity-based modularity in a graph is NP-hard, and greedy-search based approaches all suffer from the local optimum drawback [2, 9, 15, 17, 19]. To evaluate the capability of the proposed modularity function, we would like to use a global-search based approach to avoid the local optimum trap, namely a Genetic Algorithm (GA):

Encoding - For a graph with n vertices, a partition can be represented as an array of integers, for which each index corresponds to a vertex and the value to its cluster ID of this vertex. For example, the string "2 3 2 3 5 1 5 4 5" indicates that there are 5 clusters in the graph, the 1st and the 3rd vertices belong to cluster 2, the 2nd and the 4th vertices belong to cluster 3, and so on. Since each array element can have a value from 1 to n, the total search space is n^n, which is potentially very large.

Initial Population - the initial population can be generated randomly or using some task related strategies. Here we use a random initialization.

Fitness Measure - the function to be optimized. For the graph partitioning problem we use formulae (11.5) and (11.6).

Selection - some individuals are selected according to their fitness or ranks and they mate to produce offspring. We use a probabilistic selection

$$P_r(h_i) = \frac{Fitness(h_i)}{\sum\limits_{j=1}^{k} Fitness(h_j)}, \tag{11.8}$$

where k is the number of clusters.

Crossover - this operator determines how to produce offspring from parents. Either single point or multiple point crossovers can be used. An important parameter is the cross rate, which determines how many genetic elements are exchanged. We use single point crossover. The exchanged point is chosen at random and the length of the array block that is exchanged is set by the Crossover Rate parameter. There are two cases in single point crossover: with or without roll back.

a) Single point crossover (Crossover Rate = 50%, no roll back):

Parents		Children
1 1 1 3 **2 3 4 4**	⇒	1 1 1 3 **4 4 3 3**
2 2 3 1 **4 4 3 3**		2 2 3 1 **2 3 4 4**
↖	Crossover point	↗

b) Single point crossover (Crossover Rate = 50%, with roll back):

Parents		Children
1 1 1 3 2 **3 4 4**	⇒	**1** 1 1 3 4 **4 3 3**
2 2 3 1 4 **4 3 3**		**2** 2 3 1 2 **3 4 4**
↑ Roll back ↖	Crossover point	↑ ↗

Mutation - mutation adds random variation and diversity to the population. Usually the mutation rate should be kept very small.

2 2 **3** 1 2 3 4 4 ⇒ 2 2 **1** 1 2 3 4 4

Replacement - new offspring are inserted into the original population, replacing individuals with lower fitness. Usually, the replacement population size is constant.

The above procedure is repeated until a predefined number of generations have elapsed or the fitness of the population stops increasing. The string with highest fitness is the solution.

11.5. A Fast Agglomerative Algorithm

Although GA has the potential to jump out of local optimums, it has the drawback of being slow and thus may be unsuitable for large graphs. Here we extend Newman's fast hierarchical agglomerative clustering (FHAC) algorithm [1, 13] to the similarity-based modularity.

Formula (11.6) can be re-written as:

$$Q_S = \sum_{i=1}^{k} Q_{S_i} \text{ With } Q_{S_i} = \frac{IS_i}{TS} - \left(\frac{DS_i}{TS}\right)^2. \tag{11.9}$$

The algorithm is as follows:

1. Initialize by placing every vertex v_i into its own cluster c_i, and calculates Q_{S_i} for each cluster.

2. Considering merging every possible pair of cluster c_g and c_h, $(g \neq h)$ and note the change to Q_S, which is given by:

$$\Delta Q_{S_{g+h}} = Q_{S_{g+h}} - Q_{S_g} - Q_{S_h} \tag{11.10}$$

3. Choose the merge producing the largest $\Delta Q_{S_{g+h}}$, then update and record Q_S.
4. Update all affected values of $\Delta Q_{S_{g+i}}$ and $\Delta Q_{S_{h+j}}$, $(i \neq g; j \neq h)$ using formula (11.10).
5. Repeat step 3 and 4 until all vertices are grouped into one cluster.
6. Retrieve the partitioning with the highest Q_S value as the best result.

In step 4, only clusters with edges connecting them to the merged pair need to be updated. The number of clusters to be updated that are connected to cluster g is $|i|$ and the number connected to cluster h is $|j|$. There are at most $|i| + |j|$ updates, each update taking O(log k) < O(log n) time to access the data structure, where n is the number of vertices in the Graph. Thus each iteration takes O(($|i|$+$|j|$)log n) time [1].

The overall running time is the sum of all joining steps, which, as Newman indicated is at most O(md log n) [13], where m is number of edges and d is the depth of the dendrogram. More details of this algorithm may be found in Newman et al. [1, 13].

11.6. Evaluation Results

In this section, we present experimental result from analyzing synthetic graphs, detecting community structure in social networks, and clustering entities from a customer database. The size of the graphs varies from tiny to very large. Each example has a known structure that we are seeking to reproduce. We introduce a measure of "error" to gauge the performance of different clustering algorithms: if a node is classified into a cluster other than the pre-defined cluster, we count it as an error.

11.6.1. *Tests on Synthetic Graphs*

Synthetic graphs are generated based on predefined cluster structure and properties, so as to facilitate systematic study of the performance of our similarity-based modularity definition and graph partitioning algorithms. Here we use the same construction as used in several papers [6, 7, 13]. The synthetic graph consists of 128 vertices that are divided into 4 equal-size clusters. Each vertex connects to vertices within its cluster with a probability P_i, and connects to vertices outside its cluster with a probability $P_o < P_i$. On average, each vertex is connected to

$K_{out} = 96P_o$ outside vertices and to $K_{in} = 31P_i$ inside vertices. We generate several random graphs using different K_{out} and K_{in}, and then test the performance of our two algorithms. Figure 11.2 illustrates examples of these synthetic random graphs with different levels of interconnectivity [7]. Notice that these graphs belong to type I defined in Fig. 11.1.

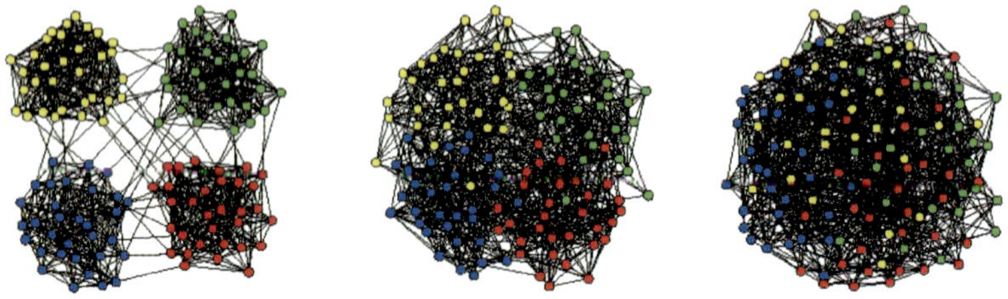

(a). $K_{in} : K_{out} = 15{:}1$ (b). $K_{in} : K_{out} = 10{:}6$ (c). $K_{in} : K_{out} = 8{:}8$

Fig. 11.2. Graphs with different $K_{in} : K_{out}$ ratio

Table 11.2. GA performances

$K_{in} : K_{out}$	Best Q_N	Errors	Best Q_S	Errors
12:4	0.499807	0	0.659431	0
11:5	0.437481	1	0.614884	0
10:6	0.376271	2	0.564478	1
9:7	0.318518	4	0.484211	6
8:8	0.261690	15	0.448903	20

Table 11.2 summarizes the results of the Genetic algorithm using Newman's modularity (Q_N) and the similarity-based modularity (Q_S). Since the Genetic algorithm falls into local minimums, we run the algorithm several times for each parameter setting and report the best result. At $K_{in} : K_{out} \geq 11{:}5$, the structures are very clear, both modularity definitions perfectly identify the structure. At $K_{in} : K_{out} = 10{:}6$, Q_N has 2 errors, but Q_S yields only one error. Since then, along with the confusion increases, Q_S begins slightly lagging from Q_N. This is due to the synthetic graphs are generated based on the ratio of $K_{in} : K_{out}$, which exactly matches with the definition of Q_N. The accuracy comparison is plotted in Fig. 11.3. We can conclude that the two modularity measures yield a comparable accuracy for the synthetic graphs.

Likewise, we tested the fast hierarchical agglomerative clustering algorithm (FHAC) to detect structure in the synthetic graphs. The results are summarized in Table 11.3, and we plot accuracy curves in Fig. 11.4.

Comparing Fig. 11.3 and 11.4, one can see that the FHAC keeps accuracy with GA when $K_{out} \leq 6$ ($K_{in} = 16 - K_{out}$), then accuracy drops significantly

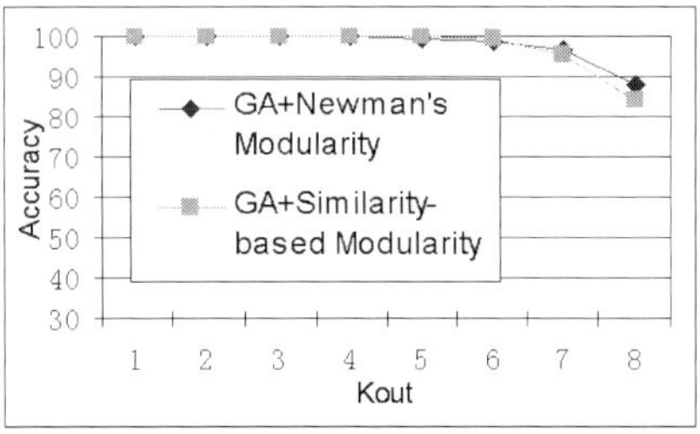

Fig. 11.3. Accuracy curves of GA using Q_N and Q_S

Table 11.3. Results of FHAC optimizing Q_S

$K_{in} : K_{out}$	Best Q_S	# of clusters	Errors
11:5	0.614885	4	0
10:6	0.565117	4	4
9:7	0.481685	5	18
8:8	0.458222	8	44

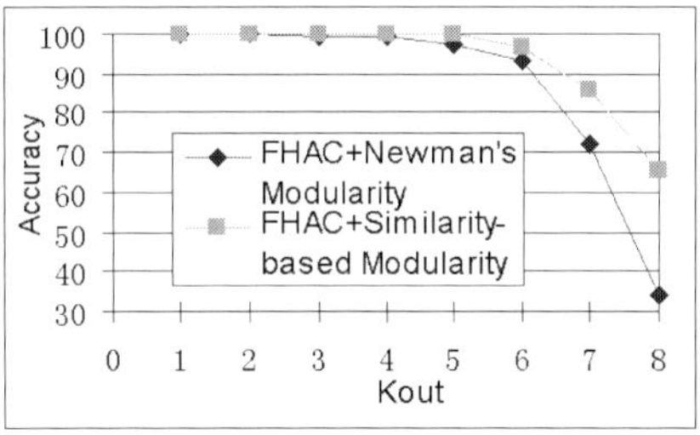

Fig. 11.4. Accuracy curves of FHAC using Q_N and Q_S

when $K_{out} > 6$, when the cluster structure becomes confused. However, one may notice that the FHAC algorithm works much better when combining with the proposed Similarity- based Modularity. It demonstrates that the proposed measurement cooperates with the greedy search algorithm FHAC much better even when the pre-defined cluster structure is not in favor of it.

While the synthetic network is a Type I network that does suffers from no hubs or outliers, it is a significant problem to study to demonstrate the stability that similarity adds. In that regard, hubs and outliers are elements that add confusion.

11.6.2. *Real Applications*

The first real application we report in our experiment is a social network - detecting communities (or conferences) of American college football teams [6, 7, 12, 13]. The NCAA divides 115 college football teams into 13 conferences. The question is how to find out the conferences from a network that represents the schedule of games played by all teams. We presume that because teams in the same conference are more likely to play each, that the conference system can be mapped as a structure despite the significant amount of inter-conference play. We analyze the graph by using both GA and FHAC. The results are reported in Table 11.4. From which, FHAC with Q_N partitions the graph into 6 conferences with 45 misclassifications [13], while FHAC with Q_S partitions to 12 conferences with only 14 errors. Again, Q_S significantly outperforms Q_N when combining with FHAC.

Table 11.4. Detecting conferences in college football teams by using Q_N and Q_S

Alg.	Best Q_N	# of clusters	Errors	Best Q_S	# of clusters	Errors
GA	0.601009	12	14	0.820668	12	14
FHAC	0.577284	6	45	0.820668	12	14

The second application is detecting the individuals from customer records coming from Acxiom Corporation, where data errors blur the boundaries between individuals with similar information (see Fig. 11.5). This example represents a Type II graph, with hubs and outliers. In this dataset, there are 3 groups of customers, a number of hubs (vertices 7, 10, 11, and 19) and a single outlier (vertex 21).

We test both GA and FHAC by using Q_N and Q_S respectively. The results are summarized in Table 11.5. Both algorithms make 3 errors by using Q_N, they misclassify hub node vertex 7 and vertex 10 into wrong cluster and fail in detecting the outlier (vertex 21). However, by using the proposed Q_S, both algorithms perfectly classify this graph, which means the Q_S has better ability to deal with hub and outlier vertices.

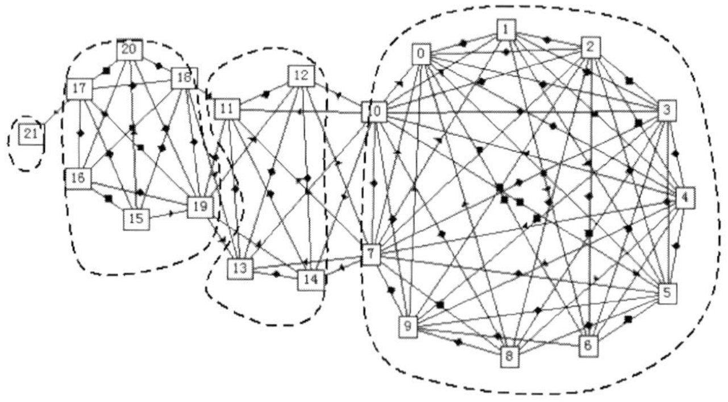

Fig. 11.5. Customer record networks

Table 11.5. Performance comparison on Acxiom dataset using Q_N and Q_S respectively

Alg.	Expected clusters	Expected Q_N	Best Q_N	clusters	errors
GA+Q_N	4	0.378893	0.399623	3	3
FHAC+Q_N	4	0.378893	0.37669	3	3
Alg.	Expected clusters	Expected Q_S	Best Q_S	clusters	errors
GA+Q_S	4	0.474738	0.474738	4	0
FHAC+Q_S	4	0.474738	0.474738	4	0

Table 11.6. FHAC running time for very large network

Vertices	Edges	FHAC+Q_N	FHAC+Q_S
52909	245300	591 Sec.	658 Sec.

The final experiment was on a very large dataset - the DBLP authors' collaboration networks of 2005 [10], which consist of 52,909 vertices (authors) and 245,300 edges (co-author relationship). The purpose of this experiment is testing the speed of FHAC on a very large network. The running time on an Intel PC with P3.2G CPU and 2GB memory are reported in Table 11.6. One can see that optimizing Q_S it runs marginally slower than optimizing Q_N, which means Q_S cooperates with the FHAC algorithm very well.

11.7. Conclusion

In this paper, we propose a novel similarity-based modularity (Q_S) to measure the quality of a graph partitioning. Through theoretical analysis and extensive experiments, we can conclude that the propose measure is significantly more accurate and robust than Newman's connection-based modularity with respect to the result of clustering. Furthermore, it has a better ability to deal with hubs and outliners. The proposed similarity-based modularity in combination with the Genetic clustering algorithm (GA) and the greedy search algorithm FHAC yields an improved accuracy for even dense, confused graphs. The FHAC often converges to the global optimal for real applications using the proposed modularity. However, in some very tough cases, such as in very confused synthetic graphs, FHAC significantly lags the global optimal obtained by GA. This suggests us to further study more powerful fast clustering algorithm in the future to exert the potential of the proposed modularity definition.

References

[1] A. Clauset, M. Newman, and C. Moore. *Finding community structure in very large networks*, *Physical Review E*, 70:066111, 2004.

[2] C. R. Dias and L. S. Ochi. Efficient evolutionary algorithms for the clustering problem in directed graphs. *The Congress on Evolutionary Computation, CEC '03*, 2003.

[3] C. H. Q. Ding, X. He, H. Zha, M. Gu, and H. Simon. A min-max cut algorithm for graph partitioning and data clustering. *Proc. of ICDM 2001*, pages 107-114, 2001.

[4] Z. Feng, X. Xu, N. Yuruk, and T. Schweiger. A novel similarity-based modularity function for graph partitioning. In *9th International Conference on Data Warehousing and Knowledge Discovery (DaWaK 2007)*, pages 385-396, Regensburg, Germany, 3-7 September 2007.

[5] L. Freeman. A set of measures of centrality based upon betweeness. *Sociometry*, 40: 35-41, 1977.

[6] M. Girvan and M. Newman. Community structure in social and biological networks. *Proc. Natl. Acad. Sci. USA*, 99: 7821-7826, 2002.

[7] R. Guimera and L. A. N. Amaral. Functional cartography of complex metabolic networks. *Nature*, 433: 895-900, 2005.

[8] L. Hegan and A. B. Kahng. New spectral methods for ratio cut partitioning and clustering. *IEEE Trans. On Computed Aided Design*, 11: 1074-1085, 1992.

[9] G. Hernadez, L. Bobadilla and Q. Sanchez. A genetic word clustering algorithm. *The Congress on Evolutionary Computation*, 2005.

[10] http://www.informatik.uni-trier.de/ ley/db/.

[11] E. A. Leicht, P. Holme, and M. E. J. Newman. Vertex similarity in networks. *Phys. Rev. E*, 73: 026120, 2006.

[12] M. Newman. Detecting community structure in networks. *Eur. Phys. J. B*, 38: 321-330, 2004.

[13] M. Newman. Fast algorithm for detecting community structure in networks. *Phys. Rev. E*, 69: 066133, 2004.

[14] M Newman. Scientific collaboration networks: II. Shortest paths, weighted networks, and centrality. *Phys. Rev. E*, 64: 015132, 2001.

[15] W. Sheng, S. Swift, L. Zhang, and X. Liu. A weighted sum validity function for clustering with a hybrid niching genetic algorithm. *IEEE Trans. On Sys., Man and Cybernetics, part B: Cybernetics*, 35(6): 1156-1167, 2005.

[16] J. Shi and J. Malik. Normalized cuts and image segmentation. *IEEE Trans. On Pattern Analysis and Machine Intelligence*, 22(8): 888-905, 2000.

[17] J. Wang, L. Xu and B. Zhang. A genetic annealing hybrid algorithm based clustering strategy in mobile ad hoc network. *Proc. on Communications, Circuits and Systems*, 2005.

[18] S. Wasserman and K. Faust. *Social Network Analysis*. Cambridge University Press, Cambridge, 1994.

[19] J. Zhang, H. Chung and B. Hu. Adaptive probabilities of crossover and mutation in genetic algorithms based on clustering technique. *The Congress on Evolutionary Computation*, 2004.

Chapter 12

Mechanism-based Clustering of Genome-wide RNA Levels: Roles of Transcription and Transcript-Degradation Rates

Sungchul Ji

Department of Pharmacology and Toxicology, Rutgers University
170 Frelinghuysen Rd., Piscataway, NJ 08855, USA
E-mail: sji@rci.rutgers.edu

W. Art Chaovalitwongse*

Department of Industrial and Systems Engineering, Rutgers University
96 Frelinghuysen Rd., Piscataway, NJ 08854, USA
Email: wchaoval@rci.rutgers.edu

Nina Fefferman

DIMACS, Rutgers University
96 Frelinghuysen Rd., Piscataway, NJ 08854, USA
Department of Public Health and Family Medicine, Tufts School of Medicine
Boston, MA, 02111, USA
Email: nina.fefferman@tufts.edu

Wonsuk Yoo

Department of Internal Medicine, Wayne State School of Medicine
4201 St. Antoine St., UHC 4H-30, Detroit, MI 48201, USA
Email: wyoo@med.wayne.edu

Jose E. Perez-Ortin

Department of Biochemistry and Molecular Biology, University of Valencia
Dr. Moliner 50, 46100 Burjassot, Spain
Email: jose.e.perez@uv.es

DNA array techniques invented over a decade ago enable biologists to measure tens of thousands of mRNA levels in cells simultaneously as functions of envi-

*Corresponding Author.

ronmental perturbations. In a few cases the same technique has been employed to measure not only genome-wide transcript levels (*TL*) but also the associated transcription rates (*TR*) simultaneously. Since *TL* is determined by the balance between two opposing processes, *i.e.*, transcription and transcript degradation, simple theoretical considerations indicate that it would be impossible to determine *TR* based on *TL* data alone. This conclusion is supported by the finding that *TL* and *TR* do not always vary in parallel. In fact, the genome-wide measurements of *TL* and *TR* in budding yeast undergoing glucose-galactose shift indicate that *TL* can decrease even though *TR* increases and *TL* can increase despite the fact that *TR* decreases. These counter-intuitive findings cannot be accounted for unless transcript-degradation rates (*TD*) are also taken into account. One of the main objectives of this contribution is to derive a mathematical equation relating *TL* to *TR* and *TD*. Based on this equation, it was predicted that there would be 9 different mechanisms by which *TL* can be altered in cells. The *TL* and *TR* data measured in budding yeast demonstrate that all of the 9 predicted mechanisms are found to be activated in budding yeast during glucose-galactose shift, except Mechanisms 5 (i.e., decreasing *TL* with no change in *TR*) and 9 (i.e., no change in *TL* nor in *TR*). It was also shown that the opposite changes in the mRNA levels of glycolytic and respiratory genes observed between 5 and 360 minutes following the glucose-galactose shift could be quantitatively accounted for in terms of what is referred to as the transcript-degradation/transcription (*D/T*) ratios calculated here for the first time. Our results suggest that the predicted 9 mechanisms of controlling *TL* may be employed to cluster the genome-wide measurements of mRNA levels as a means to characterize the functional states of both normal and diseased cells.

12.1. Introduction

The DNA array technique allows cell biologists to measure the intracellular levels of tens of thousands of different kinds of mRNA molecules in living cells simultaneously [1, 4, 22, 28, 30, 33] When mRNA levels are measured from a cell preparation as a function of time after some perturbation and the resulting data are subjected to a cluster analysis, it is frequently found that the mRNA levels can be grouped into a set of distinct clusters, each cluster exhibiting a common kinetics or a temporal pattern of the changes in mRNA levels.

It has been an almost universal practice in the field of microarray technology since its inception to interpret mRNA level changes in terms of transcription (*i.e.*, the synthesis of mRNA using DNA as template) rates only, without taking into consideration the transcript-degradation step [23]. It is well known that this transcript-degradation process can occur with rates comparable to those of transcription itself [1, 28, 29]. Ignoring the role of the transcript degradation step is tantamount to equating *transcript levels* with *transcription rates* and this has led to the following erroneous interpretations of DNA microarray data:

(i) When mRNA levels increase, it is interpreted as an indication of increased rates of transcription of the corresponding genes;
(ii) When mRNA levels undergo no change, it is taken as the evidence for unchanged transcription rates; and
(iii) When mRNA levels decrease, it is interpreted as an indication for decreased transcription rates.

For convenience, we will refer to this way of interpreting mRNA levels as the *1-to-1 interpretation*. Based on this approach, it has been widely assumed that, when a *mRNA cluster* was found by various clustering techniques, this fact can be used to infer that the underlying genes are transcribed with similar rates and hence that there exists a corresponding *gene cluster*. It is one of the main objectives of this chapter to demonstrate that this way of interpreting mRNA clusters is theoretically invalid and factually unsupported, leading to Type I and Type II errors. A Type I error (or a false positive) can arise, for example, when an increase in mRNA level is interpreted as indicating an increase in the associated transcription rate (which can be true sometimes but not always), since the level of a mRNA molecule can increase even if the associated transcription rate does not change as long as the number of mRNA molecules synthesized during the time period of observation is greater than the number of mRNA molecules degraded during the same time period. Similarly, a Type II error (or a false negative) can arise if a gene is inferred to undergo no change in its transcription rate based on the fact that its mRNA level did not change. This is because, even if a mRNA level did not change, the transcription rate could have increased (or decreased) if the changes in the rate of mRNA degradation happened to exactly counterbalance the effect of an increased transcription rate.

Direct experimental evidence for the concept that mRNA levels, also known as transcript levels (*TL*), is determined by a dynamic balance between transcription and transcript degradation have been obtained only recently when *TL* and transcription rates (*TR*) were measured simultaneously from human lung carcinoma cells [8], tobacco plant cells [19] and the budding yeast *Sacharomyces cerevisiae (S. cerevisiae)* subjected to glucose-galactose shift [10]. Here we present the results of a genome-wide analysis of the yeast *TL* and *TR* data reported in Garcia-Martinez *et al.* [10] based on a kinetic equation relating *TL* to *TR* and transcription degradation rates (*TD*), demonstrating the following points:

- *TL* and *TR* increase together (see Mechanism 2 in Table 12.1 and Fig. 12.3) in 51% of the time and decrease together (see Mechanism 6) 40% of the time.

- The opposite changes in glycolytic and respiratory *TL* (See Fig. 12.4) induced by glucose-derepression are due to opposite changes in *TD*, thus establishing an instance of degradational control in contrast to the well-known transcriptional control.
- There are 9 mechanisms underlying underlying the changes in *TL* (See Fig. 12.3).

12.2. Materials and Data Acquisition

12.2.1. *Glucose-Galactose Shift Experiments*

The *S. cerevisiae* yeast strain BQS252 was grown overnight at 28 degrees Celsius in YPD medium (2% glucose, 2% peptone, 1% yeast extract) to exponential growth phase ($OD_{600} = 0.5$) [10]. Cells were recovered by centrifugation, resuspended in YPGal medium (2% galactose, 2% peptone, 1% yeast extract), and allowed to grow in YPGal medium for 14-15 hours after the glucose-galactose shift. Cell samples were taken at 0 (denoted as t_0), 5 (t_1), 120 (t_2), 360 (t_3), 450 (t_4) and 850 (t_5) minutes after the glucose-galactose shift. The t_5 sampling time corresponds to the exponential growth phase in YPGal medium. Two different aliquots were taken from the cell culture at each sampling time. One aliquot was processed to measure *TR* according to the genomic run-on protocol (see the next section), and the other was processed to measure *TL* using the same DNA arrays recovered after *TR* measurements.

12.2.2. *Measuring Transcription Rates* (TR) *Using the Genomic Run-on* (GRO) *Method*

The experimental details of the genomic run-on procedures are given in Garcia-Martinez *et al.* [10]. This technique is a scaled-up version of the usual nuclear run-on method [13]. Lysed cells contain transcription complexes stalled on the DNA template due to lack of ribonucleotides. Transcription is re-initiated *in vitro* by adding new nucleotides, one of which is radiolabled (*e.g.*, [$\alpha-^{33}$P]UTP). After allowing transcription to finish, one can determine via autoradiography the density of RNA polymerases for each gene. Assuming a constant speed for RNA polymerase II molecules this density allows us to estimate the transcription rates in the cells of interest. By comparing the amount of gene-specific radiolabeled RNA synthesized in one nuclei preparation with another, it is possible to estimate the extent of the RNA polymerase densities and, therefore, transcription rates in the cells of interest. The *TR* data utilized in the *TL-TR* plots shown in Fig. 12.1 are available at *http://scsie.uv.es/chipsdna/chipsdna-e.html♯datos*.

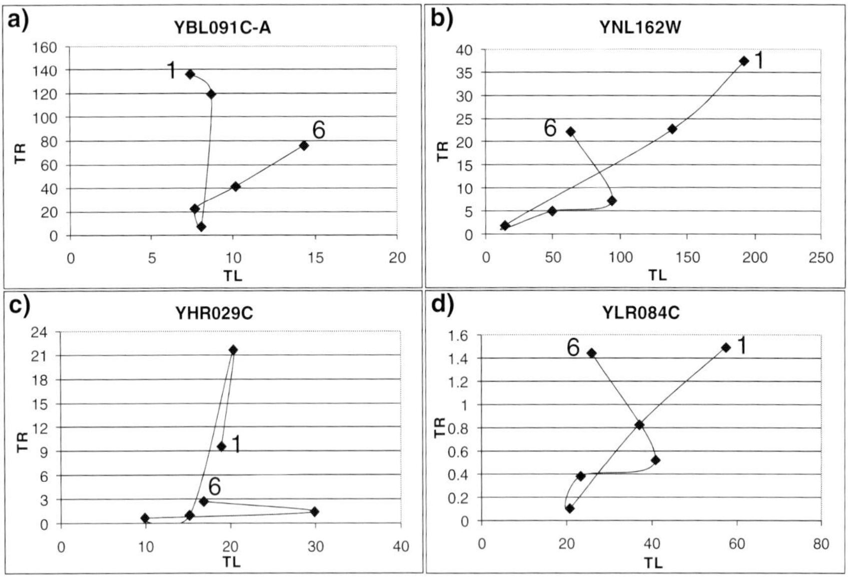

Fig. 12.1. Plots of the fold changes in transcription rates (*TR*) against those of transcript levels (*TL*) measured in budding yeast at 6 time points (1 = 0 min, 2 = 5 min, 3 = 120 min, 4 = 360 min, 5 = 450 min, and 6 = 850 min) after the glucose-galactose shift.

12.2.3. *Measuring mRNA or Transcript Levels (*TL*)*

The total RNA isolated from the same cell culture used in the previous section was reverse-transcribed into cDNA in the presence of [$\alpha-^{33}$P]dCTP [10]. The labeled cDNAs were purified and hybridized under the same conditions as described for *GRO* in order to minimize the variability due to artifacts of DNA membrane arrays. Again, the raw *TL* data in arbitrary units measured in triplicates at 6 time points are available at *http://evalga.uv.es/scsie-docs/chipsdna/chipsdna-e.html.*

12.2.4. *The* TL-TR *Plots*

The *TL* and *TR* data measured as described in the previous sections can be visualized in a 2-dimensional plane as shown in Fig. 12.1. The genes depicted in this figure were randomly chosen out of the 5,725 genes showing no missing values in their triplicate measurements of *TL* and *TR*. The notation given on the top of each figure is the name of the open reading frame (ORF) whose *TL* and TR values were measured. The trajectory of each plot can be divided into 5 segments or vectors bounded by two time points (*e.g.*, vector 1-2 between 0 to 5 min after the glucose-galactose shift and vector 2-3 between 5 to 120 min, etc.). Each vector can be characterized in terms of the angle measured counterclockwise starting from the positive x-axis (see Fig. 12.2). For example, vector 5-6 in Fig. 12.1a is approximately $45°$ and vector 1-2 in Fig. 12.1b is approximately $225°$. Thus, the angle α determining the direction of the vector from the ith point to the $(i+1)^{th}$ point in a *TL* versus *TR* plot with coordinates (x_i, y_i) and (x_{i+1}, y_{i+1}), respectively, can be calculated from the relation $\alpha = \tan^{-1}[\frac{(y_{i+1}-y_i)}{(x_{i+1}-x_i)}] + \Theta$, where $\Theta = 0°$ if both the numerator and the denominator are positive, $\Theta = 180°$ if either the numerator is positive and the denominator is negative or both the numerator and the denominator are negative, $\Theta = 360°$ if the numerator is negative but the denominator is positive.

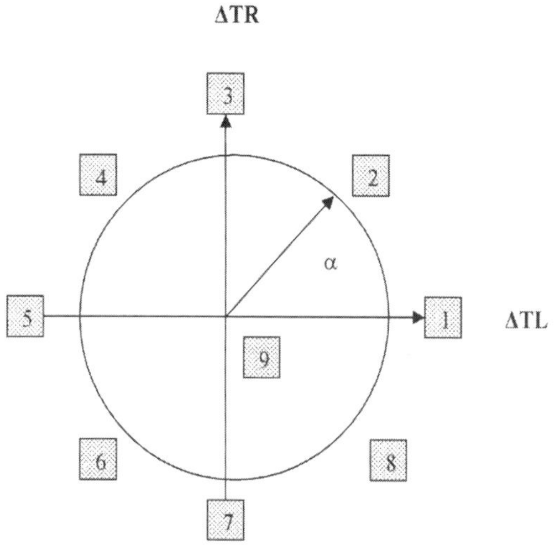

Fig. 12.2. The "unit" circle whose x-axis indicates the changes in *TL* and y-axis those in *TR* values of a trajectory in the *TL-TR* plot. The direction of the radius of the circle coincides with the direction of the component vector of a *TL-TR* trajectory. For convenience, each direction is defined by the following values of angle α: α_1 = 357 – 3; α_2 = 3 – 87; α_3 = 87 – 93; α_4 = 93 – 177; α_5 = 177 – 183; α_6 = 183 – 267; α_7 = 267 – 273; α_8 = 273 – 357. The center of the circle denoted as 9 indicates that there is no change in *TR* nor in *TL* between two time points.

12.3. Statistical Analysis

The total number of genes whose *TL* and *TR* values were measured was 5,913, of which 5,725 genes were successfully measured in triplicates without any missing values. The rest of the genes were measured in less than triplicates. A statistical analysis for the comparison between the "expected" and the "observed" distributions of genes over 8 possible mechanisms (excluding Mechanism 9) of controlling mRNA level in cells(defined in the previous section) was performed using the binomial test to make inferences about a proportion for response rate based on a series of independent observations. A hypothesis test was performed to compare the average trajectories (also called sequences, time series, or profiles) of two groups of genes coding for glycolytic and respiratory mRNA molecules. Each group generates two average profiles, labeled *TL* and *TR* (See Fig. 12.3a). Thus, the null hypothesis states that "there is no difference between the two profiles associated with glycolysis and respiration", and the alternative hypothesis states that "There exists a significant difference between the two profiles". In order to test these hypotheses, a multivariate approach for repeated measures analysis was used to calculate p-values. The p-value for the two trajectories shown in Fig. 12.3a is less than 0.0001, indicating that there are significant differences between the average glycolytic and respiratory mRNA level trajectories. The p-value for the two curves shown in Fig. 12.3b is 0.2292, indicating that there is no significant difference between the average trajectories of the glycolytic and respiratory mRNA synthesis rates.

12.3.1. *Calibration of* TL *Data*

To convert the *TL* data expressed in arbitrary unit to the corresponding values in absolute unit (mRNA molecules per cell), we utilized the reference mRNA abundance data complied by Beyer *et al.* [5] based on 36 datasets reported in the literature. Out of about 5,700 mRNA abundance data that these authors collected, we selected a total of 59 glycolytic and respiratory mRNA abundance values and plotted them against the corresponding *TL* data, leading to a linear regression line satisfying the relation, $n = 0.1114TL_0 - 7.220$, where n is the number of mRNA molecules per cell, and TL_0 is the mRNA level measured in arbitrary unit at $t = 0$. The correlation coefficient of the straight line was 0.94245. This equation was used to convert all the *TL* data (measured at five different time stamps) in arbitrary unit into the corresponding values in absolute unit.

Examples of the time courses of the average changes in *TL*, *TR*, and the transcript-degradation/transcription (*D/T*) ratios of glycolytic and respiratory genes are shown in Fig. 12.3 and 12.4. The glycolytic genes considered here include *PDC2, PFK1, PFK2, ADH5, PDC6, LAT1, PDA1, ADH2, GLK1, PGM1, ADH3, TPI1, GCR1*, and *PDB1* and the respiratory genes include *QCR6, COX13, COX9, NDI1, CYT1, COX8, COX12, SDH4, COX5A, COX4, CYC1, QCR2, QCR8*, and *COR1*.

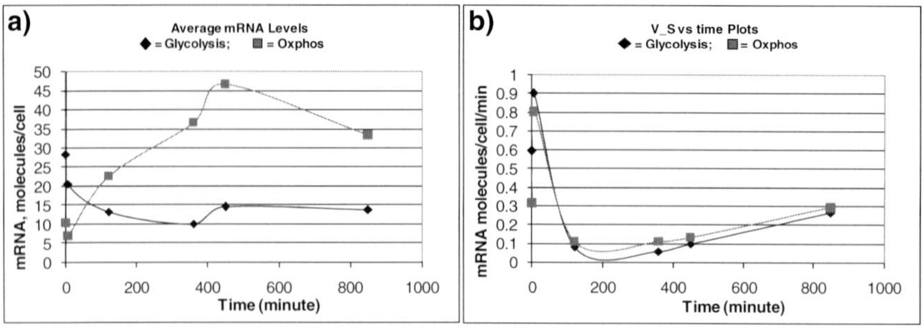

Fig. 12.3. The time courses of the average changes in the transcript level (*TL*) and transcription rate (*TR*) of 12-15 glycolytic (◆) and 14 respiratory genes (■).

12.3.2. *Calibration of* TR *Data*

Transcription rates of yeast genes at t_0 were estimated by utilizing the genome-wide mRNA decay half-lives reported by Wang *et al.* [32] The half-life data were down-loaded from their web site (*http://www-genome.stanford.edu/turnover*). The mathematical relation between transcription rate, $dn_{s,i}/dt$, where $n_{s,i}$ is the number of the ith mRNA molecules synthesized per cell during the time interval dt, and the $mRNA_i$ decay half-life, denoted by $t_{\frac{1}{2},i}$, can be derived based on the following assumptions.

(i) At t_0, budding yeast cells are at a steady state with respect to transcription and transcript degradation. In other words, at t_0, $dn_{S,i}/dt = dn_{D,i}/dt$, where $n_{D,i}$ is the number of the ith mRNA molecules per cell degraded during the time interval dt.

(ii) The decay of the ith mRNA molecules obeys a first-order rate law given by

$$-dn_i/dt = dn_{D,i}/dt = k_{D,i}[mRNA_i], \tag{12.1}$$

where $k_{D,i}$ is the first-order degradation rate constant and $[mRNA_i]$ is the concentration of the ith mRNA in the cell. Integrating Eq. (12.1) with respect to time leads to

$$[mRNA_i] = [mRNA_i]_0 e^{-k_{D,i}t}, \tag{12.2}$$

where $[mRNA_i]_0$ is the $mRNA_i$ abundance at t_0. Substituting the values, $[mRNA_i] = [mRNA_i]_0/2$ at $t = t_{\frac{1}{2},i}$, and solving the resulting equation for $k_{D,i}$ yields

$$k_{D,i} = \frac{\ln 2}{t_{\frac{1}{2},i}} = \frac{0.693}{t_{\frac{1}{2},i}}. \tag{12.3}$$

Using this relation, it is possible to convert mRNA decay half-lives measured by Wang *et al.* [32] into the corresponding transcript decay rate constants.

(iii) Since budding yeast cells are assumed to be at steady states at t_0 with respect to transcription and transcript decay (previously mentioned in the first assumption), it would follow that

$$dn_{S,i}/dt = dn_{D,i}/dt = k_{D,i}[mRNA_i]_0 = \frac{0.693[mRNA_i]_0}{t_{\frac{1}{2},i}}. \quad (12.4)$$

Thus Eq. (12.4) allows us to estimate the transcription rate, $dn_{S,i}/dt$, at t_0 from the $[mRNA_i]_0$ and $t_{\frac{1}{2},i}$ values.

(iv) The conversion factor, defined as $a_i = (dn_{S,i}/dt)/TR_i$ calculated at t_0 is assumed to apply to all the other TR_i values measured at t_1, t_2, t_3, t_4, and t_5, where TR_i denoting the *TR* values associated with the ith gene.

(v) Since the $mRNA_i$ decay measurements were made at $37°$ Celsius (C) [32], whereas the TR_i measurements were carried out at $28°$ C [10], the absolute rate values for $dn_{S,i}/dt$ calculated in (iii) were corrected for the temperature difference by dividing $dn_{S,i}/dt$ by the factor $2 \times (9/10) = 1.9$, which results from the assumption that the Q10 value (*i.e.*, the factor by which the rate increases due to a $10°$ C increase in temperature) was 2.

12.3.3. *Kinetic Analysis of the Changes in mRNA Levels*

In the absence of any exchange of mRNA molecules between budding yeast cells and their environment, it is possible to equate the rate of change of the ith mRNA molecules per cell, dn_i/dt, with the balance between the rate of $mRNA_i$ synthesis, $dn_{S,i}/dt$, and the rate of its degradation, $dn_{D,i}/dt$ given by

$$dn_i/dt = dn_{S,i}/dt - dn_{D,i}/dt \quad (12.5)$$

In general, the rate of synthesis of the ith mRNA, $dn_{S,i}/dt$, would be a function of many variables including the levels (or concentrations) of RNA polymerase II, various transcription factors (encoded by the jth gene, where $j \neq i$, except when the jth gene happens to code for a transcription factor that acts on the jth gene itself), and small molecules such as $ATP, ADP, AMP, Mg^{++}, H^+$, etc. The same would hold true for the rate of the degradation of the ith mRNA, $dn_{D,i}/dt$. A system of ordinary differential equations describing the dynamics of mRNA levels in a cell taking into account all the variables mentioned above has been derived in several previous studies [6, 26, 27]. Our model in Eq. (12.5) is similar to (but not identical with) Eq. (1) in the studies by Savageau [26, 27].

The change in the number of $mRNA_i$ molecules per cell, Δn_i, during the time interval, Δt, between two time points, t_k and t_{k+1}, can be expressed as

$$\Delta n_i = \int dn_i = \int dn_{S,i} - \int dn_{D,i} = \Delta n_{S,i} - \Delta n_{D,i}, \qquad (12.6)$$

where the integration is from t_k to t_{k+1}. The resulting equation, $\Delta n_i = \Delta n_{S,i} - \Delta n_{D,i}$, simply states that the change in the number of the ith mRNA molecules in a cell during the time interval, $\Delta t = t_{k+1} - t_k$, is determined by the balance between the number of $mRNA_i$ molecules synthesized and the number of $mRNA_i$ molecules degraded during that time interval. It should be noted here that the variables in Eq. (12.6) are in absolute units. We can estimate the variables in Eq. (12.6) as follows:

$$\Delta n_i = n_i - n_{i-1} \qquad (12.7)$$

$$\Delta n_{S,i} = AUC_S \qquad (12.8)$$

$$\Delta n_{D,i} = AUC_S - (n_i - n_{i-1}), \qquad (12.9)$$

where the subscript i refers to the ith mRNA molecule, and AUC_S is the area under the curve of the *TR*-time plots.

12.3.4. *Transcript-Degradation to Transcription* (D/T) *Ratios*

The transcript-degradation/transcription (*D/T*) ratio for the ith mRNA molecule can be defined by

$$D/T = \frac{\Delta n_{D,i}}{\Delta n_{S,i}}. \qquad (12.10)$$

Note that the subscript i is omitted from the term, D/T, for simplicity. Combining Eqs. (12.6) and (12.10) results in

$$\Delta n_i = \Delta n_{S,i}(1 - D/T). \qquad (12.11)$$

We observe that the sign of Δn_i is determined solely by the magnitude of the *D/T* ratio relative to 1. The relationship can be described by

$$\Delta n_i > 0 \text{ if } D/T < 1 \qquad (12.12)$$

$$\Delta n_i = 0 \text{ if } D/T = 1 \qquad (12.13)$$

$$\Delta n_i < 0 \text{ if } D/T > 1. \qquad (12.14)$$

The time-dependent variations of the *D/T* ratios of the glycolytic and respiratory genes were calculated as explained in the previous section. Fig. 12.4 illustrates the temporal evolution of *D/T* ratios of the glycolytic and respiratory genes given in Fig. 12.3. It should be pointed out that there are 2 less respiratory genes than the one given in Fig. 12.3b due to missing values in the degradation rates. The *D/T* ratio for the ith mRNA molecule (DTR_i) is defined as $\frac{\Delta n_{D,i}}{\Delta n_{S,i}}$ where n_i is the number of the ith mRNA molecules synthesized (S) or degraded (D) per cell over a given time period (see the previous section for more details). The calculation of $\Delta n_{S,i}$ requires integrating the TR_i versus time curves between two sampling time points, say t_1 and t_2. This is why there are only 5 $DTR's$, each of which was plotted at the mid-point of the two time points involved.

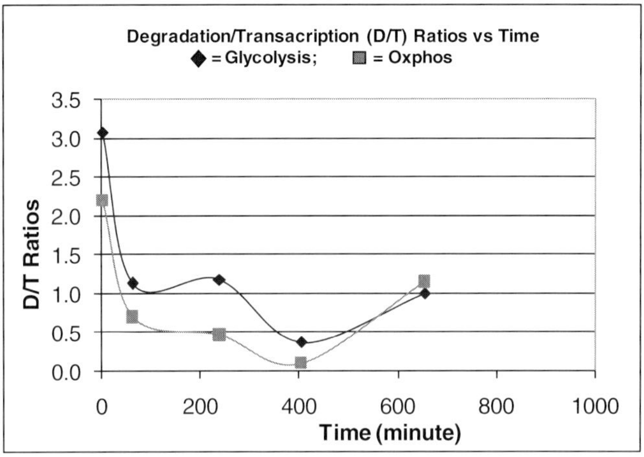

Fig. 12.4. The temporal variations of the transcript-degradation/transcription (D/T) ratios of the 15 glycolytic (◆) and 12 respiratory genes (■) given in Fig. 12.3.

12.4. Experimental Results

In our previous publication [10], we reported that, when yeast cells were grown in a glucose-containing medium to exponential growth phase, harvested by centrifugation, and resuspended in a galactose-containing medium replacing glucose, extensive metabolic changes were found to occur as reflected in *TL* and *TR* values for almost all of the 6,400 genes in the yeast genome. The *TL* trace averaged over 5,753-5,829 genes decreased by about 65% during the first 5 minutes following the glucose-galactose shift and continued to decline until 360 minutes by further 20%. There was a slight increase (17%) in *TL* between 360 and 450 minutes, which was followed by a smaller decrease (12%) between 450 and 850 minutes. In contrast, the *TR* values averaged over 5,753-5,829 genes exhibited very different kinetic behaviors: A rapid increase by about 60% during the first 5 minutes followed by a 150% decrease to the 10% level of the control by 120 minutes. This low level of *TR* was maintained until 360 minutes when it started to increase linearly up to the 45% level of the control by 850 minutes. The *TL* and *TR* values were normalized with respect to their values at t = 0, and corrected for the increasing number of cells during the observational period.

These observations clearly indicate that the average *TL* and *TR* values do not change in parallel (as would have been expected if *TL* reflected the rates of gene expression, namely *TR*) but exhibit seemingly *independent* kinetic behaviors, especially during the first (0-5 minutes) and the last (450 - 850 minutes) periods of observation. The relation between *TL* and *TR* can be visualized by plotting *TL* against *TR* on a 2-dimensional plane as shown in Fig. 12.1a through 12.1d. A trajectory in the *TR* versus *TL* plane consists of a set of 5 vectors, each representing the change in *TL* and *TR* over a time interval which varies from 5 minutes to 400 minutes. The trajectory associated with the gene YBL091C-A consists of a series of five connected vectors starting from time point 1 (0 minute) and ending at time point 6 (850 minutes) having approximate angles of $315°$, $270°$, $135°$, $45°$, and $45°$. A given vector is associated with a mechanism of RNA metabolism expressed in terms of *TL* and *TR*. Specifically, when $\alpha = 315°$, it indicates that, during the first time interval (*i.e.*, from 0 to 5 minutes), the transcript level *TL* of gene TBL091C-A is increased but its transcription rate *TR* is decreased; when $\alpha = 270°$, it indicates that the same gene experienced no change in *TL* but a decrease in *TR*, *etc*. It is important to note that this gene experienced an increase in both *TL* and *TR* only during the last two time intervals, since their associated α values are approximately $45°$. The 9 mechanisms of RNA metabolism depicted in Fig. 12.2 are defined as follows:

(1) **Mechanism 1** is defined by the α values ranging from $357°$ (or $-3°$) to $3°$ and indicates that *TL* increases without any change in *TR*;
(2) **Mechanism 2** is defined by the α values ranging from $3°$ to $87°$ and indicates that both *TL* and *TR* increase;
(3) **Mechanism 3** is defined by the α values ranging from $87°$ to $93°$ and indicates that *TL* does not change although *TR* increases;
(4) **Mechanism 4** is defined by the α values ranging from $93°$ to $177°$ and indicates that *TL* decreases even though *TR* increases;
(5) **Mechanism 5** is defined by the α values ranging from $177°$ to $183°$ and indicates that *TL* decreases without any change in *TR*;

(6) **Mechanism 6** is defined by the α values ranging from 183° to 267° and indicates that both *TL* and *TR* decreases;
(7) **Mechanism 7** is defined by the α values ranging from 267° to 273° and indicates that *TL* does not change although *TR* decreases; and
(8) **Mechanism 8** is defined by the α values ranging from 273° to 357° and indicates that *TL* increases even though *TR* decreases.

Note that **Mechanisms 9** is not shown above because it is defined not by α but by the values of ΔTL and ΔTR both being zero.

Since there are 5 vectors per *TL-TR* trajectory, there are a total of $5 \times 5,725 = 28,625$ values of α to be calculated. These values are calculated using the formula given in the previous section and displayed in Table 12.1. The rows numbered 1 through 5 refer to the 5 time intervals and the columns numbered 1 through 8 refer to the mechanisms of controlling mRNA levels. Due to a large number of genes involved, we used the normal approximation approach, leading to the results that the observed proportions for Mechanisms 1, 2, 3, 4, 6, 7, and 8 are significantly different from the expected except for Mechanism 5. In other words, all of the 8 mechanisms are observed to occur in budding yeast cells under the experimental conditions employed, except Mechanisms 5 and 9.

Table 12.1. The frequency distributions of the 8 modules of RNA metabolism (defined in the legend to Fig. 12.2) as the functions of the 5 time periods following the glucose-galactose shift.If the angles are randomly distributed over 360°, the expected distributions can be calculated as shown in the 8th row. The p-values for the difference between the observed and the expected distributions are given in the last row. The differences are all significant, except for Mechanism 5.

	Mechanism								
Vector	1	2	3	4	5	6	7	8	Total
1	0	142	234	3470	96	1732	12	39	5725
2	14	18	3	37	5	3729	617	1302	5725
3	340	1914	52	638	314	1471	28	968	5725
4	477	4237	21	151	61	143	19	616	5725
5	12	1151	238	4213	38	56	4	13	5725
Total Observed	843	7462	548	8509	514	7131	680	2938	28625
Total Expected	477	6678	477	6678	477	6678	477	6678	28625
p-value	< 0.0001	< 0.0001	0.001	< 0.0001	0.092	< 0.0001	< 0.0001	< 0.0001	

The numbers in Table 12.1 represent the frequencies of the different mechanisms that occur in budding yeast during one of the five time intervals. Thus, during the first time interval (*i.e.*, from 0 to 5 minutes), no gene experienced (or exhibited) Mechanism 1; 142 genes experienced Mechanism 2; 234 genes exhibited Mechanism 3; 3,470 genes experienced Mechanism 4; 96 experienced Mechanism 5; 1,732 genes experienced Mechanism 6; 12 genes experienced Mechanism 7; and 39 experienced Mechanism 8. If the various mechanisms occur randomly during this time interval, their frequency of distribution would be expected to be proportional to the magnitude of the angle α associated with the vector spanning the time intervals involved, *i.e.*, 6° for Mechanisms 1, 3, 5, and 7, and 84° for Mechanisms 2, 4, 6 and 8. The theoretically predicted distributions of the mechanisms based on the angular sizes are given in the 9th row in Table 12.1. Comparing Rows 8 and 9, it is clear that all of the 8 mechanisms occur with frequencies different from those expected on the basis of random distributions, except Mechanism 5, as evidenced by the fact that the associated p-values are all less than 0.001 except that associated with Mechanism 5.

In 2002, Gorospe and her group measured for the first time the *TL* and *TR* data for about 2,000 genes from non-small cell human lung carcinoma H1299 [8] and found that *TL* could increase or decrease without any changes in *TR* (see Groups *IV* and *V* in Table 1 of [8]), from which they concluded that transcript degradation played a critical role in determining mRNA levels. The first genome-wide measurements of *TL* and *TR* in budding yeast *S. cerevisiae* were reported by Garcia-Martinez *et al.* [10], whose results also indicated that there were no *1-to-1* correlation between *TL* and *TR*. However, neither of these publications included any mathematical equation relating *TL, TR,* and *TD*. One of the main objectives of this paper is to fill this gap in our knowledge and use the derived equation to analyze the *TL* and *TR* data of functionally well-defined groups of mRNAs in order to investigate the possible functional roles of mRNA levels in cell biology. For this purpose, we chose the glycolytic and respiratory mRNA molecules for a detailed analysis because the biochemistry of glycolysis and respiration (leading to oxidative phosphorylation) and their antagonistic interactions are well known in *S. cerevisiae* during glucose-galsctose shift [2, 7, 15, 16, 18].

The unicellular organism *S. cerevisiae* (also known as budding yeast, baker's yeast, or wine yeast) has the capacity to metabolize glucose and galactose but prefers the former as the carbon and energy sources when both nutrients are present in its environment. In the presence of glucose, the organism turns on those genes coding for the enzymes needed to convert glucose to ethanol (which phenomenon is known as *glucose induction*) and turns off those genes needed for galactose metabolism (which phenomenon is known as *glucose repression*) [2, 7, 15, 16, 18]. The detailed molecular mechanisms underlying these phenomena (called *diauxic shift*) are incompletely understood at present and are under intensive studies [11, 25, 31]. When glucose is depleted, *S. cerevisiae* increases its rate of metabolism of ethanol to produce ATP via the Krebs cycle and mitochondrial respiration [11, 25]. This metabolic control is exerted by reversing the glucose repression of the genes encoding the enzymes required for respiration (*i.e.*, oxidative phosphorylation) – the process referred to as *glucose de-repression* [11].

12.5. Conclusion and Discussion

Fig. 12.3a shows the time courses of the average levels of 14 each of the glycolytic and respiratory mRNA molecules during the 850 minutes of observation after shifting glucose to galactose. The time course of the average transcription rates of the same sets of glycolytic and respiratory genes are displayed in Fig. 12.3b. The most striking feature of these figures is that, despite the similarity between the time courses of the transcription rates (*TR*) of glycolytic and respiratory genes, those of the corresponding transcript levels (*TL*) are quite different. In fact, the *TL* trajectories of glycolytic and respiratory genes change in opposite directions during the period between 5 and 360 minutes after glucose-galactose shift, whereas the corresponding *TR* trajectories almost coincide. As shown in Fig. 12.4, these opposite changes in *TL* appear to be the consequences of the opposite changes in the degradation rates of glycolytic and respiratory mRNA molecules.

The qualitative features of the temporal behaviors of *TL* and *TR* changes displayed in Fig. 12.3a and 12.3b are summarized in Table 12.2. As indicated in the first two rows, the total observational period of 850 minutes are broken down to 5 phases, labeled *I* through *V*. During Phase I, the transcript levels of both glycolytic and respiratory genes decrease precipitously although the corresponding transcription rates increase, most likely because the stress induced by glucose-galactose shift increase transcript degradation rates more than can be compensated for by increased transcription. This interpretation is supported by the transcription/degradation ratio of 0.5 calculated for Phase I (see Fig. 12.4). During Phases II and III, the glycolytic transcript levels decrease by 2 fold, whereas the respiratory transcript levels increase by 4 fold. Since the corresponding transcription rates of both the glycolytic and respiratory genes decline rapidly followed by a plateau, the increased respiratory mRNA levels cannot be accounted for in terms of transcriptional control but must implicate degradational control. That is, just as the removal of glucose "de-induces" glycolytic mRNA molecules (leading to the declining *TL* and *TR* trajectories for glycolysis in Fig. 12.3a and 12.3b), so it might repress the degradation of respiratory mRNA molecules, leading to a rise in respiratory mRNA levels as seen in Fig. 12.3a between the second and fourth time points. This phenomenon may be referred to as "glucose de-induction" in analogy to glucose induction [7, 18, 20]. If this interpretation is correct, one intriguing hypothesis suggests itself that glucose normally keeps the respiratory mRNA levels low by both enhancing the degradation and repressing the synthesis of respiratory mRNA molecules. During Phase IV, both *TL* and *TR* for glycolytic and respiratory genes increase, and this may be attributed to galactose induction [11, 18, 34]. In support of this interpretation, it was found that glucose-galactose shift induced an increase in both *TL* and *TR* of the *Leloir* genes (GAL 1, 2, 3, 7 and 10) between 120 and 450 minutes by more than 10 folds (data not shown). The *Leloir* genes code for the enzymes and transport proteins that are involved in converting extracellular galactose to intracellular glucose-1-phosphate [9], which is then metabolized via the glycolytic and respiratory pathways. Finally, during Phase V, the glycolytic mRNA levels remain constant while the respiratory mRNA levels decline slightly, the latter probably due to galactose repression (in analogy to the well-known glucose repression [18]) of respiration following the formation of *glucose-1-phosphate* via the Leloir pathway [9]. The transcription rate of glycolytic genes continue to increase during Phase V probably due to galactose induction [21, 25], although the corresponding transcript levels remain unchanged, which may also indicate the degradational control of glycolytic mRNA molecules during this time period. That is, budding yeast seems able to keep glycolytic *TL* constant in the face of increasing *TR*, by increasing *TD* - the transcript degradation rate. The *TR* trajectory of respiratory genes also continue to increase during Phase V despite the fact that their *TL* trajectory decline, which can be best explained in terms of the hypothesis that that respiratory mRNA levels are controlled by transcript degradation. It is quite evident that the *TL* and *TR* data presented in Fig. 12.3a and 12.3b cannot be accounted for in terms of *TR* alone but requires taking into account both *TR* and *TD* on an equal footing for their logically consistent explications, which is tantamount to the conclusion that *TL* is determined by the *D/T* ratio (see Fig. 12.3a and 12.4).

Table 12.2. A summary of the kinetics of the *TL* and *TR* changes depicted in Fig. 12.3a and 12.3b. The upward and downward arrows indicate an increase and decrease, respectively.

Time (min)		0-5	5-120	120-360	360-450	450-850
Phase (or Time Period)		I	II	III	IV	V
Transcript Level (TL)	Glycolysis	↓	↓	↓	↑	No Change
	Oxidative Phosphorylation	↓	↑	↑	↑	↓
Transcription Rate (TR)	Glycolysis	↑	↓	↓	↑	↑
	Oxidative Phosphorylation	↑	↓	↓	↑	↑

In conclusion, the genome-wide *TL* and *TR* data of *S. cerevisiae* measured by Garcia-Martinez *et al.* [10] have provided us with a concrete experimental basis to establish the concept that mRNA levels measured with cDNA arrays cannot be interpreted in terms other than what is here called the transcription/degradation (D/T) ratios. These ratios have been found useful in characterizing the temporal evolution of the molecular mechanisms underlying mRNA level changes induced by glucose-galactose shift in this microorganism, which may be extended to other cellular systems for similar determinations. Since these mRNA level changes reflect the dynamic metabolic states of cells (*i.e.*, cell states) supported by dissipation of free energy, they qualify as examples of what was referred to as intracellular dissipative structures (*IDSs*), an example of the application of Prigogine's dissipative structure concept to molecular cell biology [3, 14, 17]. It was postulated that *IDSs* serve as the immediate driving forces for all cell functions and reflect the functional states of the cell. If this interpretation turns out to be correct upon further investigations, the DNA array technique, due to its ability to measure D/T ratios as demonstrated here, may prove to be an invaluable experimental tool to characterize and investigate *IDSs* and their biological functions, leading to numerous applications in basic cell biology, biotechnology, and medicine, including developments of diagnostic procedures to recognize cancer cells in their early developmental stages and testing drug candidates for their ability to reverse such pathological cell states.

Acknowledgments

This material is based upon work supported by the National Science Foundation under Grant No. 0546574. The first author thanks R. Miura and S. Dhar of NJIT for their valuable contributions in the early phase of the research program described here.

References

[1] U. Alon, N. Barkai, D. A. Notterman, K. Gish, S. Ybarra, D. Mack, and A.J. Levine. Broad patterns of gene expression revealed by clustering analysis of tumor and normal colon tissues probed by oligonucleotide arrays. *Proc. Nat. Acad. Sci. USA*, 96: 6745-6750, 1999.

[2] M.P. Ashe, S.K. De Long, and A.B. Sachs. Glucose depletion rapidly inhibits translation initiation in yeast. *Mol. Biol. Cell*, 11: 833-848, 2000.

[3] A. Babloyantz. *Molecules, Dynamics and Life: An Introduction to Self-Organization of Matter*. Wiley-Interscience, New York, 1986.

[4] J.M. Berg, J.L. Tymoczko, and L. Stryer. *Biochemistry, Fifth Edition*. W. H. Freeman and Company, New York, 2002.

[5] A. Beyer, J. Hollunder, H.-P. Nasheuer, and T. Wilhelm. Post-transcriptional expression regulation in the yeast saccharomyces cerevisiae on a genomic scale. *Mol. Cell. Proteomics*, 3: 1083-1092, 2004.

[6] T. Chen, H. L. He, and G. M. Church. Modeling gene expression with differential equations. In R.B. Altman, A.K. Dunker, L. Hunter, T.E. Klein, and K. Lauderdale

(editors), *Pacific Symposium on Biocomputing*, World Scientific, Singapore, pages 29-40, 1999.

[7] J.L. DeRisi, V. R. Iyer, and P. O. Brown. Exploring the metabolic and genetic control of gene expression on a genomic scale. *Science*, 278: 680-686, 1997.

[8] J. Fan, X. Yang, W. Wang, W. H. Wood, K. G. Becker, and M. Gorospe. Global analysis of stress-regulated mRNA turnover by using cDNA arrays. *Proc. Nat. Acad. Sci.*, 99(16): 10611-10616, 2002.

[9] L. Fu, P. Bounelis, N. Dey, B. L. Browne, R. B. Marchase, and D. M. Bedwell. The posttranslational modification of phosphoglucomutase is regulated by galactose induction and glucose repression in Saccharomyces cerevisiae. *J. Bacteriol.*, 7(11): 3087-3094, 1995.

[10] J. Garcia-Martinez, A. Aranda, and J. E. Perez-Ortin. Genomic run-on evaluates transcription rates for all yeast genes and identifies gene regulatory mechanisms. *Mol. Cell*, 15: 303-313, 2004.

[11] A. P. Gasch. The environmental stress response: a common yeast response to diverse environmental stresses. In S. Hohmann and W.H. Mager (editors), *Yeast Stress Responses*, Springer, Berlin, pages 11-70, 2003.

[12] J. L. Hargrove and F. H. Schmidt. The role of mRNA and protein stability in gene expression. *FASEB J.*, 3: 2360-2370, 1989.

[13] K. Hirayoshi and J. T. Lis. Nuclear run-on assays: assessing transcription by measuring density of engaged RNA polymerases. *Methods in Enzymology*, 304: 351-362, 1999.

[14] S. Ji. The Bhopalator: a molecular model of the living cell based on the concepts of conformons and dissipative structures. *J. theoret. Biol.*, 116: 399-426, 1985.

[15] G. Jona, M. Choder, and O. Gileadi. Glucose starvation induces a drastic reduction in the rates of both transcription and degradation of mRNA in yeast. *Biochim. Biophys. Acta*, 1491: 37-48, 2000.

[16] M. Johnnston. Feasting, fasting and fermenting: glucose sensing in yeast and other cells. *Trends Genetics*, 15(1): 29-33, 1999.

[17] D. Kondepudi and I. Prigogine. *Modern Thermodynamics: From Heat Engine to Dissipative Structures*. John Wiley and Sons, Inc., Chichester, 1998.

[18] K. M. Kuhn, J. L. DeRisi, P. O. Brown, and P. Sarnow. Global and specific translational regulation in the genomic response of saccharomyces cerevisiae to nonfermental carnbon source. *Mol. Cell. Biol.*, 21(3): 916-927, 2001.

[19] J. Legen, S. Kemp, K. Krause, B. Profanter, R. G. Hermann, and R. M. Maier. Comparative analysis of plastid transcription profiles of entire plastid chromosomes from tobacco attributed to wild-type and PEP-deficient transcription machineries. *Plant J.*, 31: 171-188, 2002.

[20] K. K. Leuther and S. A. Johnston. Nondissociation of GAL4 and GAL80 in vivo after galactose induction. *Science*, 256(5061): 33-1335, 1992.

[21] A. L. Mosley, J. Lakshmann, B. K. Aryal, and S. Özcan. Glucose-mediated phosphorylation converts the transcription factor Rgt1 from a repressor to an activator. *J. Biol. Chem.*, 278 (12): 10322-10327, 2003.

[22] A. C. Pease, D. Solas, E. J. Sullivan, M. T. Cronin, C. P. Holmes, and P. A. Fodor. Light-generated oligonucleotide arrays for rapid DNA sequence analysis. *Proc. Nat. Acad. Sci. USA*, 91: 5022-5026, 1994.

[23] I. Prigogine. Time, structure, and fluctuations. *Science*, 201: 777-785, 1978.

[24] D. R. Rhodes and A. M. Chinnaiyan. Bioinformatics strategies for translatingh genome-wide expression analyses into clinically useful cancer markers. *Ann. N.Y. Acad. Sci.*, 1020: 32-40, 2004.

[25] H. Ronne. Glucose repression in fungi. *Trends Genetics*, 11(1): 12-17, 1995.

[26] M. A. Savageau. Biochemcial Systems Analysis I: Some mathematical properties of the rate law for the component enzymatic reactions. *J. theoret. Biol.*, 25: 365-369, 1969.

[27] M. A. Savageau. Biochemcial Systems Analysis II: The steady-state solutions for an n-pool system using a power-law approximation. *J. theoret. Biol.*, 25: 370-379, 1969.

[28] S. M. Schena, D. Shalon, D. R. Davis, and P. O. Brown. Quantitative monitoring of gene expression patterns with a complementary DNA microarray. *Science*, 91: 467-470, 1995.

[29] D. J. Shapiro, J. E. Blume, and D. A. Nielsen. Regulation of messenger RNA stability in eukaryotic cells. *BioEssays*, 6(5): 221-226, 1986.

[30] M. A. Troester, K. A. Hoadley, T. Sorlie, B.-S. Herbert, A.-L. Borresen-Dale, P. E. Lonning, J. W. Shay, W. K. Kaufmann, and C. M. Perou. Cell-type-specific responses to chemotherapeutics in breast cancer. *Cancer Research*, 64: 4218-4226, 2004.

[31] B. P. Tu, A. Kudlicki, M. Towicka, and S. L. McKnight. Logic of the yeast metabolic cycle: temporal compartmentalization of cellular processes. *Science*, 310: 1151-1158, 2005.

[32] Y. Wang, C. L. Liu, J. D. Storey, R. J. Tibshirani, D. Herschlag, and P. O. Brown. Precision and functional specificity in mRNA cay. *PNAS*, 99(9): 5860-5865, 2002.

[33] S. J. Watson and U. Akil. Gene chips and arrays revealed: A primer on their power and their uses. *Biol. Psychiatry*, 45: 533-543, 1999.

[34] J. Winderickx, I. Holsbeeks, O. Lagatie, F. Giots, J. Thevelein, and H. de Winde. From feast to famine; adaptation to nutrient availability in yeast. In S. Hohmann and W.H. Mager (editors). *Yeast Stress Responses*, Springer, Berlin, pages 305-386, 2003.

[35] E. Yang, E. van Nimwegen, M. Zavolan, N. Rajewsky, M. Schoeder, M. Magnasco, and J. E. Darnell, Jr. Decay rates of human mRNAs: Correlation with functional characteristics and sequence attributes. *Genome Research*, 13: 1863-1872, 2003.

Chapter 13

The Complexity of Feature Selection for Consistent Biclustering

O. Erhun Kundakcioglu and Panos M. Pardalos

Department of Industrial and Systems Engineering
University of Florida
303 Weil Hall
Gainesville, FL, 32611, USA
{erhun, pardalos}@ufl.edu

Biclustering is simultaneous classification of the samples and features in a way that samples from the same class have similar values for that class' characteristic features. A biclustering is *consistent* if in each sample (feature) from any set, the average expression of features (samples) that belong to the same class is greater than the average expression of features (samples) from other classes. Supervised biclustering uses a training set to classify features whose consistency is achieved by *feature selection*. The worst case complexity of this feature selection process is studied.

13.1. Introduction

Biclustering is a methodology allowing simultaneous partitioning of a set of samples and their features into classes. Samples and features classified together are supposed to have a high relevance with each other which can be observed by intensity of their expressions. The notion of consistency for biclustering is defined using interrelation between centroids of sample and feature classes. Previous works on biclustering concentrated on unsupervised learning and did not consider employing a training set, whose classification is given. However, with the introduction of consistent biclustering, significant progress has been made in supervised learning as well.

A data set (e.g., from microarray experiments) is normally given as a rectangular $m \times n$ matrix A, where each column represents a data sample (e.g., patient) and each row represents a feature (e.g., gene)

$$A = (a_{ij})_{m \times n}$$

where a_{ij} is the expression of ith feature in j^{th} sample.

Biclustering is applied by simultaneous classification of the samples and features (i.e., columns and rows of matrix A, respectively) into k classes. Let $S_1, S_2, \ldots, S_k$ denote the classes of the samples (columns) and $F_1, F_2, \ldots, F_k$ denote the classes of features (rows). Formally biclustering can be defined as a collection of pairs of sample and feature subsets $\mathcal{B} = \{(S_1, F_1), (S_2, F_2), \ldots, (S_k, F_k)\}$ such that

$$S_1, S_2, \ldots, S_k \subseteq \{a^j\}_{j=1,\ldots,n},$$

$$\bigcup_{r=1}^{k} S_r = \{a^j\}_{j=1,\ldots,n},$$

$$S_\zeta \bigcap S_\xi = \emptyset \Leftrightarrow \zeta \neq \xi,$$

$$F_1, F_2, \ldots, F_k \subseteq \{a_i\}_{i=1,\ldots,m},$$

$$\bigcup_{r=1}^{k} F_r = \{a_i\}_{i=1,\ldots,m},$$

$$F_\zeta \bigcap F_\xi = \emptyset \Leftrightarrow \zeta \neq \xi,$$

where $\{a^j\}_{j=1,\ldots,n}$ and $\{a_i\}_{i=1,\ldots,m}$ denote the set of columns and rows of the matrix A, respectively.

The ultimate goal in a biclustering problem is to find a classification for which samples from the same class have *similar* values for that class' characteristic features. The visualization of a reasonable classification should reveal a block-diagonal or "checkerboard" pattern. A detailed survey on biclustering techniques can be found in [5] and [8].

The concept of *consistent biclustering* is introducted in [3]. Formally, a biclustering $\mathcal{B}$ is consistent if in each sample (feature) from any set S_r (set F_r), the average expression of features (samples) that belong to the same class r is greater than the average expression of features (samples) from other classes. The model for supervised biclustering involves solution of a special case of fractional 0-1 programming problem whose consistency is achieved by feature selection. Computational results on microarray data mining problems are obtained by reformulating the problem as a linear mixed 0-1 programming problem.

An improved heuristic procedure is proposed in [9], where a linear programming problem with continuous variables is solved at each iteration. Numerical experiments on the data, which consists of samples from patients diagnosed with *acute lymphoblastic leukemia (ALL)* or *acute myeloid leukemia (AML)* diseases (see [1, 2, 7, 12, 13]), confirm that the algorithm outperforms the previous results in the quality of solution as well as computation time. Consistent biclustering is also used to analyze scalp EEG data obtained from epileptic patients undergoing treatment with a vagus nerve stimulator (VNS) (see [4]).

The rest of the chapter is organized as follows: We first introduce mathematical formulations for consistent biclustering and introduce the application of feature selection for consistent biclustering. Next, the complexity results are shown and the chapter is concluded with closing remarks.

13.2. Consistent Biclustering

Given a classification of the samples, S_r, let $S = (s_{jr})_{n\times k}$ denote a 0-1 matrix where $s_{jr} = 1$ if sample j is classified as a member of the class r (i.e., $a^j \in S_r$), and $s_{jr} = 0$ otherwise. Similarly, given a classification of the features, F_r, let $F = (f_{ir})_{m\times k}$ denote a 0-1 matrix where $f_{ir} = 1$ if feature i belongs to class r (i.e., $a_i \in F_r$), and $f_{ir} = 0$ otherwise. Construct corresponding *centroids* for the samples and features using these matrices as follows

$$C_S = AS(S^TS)^{-1} = (c^S_{i\xi})_{m\times r} \tag{13.1}$$

$$C_F = A^TF(F^TF)^{-1} = (c^F_{j\xi})_{n\times r} \tag{13.2}$$

The elements of the matrices, $c^S_{i\xi}$ and $c^F_{j\xi}$, represent the average expression of the corresponding sample and feature in class ξ, respectively. In particular,

$$c^S_{i\xi} = \frac{\sum_{j=1}^n a_{ij}s_{j\xi}}{\sum_{j=1}^n s_{j\xi}} = \frac{\sum_{j|a^j\in S_\xi} a_{ij}}{|S_\xi|},$$

and

$$c^F_{j\xi} = \frac{\sum_{i=1}^m a_{ij}f_{i\xi}}{\sum_{i=1}^m f_{i\xi}} = \frac{\sum_{i|a_i\in F_\xi} a_{ij}}{|F_\xi|}.$$

Using the elements of matrix C_s, one can assign a feature to a class where it is over-expressed. Therefore feature i is assigned to class $\hat{r}$ if $c^S_{i\hat{r}} = \max_\xi\{c^S_{i\xi}\}$, i.e.,

$$a_i \in \hat{F}_{\hat{r}} \Longrightarrow c^S_{i\hat{r}} > c^S_{i\xi}, \qquad \forall \xi, \xi \neq \hat{r}. \tag{13.3}$$

Note that the constructed classification of the features, $\hat{F}_r$, is not necessarily the same as classification F_r. Similarly, one can use the elements of matrix C_F to classify the samples. Sample j is assigned to class $\hat{r}$ if $c^F_{j\hat{r}} = \max_\xi\{c^F_{j\xi}\}$, i.e.,

$$a^j \in \hat{S}_{\hat{r}} \Longrightarrow c^F_{j\hat{r}} > c^F_{j\xi}, \qquad \forall \xi, \xi \neq \hat{r}. \tag{13.4}$$

As before, the obtained classification $\hat{S}_r$ does not necessarily coincide with classification S_r.

Biclustering $\mathcal{B}$ is referred to as a *consistent biclustering* if relations (13.3) and (13.4) hold for all elements of the corresponding classes, where matrices C_S and C_F are defined according to (13.1) and (13.2), respectively.

A data set is *biclustering-admitting* if some consistent biclustering for it exists. Furthermore, the data set is called *conditionally biclustering-admitting* with respect to a given (partial) classification of some samples and/or features if there exists a consistent biclustering preserving the given (partial) classification.

Theorem 13.1. *Let $\mathcal{B}$ be a consistent biclustering. Then there exist convex cones $P_1, P_2, \ldots, P_k \subseteq \mathbb{R}^m$ such that only samples from S_r belong to the corresponding cone P_r, $r = 1, \ldots, k$. Similarly, there exist convex cones $Q_1, Q_2, \ldots, Q_k \subseteq \mathbb{R}^n$ such that only features from class F_r belong to the corresponding cone Q_r, $r = 1, \ldots, k$.*

See [3] for the proof of Theorem 13.1. It also follows from the proven conic separability that convex hulls of classes do not intersect.

By definition, a biclustering is consistent if $F_r = \hat{F}_r$ and $S_r = \hat{S}_r$. However, a given data set might not have these properties. The features and/or samples in the data set might not clearly belong to any of the classes and hence a consistent biclustering might not be constructed. In such cases, one can remove a set of features and/or samples from the data set so that there is a consistent biclustering for the truncated data. Selection of a representative set of features that satisfies certain properties is a widely used technique in data mining applications. This feature selection process may incorporate various objective functions depending on the desirable properties of the selected features, but one general choice is to select the maximal possible number of features in order to lose minimal amount of information provided by the training set.

A problem with selecting the most representative features is the following. Assume that there is a consistent biclustering for a given data set, and there is a feature, i, such that the difference between the two largest values of c^S_{ir} is negligible, i.e.,

$$\min_{\xi \neq \hat{r}}\{c^S_{i\hat{r}} - c^S_{i\xi}\} \leq \alpha,$$

where α is a small positive number. Although this particular feature is classified as a member of class $\hat{r}$ (i.e., $a_i \in F_{\hat{r}}$), the corresponding relation (13.3) can be violated by adding a slightly different sample to the data set. In other words, if α is a relatively small number, then it is not statistically evident that $a_i \in F_{\hat{r}}$, and feature i cannot be used to classify the samples. The significance in choosing the most representative features and samples comes with the difficulty of problems that require feature tests and large amounts of samples that are expensive and time consuming. Some stronger additive and multiplicative consistent biclusterings can replace the weaker consistent biclustering. *Additive consistent biclustering* is introduced in [9] by relaxing (13.3) and (13.4) as

$$a_i \in F_{\hat{r}} \Longrightarrow c^S_{i\hat{r}} > \alpha^S_i + c^S_{i\xi}, \qquad \forall \xi, \xi \neq \hat{r}, \tag{13.5}$$

and

$$a^j \in S_{\hat{r}} \Longrightarrow c^F_{j\hat{r}} > \alpha^F_j + c^F_{j\xi}, \qquad \forall \xi, \xi \neq \hat{r}, \tag{13.6}$$

respectively, where $\alpha^F_j > 0$ and $\alpha^S_i > 0$.

Another relaxation in [9] is *multiplicative consistent biclustering* where (13.3) and (13.4) are replaced with

$$a_i \in F_{\hat{r}} \Longrightarrow c^S_{i\hat{r}} > \beta^S_i c^S_{i\xi}, \qquad \forall \xi, \xi \neq \hat{r}, \tag{13.7}$$

and

$$a^j \in S_{\hat{r}} \Longrightarrow c^F_{j\hat{r}} > \beta^F_j c^F_{j\xi}, \qquad \forall \xi, \xi \neq \hat{r}, \tag{13.8}$$

respectively, where $\beta^F_j > 1$ and $\beta^S_i > 1$.

Supervised biclustering uses accurate data sets that are called the *training set* to classify features to formulate consistent, α-consistent and β-consistent biclustering problems. Then, the information obtained from these solutions can be used to classify additional samples that are known as the *test set*. This information is also useful for adjusting the values of vectors α and β to produce more characteristic features and decrease the number of misclassifications.

Given a set of training data, construct matrix S and compute the values of $c^S_{i\xi}$ using (13.1). Classify the features according to the following rule: feature i belongs to class $\hat{r}$ (i.e., $a_i \in F_{\hat{r}}$), if $c^S_{i\hat{r}} > c^S_{i\xi}$, $\forall \xi \neq \hat{r}$. Finally, construct matrix F using the obtained classification. Let x_i denote a binary variable, which is one if feature i is included in the computations and zero otherwise. Consistent, α-consistent and β-consistent biclustering problems are formulated as follows.

CB:

$$\max_x \quad \sum_{i=1}^{m} x_i \tag{13.9a}$$

$$\text{subject to} \quad \frac{\sum_{i=1}^{m} a_{ij} f_{i\hat{r}} x_i}{\sum_{i=1}^{m} f_{i\hat{r}} x_i} > \frac{\sum_{i=1}^{m} a_{ij} f_{i\xi} x_i}{\sum_{i=1}^{m} f_{i\xi} x_i}, \quad \forall \hat{r}, \xi \in \{1, \ldots, k\}, \hat{r} \neq \xi, j \in S_{\hat{r}} \tag{13.9b}$$

$$x_i \in \{0, 1\}, \quad \forall i \in \{1, \ldots, m\} \tag{13.9c}$$

α-CB:

$$\max_x \quad \sum_{i=1}^{m} x_i \tag{13.10a}$$

$$\text{subject to} \quad \frac{\sum_{i=1}^{m} a_{ij} f_{i\hat{r}} x_i}{\sum_{i=1}^{m} f_{i\hat{r}} x_i} > \alpha_j + \frac{\sum_{i=1}^{m} a_{ij} f_{i\xi} x_i}{\sum_{i=1}^{m} f_{i\xi} x_i}, \quad \forall \hat{r}, \xi \in \{1, \ldots, k\}, \hat{r} \neq \xi, j \in S_{\hat{r}} \tag{13.10b}$$

$$x_i \in \{0, 1\}, \quad \forall i \in \{1, \ldots, m\} \tag{13.10c}$$

β-CB:

$$\max_x \quad \sum_{i=1}^{m} x_i \tag{13.11a}$$

$$\text{subject to} \quad \frac{\sum_{i=1}^{m} a_{ij} f_{i\hat{r}} x_i}{\sum_{i=1}^{m} f_{i\hat{r}} x_i} > \beta_j \frac{\sum_{i=1}^{m} a_{ij} f_{i\xi} x_i}{\sum_{i=1}^{m} f_{i\xi} x_i}, \quad \forall \hat{r}, \xi \in \{1, \ldots, k\}, \hat{r} \neq \xi, j \in S_{\hat{r}} \tag{13.11b}$$

$$x_i \in \{0, 1\}, \quad \forall i \in \{1, \ldots, m\} \tag{13.11c}$$

The goal in the CB problem is to find the largest set of features that can be used to construct a consistent biclustering*. The α-CB and β-CB problems are similar to the original CB problem but the aim is to select features that can be used to construct α-consistent and β-consistent biclusterings, respectively.

In (13.9), x_i, $i = 1, \ldots m$ are the decision variables. $x_i = 1$ if i-th feature is selected, and $x_i = 0$ otherwise. $f_{ik} = 1$ if feature i belongs to class k, and $f_{ik} = 0$ otherwise. The objective is to maximize the number of features selected and (13.9b) ensures that the biclustering is consistent with respect to the selected features.

*Note that the number of selected features is the most commonly used objective function. Other objectives such as maximizing the weighted sum of selected features can also be considered.

13.3. Complexity Results

The optimization problem (13.9) is a specific type of *fractional 0-1 programming problem* which is defined as

$$\max \quad \sum_{i=1}^{m} w_i x_i \tag{13.12a}$$

$$\text{subject to} \quad \sum_{j=1}^{n_s} \frac{\alpha_{j0}^s + \sum_{i=1}^m \alpha_{ji}^s x_i}{\beta_{j0}^s + \sum_{i=1}^m \beta_{ji}^s x_i} \geq p_s, \qquad s = 1, \dots, S \tag{13.12b}$$

This problem is $\mathcal{NP}$-hard since linear 0-1 programming is a special class of Problem (13.12) when $\beta_{ji}^s = 0$ and $\beta_{j0}^s = 1$ for $j = 1, \dots, n_s$, $i = 1, \dots m$ and $s = 1 \dots, S$. A typical way to solve a fractional 0-1 programming problem is to reformulate it as a linear mixed 0-1 programming problem, and solve new problem using standard linear programming solvers (see [10, 11]).

In [3], a linearization technique for a generalized $\mathcal{NP}$-hard formulation (13.12) is applied to solve (13.9). In [9] heuristics are proposed for (13.9) and generalizations. These attempts are appropriate if the problem is $\mathcal{NP}$-hard. However, whether (13.9) itself is $\mathcal{NP}$-hard or not was an open question. This chapter intents to fill this gap by proving the $\mathcal{NP}$-hardness of (13.9).

Theorem 13.2. *Feature selection for consistent biclustering (i.e. (13.9)) is $\mathcal{NP}$-hard.*

Proof. To prove that the problem is $\mathcal{NP}$-hard, a special case of the problem is proven to be $\mathcal{NP}$-hard. In the case considered, there are 2 samples and m features. Suppose that there are two classes and all but one of the features belong to the same class. Without loss of generality, assume that m-th feature belongs to one class alone and hence it is selected in the optimal solution unless the problem is infeasible (i.e., $x_m = 1$). Then (13.9b) becomes

$$\frac{\sum_{i=1}^{m-1} a_{i1} x_i}{\sum_{i=1}^{m-1} x_i} > a_{m1} \tag{13.13}$$

$$\frac{\sum_{i=1}^{m-1} a_{i2} x_i}{\sum_{i=1}^{m-1} x_i} < a_{m2} \tag{13.14}$$

It has to be proven that the decision problem is $\mathcal{NP}$-complete in order to prove that the corresponding optimization problem is $\mathcal{NP}$-hard (see [6]). The decision version of feature selection for consistent biclustering problem is

D-CB: Is there a set of features that ensures biclustering is consistent, i.e., satisfies (13.13)-(13.14)?

Clearly, D-CB is in NP since the answer can be checked in $O(m)$ time for a given set of features.

Next, the KNAPSACK problem will be reduced to D-CB in polynomial time to complete the proof.

In a knapsack instance, a finite set U_1, a size $s(u) \in Z^+$ and a value $v(u) \in Z^+$ for each $u \in U_1$, a size constraint $B \in Z^+$, and a value goal $K \in Z^+$ are given. The question is

KNAPSACK: Is there a subset $U' \subseteq U_1$ such that $\sum_{u \in U'} s(u) \leq B$ and $\sum_{u \in U'} v(u) \geq K$?

We can modify the knapsack problem as

Π: Is there a subset $U' \subseteq U$ such that

$$\sum_{u \in U'} s(u) \leq 0 \tag{13.15}$$

$$\sum_{u \in U'} v(u) \geq 0\,? \tag{13.16}$$

Obviously, Π remains $\mathcal{NP}$-complete, since KNAPSACK can be reduced to its modified variant if we define $U = U_1 \cup t$, $s(t) = -B$, and $v(t) = -K$.

Defining $s'(u) = s(u) + \alpha$, $v'(u) = v(u) + \beta$ for each $u \in U$ and it can easily be seen that

$$\sum_{u \in U'} s(u) \leq 0 \Leftrightarrow \frac{\sum_{u \in U'} s'(u)}{|U'|} \leq \alpha \tag{13.17}$$

$$\sum_{u \in U'} v(u) \geq 0 \Leftrightarrow \frac{\sum_{u \in U'} v'(u)}{|U'|} \geq \beta \tag{13.18}$$

In microarray data sets, negative a_{ij} values usually correspond to "bad" data points. Note that, by selecting sufficiently large α and β values (i.e., $\alpha > B$ and $\beta > K$), the reduction is valid for the case where a_{ij} are nonnegative.

The inequality signs in (13.17)-(13.18) can be changed to strong inequality as follows

$$\frac{\sum_{u \in U'} s'(u)}{|U'|} \leq \alpha \Leftrightarrow \frac{\sum_{u \in U'} s'(u)}{|U'|} < \alpha + \epsilon_1 \tag{13.19}$$

$$\frac{\sum_{u \in U'} v'(u)}{|U'|} \geq \beta \Leftrightarrow \frac{\sum_{u \in U'} v'(u)}{|U'|} > \beta - \epsilon_2 \tag{13.20}$$

where $0 < \epsilon_1 < \min_{u,w \in U, s'(u) \neq s'(w)} \{|s'(u) - s'(w)|\}/|U|$ and $0 < \epsilon_2 < \min_{u,w \in U, v'(u) \neq v'(w)} \{|v'(u) - v'(w)|\}/|U|$. Note that, another upper bound on ϵ_2 is β to ensure that the resulting problem has nonnegative a_{ij} values.

As a result, the problem is reduced to selecting a subset $U' \subseteq U$ such that

$$\frac{\sum_{u \in U'} s'(u)}{|U'|} < \alpha + \epsilon_1 \tag{13.21}$$

$$\frac{\sum_{u \in U'} v'(u)}{|U'|} > \beta - \epsilon_2 \tag{13.22}$$

which is in the form of (13.13)-(13.14). The reduction is polynomial and (13.21-13.22) holds true if and only if (13.15-13.16) holds true. Thus D-CB is $\mathcal{NP}$-complete and the proof is complete. □

Corollary 13.1. *Problems (13.10) and (13.11) are $\mathcal{NP}$-hard.*

Proof. Problem (13.9) is a special class of Problem (13.10) when $\alpha_j = 0$ for $j \in S_{\hat{r}}$. Similarly Problem (13.9) is a special class of Problem (13.11) when $\beta_j = 1$ for all $j \in S_{\hat{r}}$. Hence both (13.10) and (13.11) are $\mathcal{NP}$-hard. □

13.4. Closing Remarks

The concept of feature selection for consistent biclustering is discussed. The aim in this setting is to select a subset of features in the original data set such that the obtained subset of data becomes conditionally biclustering-admitting with respect to the given classification of training samples. The additive and multiplicative variations of the problem are considered to extend the possibilities of choosing the most representative set of features. It is shown that the feature selection for consistent biclustering is $\mathcal{NP}$-hard.

References

[1] A. Ben-Dor, L. Bruhn, N. Friedman, I. Nachman, M. Schummer, and Z. Yakhini. Tissue classification with gene expression profiles. In *RECOMB '00: Proceedings of the fourth annual international conference on Computational molecular biology*, pages 54–64, New York, NY, USA, 2000. ACM Press.

[2] A. Ben-Dor, N. Friedman, and Z. Yakhini. Class discovery in gene expression data. In *RECOMB '01: Proceedings of the fifth annual international conference on Computational biology*, pages 31–38, New York, NY, USA, 2001. ACM Press.

[3] S. Busygin, O. A. Prokopyev, and P. M. Pardalos. Feature selection for consistent biclustering. *Journal of Combinatorial Optimization*, 10:7–21, 2005.

[4] S. Busygin, N. Boyko, P.M. Pardalos, M. Bewernitz, and G. Ghacibeh. Biclustering EEG data from epileptic patients treated with vagus nerve stimulation. In Onur Seref, O. Erhun Kundakcioglu, and Panos M. Pardalos, editors, *Data mining, systems analysis and optimization in biomedicine*, volume 953, pages 220–231. American Institute of Physics, 2007.

[5] S. Busygin, O. Prokopyev, and P. M. Pardalos. Biclustering in data mining. *Computers & Operations Research*, 35(9): 2964-2987, 2008.

[6] M. R. Garey and D. S. Johnson. *Computers and Intractability: A Guide to the Theory of NP-Completeness*. W. H. Freeman & Co., New York, NY, USA, 1979.

[7] T. R. Golub, D. K. Slonim, P. Tamayo, C. Huard, M. Gaasenbeek, J. P. Mesirov, H. Coller, M. L. Loh, J. R. Downing, M. A. Caligiuri, C. D. Bloomfield, and E. S. Lander. Molecular classification of cancer: Class discovery and class prodiction by gene expression monitoring. *Science*, 286(5439):531–537, 1999.

[8] S. Madeira and A. Oliveira. Biclustering algorithms for biological data analysis: a survey. *IEEE/ACM Transactions on Computational Biology and Bioinformatics*, 1(1):24–45, 2004.

[9] A. Nahapetyan, S. Busygin, and P. M. Pardalos. *Mathematical Modelling of Biosystems*, chapter An improved heuristic for consistent biclustering problems, pages 185–198. Applied Optimization. Springer, 2008.

[10] T.-H.Wu. A note on a global approach for general 0-1 fractional programming. *European Journal Of Operational Research*, 16:220–223, 1997.

[11] M. Tawarmalani, S. Ahmed, and N. V. Sahinidis. Global optimization of 0-1 hyperbolic programs. *J. of Global Optimization*, 24(4):385–416, 2002.

[12] J. Weston, S. Mukherjee, O. Chapelle, M. Pontil, T. Poggio, and V. Vapnik. Feature selection for svms. In *NIPS*, pages 668–674, 2000.

[13] E. P. Xing and R. M. Karp. Cliff: Clustering of high-dimensional microarray data via iterative feature filtering using normilized cuts. *Bioinformatics Discovery Note*, 17:306–315, 2001.

Chapter 14

Clustering Electroencephalogram Recordings to Study Mesial Temporal Lobe Epilepsy

Chang-Chia Liu

Department of Industrial and Systems Engineering
Department of Biomedical Engineering, University of Florida, USA
iamjeff@ufl.edu

Wichai Suharitdamrong

Department of Industrial and Systems Engineering, University of Florida, USA
wichais@ufl.edu

W. Art Chaovalitwongse

Department of Industrial and Systems Engineering, Rutgers University, USA
wchaoval@rci.rutgers.edu

Georges A. Ghacibeh

Northeast Regional Epilepsy Group, 20 Prospect Ave, Suite 800, Hackensack NJ, 07601, USA
gghacibeh@gmail.com

Panos M. Pardalos

Department of Industrial and Systems Engineering
Department of Biomedical Engineering, University of Florida, USA
pardalos@ufl.edu

The brain connectivity is known to have substantial influences over the brain function and its underlying information processes. In this chapter, a novel graph-theoretic approach is introduced to investigate the connectivity among brain regions through electroencephalogram (EEG) recordings acquired from a patient with mesial temporal lobe epilepsy (MTLE). The first step of the proposed approach is to transform the brain connectivity behavior into a complete graph. The connectivity for each pair of the brain regions is first quantified by the cross mutual information (CMI) measure, and then the maximum clique algorithm is subsequently applied to find the clique that contained a group of highly connected brain regions that is represented by a clique with maximum size. The CMI is known to have the ability to capture the connectivity between EEG signals. The adopted maximum clique algorithm can reduce the complexity of the clustering procedure for finding the maximum connected brain regions. The proposed graph-theoretic approach offers better assessments to visualize the structure of the brain connectivity over time. The results indicate that the maximum connected brain regions prior to seizure onsets were where the impending seizure was initiated. Furthermore, the proposed approach may be used to improve the outcome of the epilepsy surgery by identifying the seizure onset region(s) correctly.

14.1. Introduction

Neural activity is manifested by electrical signals known as graded and action potentials. Berger's [3] demonstration in 1929 has shown that it is possible to record the electrical activity from the human brain, particularly the neurons located near the surface of the brain. While we often think of electrical activity in neurons in terms of action potentials, the action potentials do not usually contribute directly to the electroencephalogram (EEG) recordings. In fact, for scalp EEG recordings, the EEG patterns are mainly the graded potentials accumulated from hundreds of thousands of neurons. The EEG patterns vary greatly in both amplitude and frequency. The amplitude of the EEG reflects the degree of synchronous firing of the neurons located around the recording electrodes. In general, the high EEG amplitude indicates that neurons are activated simultaneously. Low EEG frequency indicates less responses of the brain, such as sleep, whereas higher EEG frequency implies the increased alertness. Given the above descriptions, an acquired EEG time series can be defined as a record of the fluctuating brain activity measured at different times and spaces. The high degree of synchronicity for two different brain regions implies strong connectivity among them and vice versa. We will interchangeably use the terms synchronicity and connectivity for rest of the chapter.

Epileptic seizures involve the synchronization of large populations of neurons [13]. Measuring the connectivity and synchronicity among different brain regions through EEG recordings has been well documented [21, 22, 26]. The structures and the behaviors of the brain connectivity have been shown to contain rich information related to the functionality of the brain [4, 16, 30]. More recently, the mathematical principles derived from information theory and nonlinear dynamical systems have allowed us to investigate the synchronization phenomena in highly non-stationary EEG recordings. For example, a number of synchronization measures were used for analyzing the epileptic EEG recordings to reach the goals of localizing the epileptogenic zones and predicting the impending epileptic seizures [1, 14, 20, 26, 29]. These studies also suggest that epilepsy is a dynamical brain disorder in which the interactions among neuron or groups of neurons in the brain alter abruptly. Moreover, the characteristic changes in the EEG recordings have been shown to have clear associations with the synchronization phenomena among epileptogenic and other brain regions. When the conductivities between two or among multiple brain regions are simultaneously considered, the univariate analysis alone will not be able to carry out such a task. Therefore it is appropriate to utilize multivariate analysis. Multivariate analysis has been widely used in the field of neuroscience to study the relationships among sources obtained simultaneously. In this study, the cross mutual information (CMI) approach is applied to measure the connectivity among brain regions [9]. The CMI approach is a bivariate measure and has been shown to have ability for quantifying the connectivity of the EEG signals [8, 21, 24, 25]. The brain connectivity graph is then constructed where vertices in the graph represent the EEG electrodes.

Every distinct pair of vertices is connected by an arc with the length equal to the connectivity quantified by CMI. After constructing a brain connectivity graph, which is a complete graph, we then remove arcs of connectivity below a specified threshold value to preserve only strong couplings of electrode pairs. Finally, we employ a maximum clique algorithm to find a maximum clique in which the brain regions are strongly connected (See Section 14.4.2). The maximum clique size can be, in turn, used to represent the amount of largest connected regions in the brain. The maximum clique algorithm reduces the computational effort for searching in the constructed brain connectivity graph. The proposed graph-theoretic approach offers an easy protocol for inspecting the structures of the brain connectivity over time and possibly identifying the brain regions where seizures are initiated.

14.2. Epilepsy as a Dynamical Brain Disorder

Epilepsy can be caused by multiple factors. Some people may even begin having seizures from their childhood. Epilepsy in children can result from almost everything related to the brain development or function. Lack of oxygen supply can cause cerebral palsy and seizure for example new born infants that suffer a lack of oxygen supply to the brain before or during birth have higher risks for developing epilepsy in their lives [31]. Epilepsy can also occur in adult subjects that have bleeding in the brain as a result of prematurity or defective blood vessels in the brain. Some studies have also reported epilepsy can be induced by genetic changes. Some patients are born with genes related to epilepsy that can cause them to develop epilepsy for example the Unverricht-Lundborg disease. The majority of patients who have seizures are first treated with anti-epileptic drugs (AEDs). About 70% to 80% of these patients will become seizure free after the AED treatment. The choice of the AEDs depends on several factors, including the type of seizures, the age of the subject, and the potential side effects of the medicine. Some patients, however, do not respond to the usual pharmacological treatments and will then consider undergoing the epilepsy surgery.

14.3. Data Information

In the study, the EEG recordings (Table 14.1) were obtained using bilaterally, surgically implanted electrodes in the hippocampus, temporal and frontal lobe cortexes of the brain. The subject in this study was determined to have both left and right hippocampal regions (bi-lateral) as epileptic seizure onset zones by neurologists. The EEG recordings were meant for pre-surgical clinical evaluation for possible surgical treatment of intractable temporal lobe epilepsy (TLE). The recordings were obtained using the Nicolet BMSI 4000 with amplifiers of an input range of 0.6 mV, sampling rate of 200 Hz and filters with a frequency range of a 0.5–70 Hz. The recording included a total number of 30 intracranial electrodes (8 subdural and 6 hippocampal depth electrodes for each cerebral hemisphere, and a strip of 2 additional electrodes, see Fig. 14.1). The recorded EEG signals were digitized and stored on magnetic media for subsequent off-line analysis.

Table 14.1. EEG Data information

Patient No.	Gender	Age	Number of Electrodes	Seizure Onset Zone	Duration of EEG Recordings (hours)
1	Male	37	30	L./R. Hippocampus	22

14.4. Graph-Theoretic Modeling for Brain Connectivity

Although the underlying mechanism of the transition from normal to seizure onset for human brains is still largely unknown, the studies have reported that certain type of temporal lobe seizures (TLE) is initiated by specific brain connectivity patterns [25, 27]. The brain connectivity can be studied using a broad range of network approaches. Several studies have attempted to use graph-theoretic methods to model the brain connectivity [2, 15, 18], especially the theory of directed graphs is of special interest as it applies to structural and functional brain connectivity at all levels. Applying the graph-theoretic models for the EEG signals, the brain connectivity is presented using a complete graph $G(V, E)$, where V is

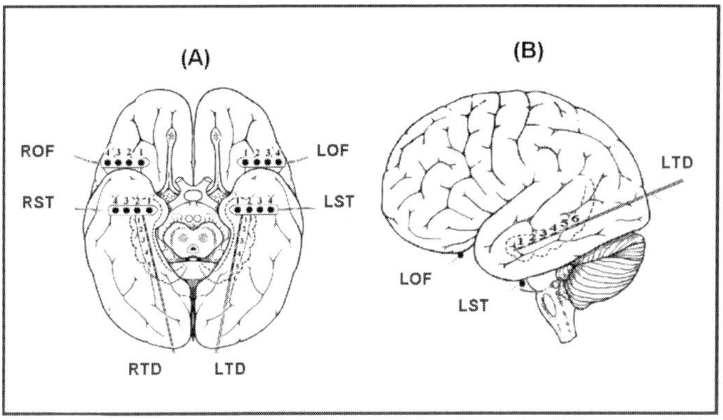

Fig. 14.1. Electrode placement (30 electrodes)

the set of vertices and E is the set of arcs. The vertex set represents EEG recording electrodes, an arc in the graph represents the synchronicity and connectivity between pair of EEG recording electrodes. Here we focus on the structure of the connectivity behavior in the construct graph which may relate to epileptic seizures.

14.4.1. *Cross–Mutual Information (CMI)*

The concept of CMI dates back to the work of Shannon in 1948 [28]. Generally, CMI measures the information obtained from observations of one random event for the other. It is known that CMI has the capability to capture both linear and nonlinear relationships between two random variables since both linear and nonlinear relationships can be described through probabilistic theories. Here in our model, the CMI measures how much information of EEG time series acquired from electrode x is presented by electrode y and vice versa. Let X be the set of data points where its possible realizations are $x_1, x_2, x_3, ..., x_n$ with probabilities $P(x_1), P(x_2), P(x_3), ...$ The Shannon entropy $H(X)$ of X is defined as

$$H(X) = -\sum_{i=1} p_i \ln p_i. \tag{14.1}$$

Shannon entropy measures the uncertainty content of X. It is always positive and measured in bits, if the logarithm is taken with base 2. Now let us consider another set of data points Y, where all possible realizations of Y are $y_1, y_2, y_3, ... y_n$ with probabilities $P(y_1), P(y_2), P(y_3), ...$ .. The degree of synchronicity and connectivity between X and Y can be measured by the joint entropy of X and Y, defined as

$$H(X,Y) = -\sum_{i,j} p_{ij}^{XY} \ln p_{ij}^{XY}. \tag{14.2}$$

where p_{ij}^{XY} which is the joint probability of $X = X_i$ and $Y = Y_j$. The cross information between X and Y, $CMI(X,Y)$, is then given by

$$CMI(X,Y) = H(Y) - H(X|Y) = H(X) - H(Y|X) \tag{14.3}$$

$$= H(X) + H(Y) - H(X,Y) \tag{14.4}$$

$$= \int\int f_{XY}(x,y) \log_2 \frac{f_{XY}(x,y)}{f_X(x)f_Y(y)} dxdy. \tag{14.5}$$

The cross mutual information is nonnegative. If these two random variables X, Y are independent, $f_{XY}(x,y) = f_X(x)f_Y(y)$, then $CMI(X,Y) = 0$, which implies that there is no relationship or correlation between X and Y. The probabilities are estimated using the histogram based box counting method. The random variables representing the observed number of pairs of point measurements in histogram cell (i,j), row i and column j,are respectively $k_{ij}, k_{i.}$ and $k_{.j}$. Here, we assume the probability of a pair of point measurements outside the area covered by histogram is negligible, therefore $\sum_{i,j} P_{ij} = 1$ [9, 10, 19].

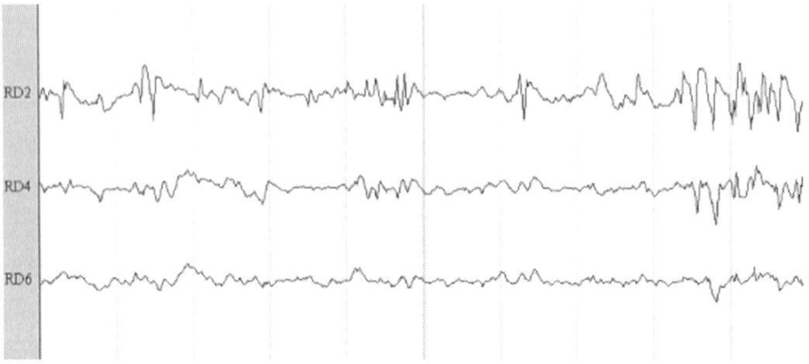

Fig. 14.2. A 10 sec. EEG epoch for $RTD2$, $RTD4$ and $RTD6$

Figure 14.2 illustrates an example of 10-second epochs recorded from the right mesial temporal depth (R(T)D) region. Figure 14.3 displays the scatter plots for pair-wise EEG signals shown in Fig. 14.2. Figure 14.4 shows the CMI values measured from EEG electrode pairs from Fig. 14.2. From the scatter plot, it is clear that EEG recordings between $R(T)D2$ and $R(T)D4$ have weak linear correlation which have also yielded lower CMI values in Fig. 14.4. The stronger linear relationship is discovered between $R(T)D4$ and $R(T)D6$ and this linear correlation pattern has resulted in higher CMI values in Fig. 14.4. Before measuring the

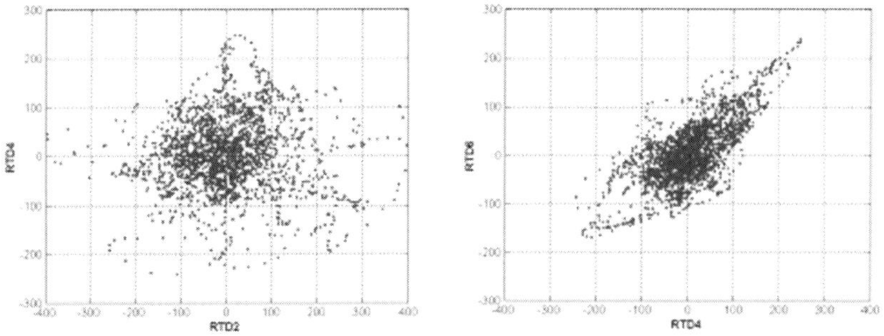

Fig. 14.3. Scatter plot for EEG epoch (10 seconds) of $RTD2$ vs. $RTD4$ and $RTD4$ vs. $RTD6$

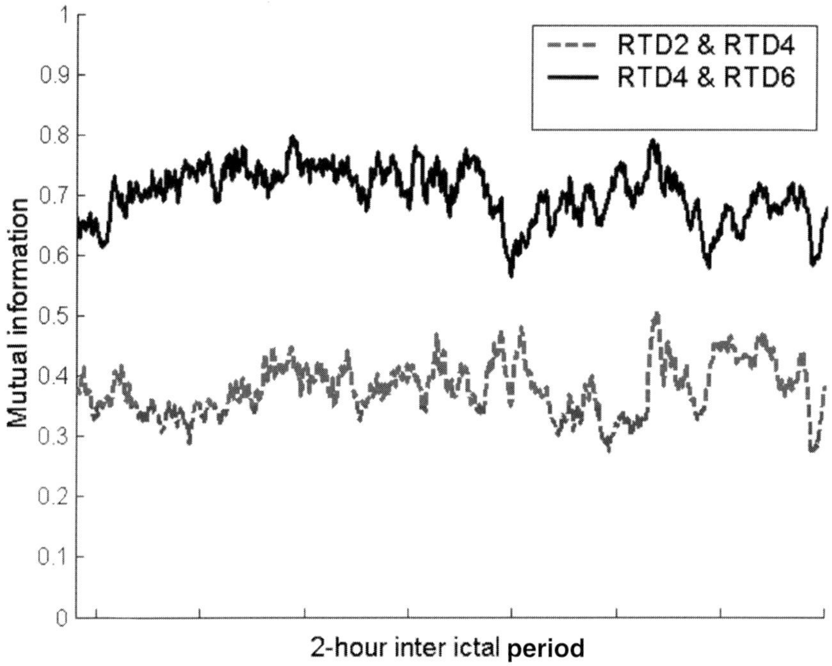

Fig. 14.4. Cross-mutual information for $RTD4$ vs. $RTD6$ and $RTD2$ vs. $RTD4$

CMI, we first divided EEG recordings into smaller non-overlapping EEG epochs. The segmentation procedure is widely utilized to subdue the non-stationary nature of the EEG recordings. The changes of EEG pattern tend to appear very briefly, examples include sharp wave transients, spikes, spike-wave complexes, and spindles. Working on shorter EEG epochs will insure the stationarity for the underlying processes and thus any change in the connectivity can be detected. Therefore, a proper length of EEG epochs has to be determined for measuring the connectivity among EEG recordings. We chose the length of the EEG epochs equal to 10.24 seconds (2048 points), which has also been utilized in many pervious EEG research studies [11, 12]. The brain connectivity measured using CMI form the complete graph, in which each node has an arc to every other adjacent vertex. In the procedure for removing the insignificant arcs (weak connection between brain regions), we first estimated an appropriate threshold value by utilizing the statistical tests. We determined this threshold by observing the statical significance over the complete connectivity graph, this threshold value was set to be a value where the small noise is eliminated, but yet the real signal is not deleted [5]. Figure 14.5 shows an example of a complete graph and a correspondent graph after edges with weak connectivity were removed.

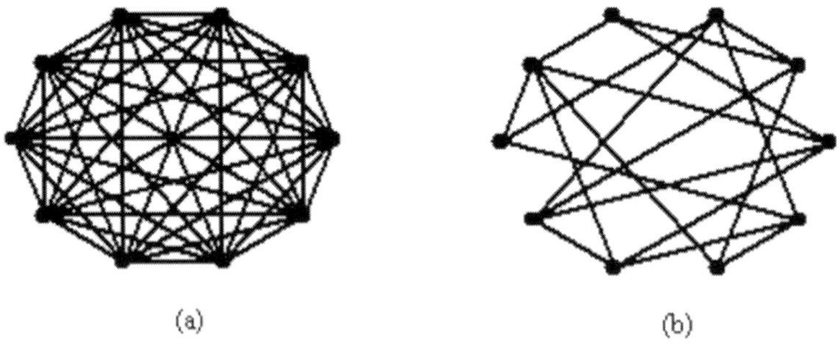

Fig. 14.5. (a) A complete connectivity graph (b) After applying the threshold and removing the arcs with insignificant connectivity

14.4.2. *Maximum Clique Algorithm*

We adopted the algorithm to find a maximum clique in the brain connectivity graph after deleting the insignificant arcs in the original complete graph as follows: Let $G = G(V, E)$ be a simple, undirected graph where $V = \{1, \ldots, n\}$ is the set of vertices (nodes), and E denotes the set of arcs. Assume that there is no parallel arcs (and no self-loops joining the same vertex) in G. Denote an arc joining vertex i and j by (i, j). We define a clique of G as a subset C of vertices with the property that every pair of vertices in C is connected by an arc; that is, C is a clique if the subgraph $G(C)$ induced by C is complete. Then, the maximum clique problem is to find a clique C with maximum cardinality (size) $|C|$. The maximum clique problem can be represented in many equivalent formulations (e.g., an integer programming problem, a continuous global optimization problem, and an indefinite quadratic programming). In this paper, we represent it in a simple integer programming form given by

$$\max x_i \tag{14.6}$$

$$\text{s.t.} \quad x_i + x_j \leq 1, \quad \text{where } (i,j) \notin E' \tag{14.7}$$

$$x_i \in \{0,1\}, \tag{14.8}$$

where x_i is binary variable indicating if electrode i is a member of the maximum clique. In finding the maximum clique in the brain networks, we applied the Carraghan-Pardalos maximum clique algorithm [6]. A pseudocode for this algorithm is provided in Fig. 14.6.

```
algorithm:    maximum clique
begin
              sort all nodes based on vertex ordering
              LIST = ordered nodes
              cbc = 0 current best clique size
              depth = 0 current depth level
              enter-next-depth(LIST,depth)
end
procedure:    enter-next-depth(LIST,depth)
begin
              m = the number of nodes in the LIST
              depth=depth+1
              for a node in position i in the LIST
                 if depth+(m-i)≤ cbc then
                     return prune the search
                  else
                     mark node i
                     if no adjacent node then
                        cbc=depth (maximum clique found)
                     else
                        enter to next depth (adjacent node of i, depth)
                     end
                 end
                  unmark node i
                 if depth=1
                     delete node i from LIST
                 end
              end
end
```

Fig. 14.6. The maximum clique algorithm

14.5. Results

In this section, we present the computational results from a patient with epilepsy. Figures 14.7, 14.8 and 14.9 show the connectivity among brain regions clustered by the proposed graph-theoretic approach. Figures 14.7 and 14.8 manifest the same pattern that the $L(T)D$ regions consistently exhibit high connectivity prior to the seizure onset over a 2-hour time horizon. With a proper threshold for constructing the connectivity graph, the uniqueness maximum clique is generated during interictal state of the subject. Moreover, the maximum clique size increased and covered almost all the brain regions during the epileptic seizure. It is known during the epileptic seizure, the abnormal discharge from epileptic brain regions may spread to the other brain regions and thus results the increase of the brain connectivity. Thus the maximum clique size reaching the maximum possible size (i.e., equal to the network size) was observed during an epileptic seizure. Through visual inspection on raw EEG recordings, the $L(T)D$ region is the area where the seizure was initiated. The EEG recordings were reviewed by two separated board certificated physicians to identify the location. In Fig. 14.9, the brain regions clustered by our approach were both from left and right orbitofrontal (i.e., $R(O)F$ and $L(O)F$) regions whereas the right hippocampus appeared to have

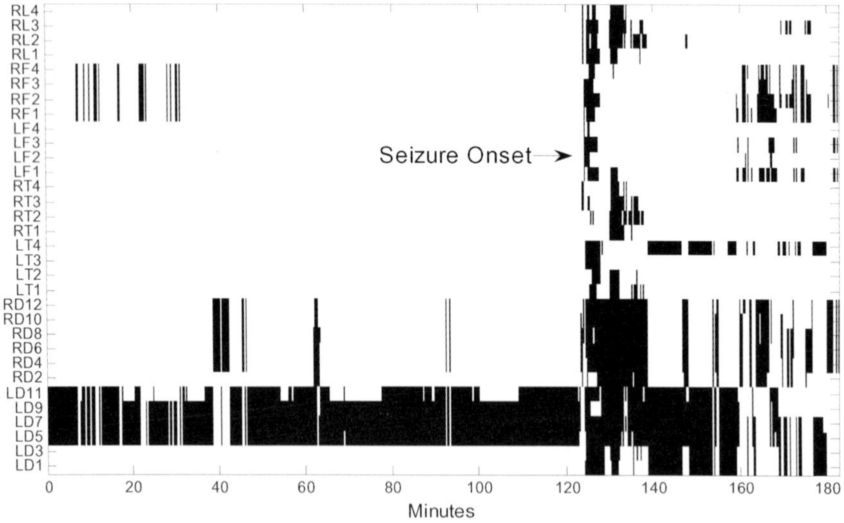

Fig. 14.7. A plot shows the brain regions clustered by graph-theoretic approach in the constructed brain connectivity graph. The highlighted areas represent the selected brain regions in the maximum clique. The $L(T)D$ areas tend to have strong connectivity prior to the actual seizure onset.

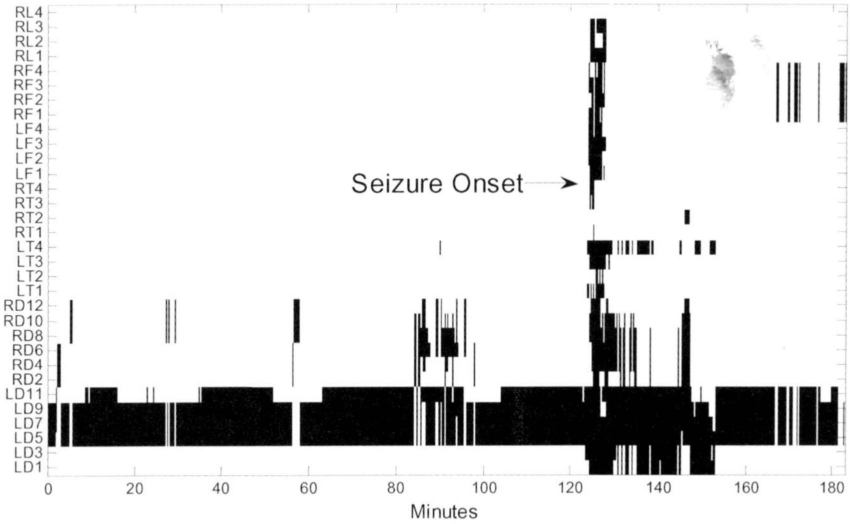

Fig. 14.8. A plot shows the brain regions clustered by graph-theoretic approach in the constructed brain connectivity graph. The highlighted areas represent the selected brain regions in the maximum clique. This figure has very similar connectivity structure as Figure 14.7 also manifests strong connectivity in the $L(T)D$ regions prior to the actual seizure onset.

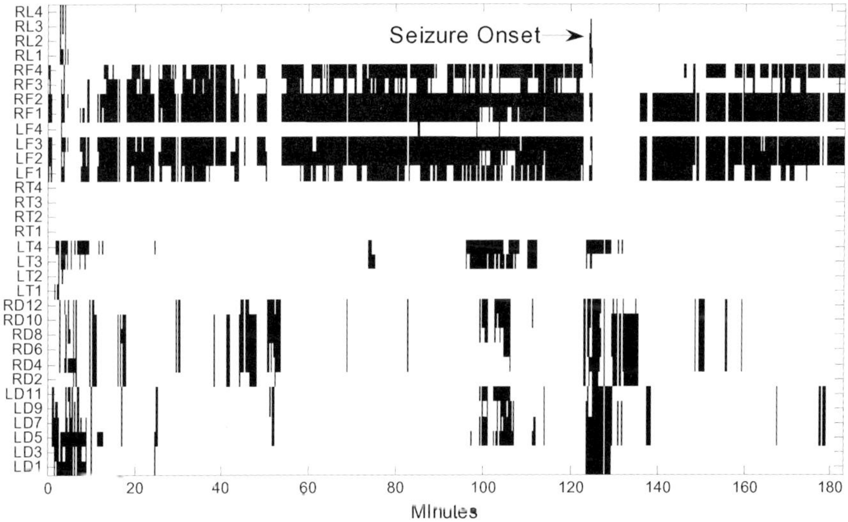

Fig. 14.9. A plot shows the brain regions clustered by graph-theoretic approach in the constructed brain connectivity graph. The highlighted areas represent the selected brain regions in the maximum clique. This plot illustrates the strong connectivity in the orbitofrontal regions (both left and right) prior to the actual seizure onset.

stronger connectivity than left hippocampus. The raw EEG recordings revealed that the epileptic seizure originated from $R(O)F$ regions. The above results suggest that the brain regions selected in a maximum clique prior to seizure onset, are indeed the regions where the seizure is initiated.

14.6. Conclusion and Discussion

In conclusion, we have presented a new approach for modeling and analyzing the brain connectivity. This approach successfully localized the epileptic focus (foci), the strongest connectivity were found in the regions where epileptic seizures are initiated. Previous studies also reported that the EEG recording acquired from epileptic regions contained higher nonlinear information than other brain regions [17]. The clustering result can be very useful for the current clinical environment, since about 20% of epilepsy patients will undergo the epilepsy surgery seeking better seizure control or the possibility of becoming seizure free. The outcome of the epileptic surgery depends largely on the accuracy of the focus localization. From the results, the proposed approach can serve as a tool for focus localization in pre-surgical evaluation stage. To further confirm our findings, a larger number of patients with different type of epileptic seizures is required. Moreover, our result also agrees with the fact that the connectivity pattern plays an important role for the brain function. Although these observations could lead us toward the understanding of the information process in temporal lobe epilepsy, however, the questions such as how the alteration of connectivity occurs and how it develops prior to a seizure still remain open. The understanding of the mechanisms of seizure development is the key to develop a reliable epilepsy seizure detector and to control or prevent an impending seizure well before its actual onset.

Acknowledgment

This material is based upon work supported by the National Science Foundation under Grant No. 0546574. The authors thank Dr. Chris Sackellares and Dr. Deng-Shan Shiau for their fruitful knowledge about epilepsy.

References

[1] J. Arnhold, P. Grassberger, K. Lehnertz, and C. E. Elger. A robust method for detecting interdependences: Application to intracranially recorded EEG. *Physica D*, 134(4):419–430, 1999.

[2] L. A. Baccala, M. Y. Alvarenga, K. Sameshima, C. L. Jorge, and L. H. Castro. Graph theoretical characterization and tracking of the effective neural connectivity during episodes of mesial temporal epileptic seizure. *J. Integr. Neurosci.*, 3(4):379–95, 2004.

[3] H. Berger. Ueber das elektrenkephalogramm des menschen. *Arch. Psychiatr. Nervenkr*, 87:527–570, 1929.

[4] A. Brovelli, M. Ding, A. Ledberg, Y. Chen, R. Nakamura, and S. L. Bressler. Beta oscillations in a large-scale sensorimotor cortical network: Directional influences revealed by granger causality. *Hum Brain Mapping*, 101:9849–9854, 2004.

[5] A. J. Butte and I. S. Kohane. Mutual information relevance networks: functional genomic clustering using pairwise entropy measurements. *Pac. Symp. Biocomput.*, pages 418–29, 2000.

[6] R. Carraghan and P. M. Pardalos. An exact algorithm for the maximum clique problem. *Operations Research Letters*, 9:375–382, 1990.

[7] T. M. Cover and J. A. Thomas. *Elements of Information Theory*. Wiley, 1991.

[8] R. B. Duckrow and A. M. Albano. Comment on performance of different synchronization measures in real data: A case study on electroencephalographic signals. *Phys. Rev. E*, 67(6):063901, Jun 2003.

[9] A. M. Fraser and H. L. Swinney. Independent coordinates for strange attractors from mutual information. *Phys Rev A*, 33:1134–1140, 1986.

[10] P. Grassberger. Generalized dimensions of strange attractors. *Phys. Lett*, 97A:227–230, 1983.

[11] L. D. Iasemidis, J. C. Principe, and J. C. Sackellares. Measurement and quantification of spatiotemporal dynamics of human epileptic seizures. In M. Akay, editor, *Nonlinear biomedical signal processing*, pages 294–318. Wiley–IEEE Press, vol. II, 2000.

[12] L. D. Iasemidis, D.-S. Shiau, W. A. Chaovalitwongse, J. C. Sackellares, P. M. Pardalos, P. R. Carney, J. C. Principe, A. Prasad, B. Veeramani, and K. Tsakalis. Adaptive epileptic seizure prediction system. *IEEE Transactions on Biomedical Engineering*, 5(5):616–627, 2003.

[13] H. H. Jasper. Mechanisms of propagation: Extracellular studies. In H.H. Jasper, A. A. Ward, and A. Pope, editors, *Basic mechanisms of the epilepsies*, pages 421–440, Boston, 1969. Little Brown.

[14] S. Ken, G. Di Gennaro, G. Giulietti, F. Sebastiano, D. De Carli, G. Garreffa, C. Colonnese, R. Passariello, J. Lotterie, and B. Maraviglia. Quantitative evaluation for brain ct/mri coregistration based on maximization of mutual information in patients with focal epilepsy investigated with subdural electrodes. *Magn Reson Imaging*, 25(6): 883–8, 2007.

[15] A. Klatchko, G. Raviv, W. R. Webber, and R. P. Lesser. Enhancing the detection of seizures with a clustering algorithm. *Electroencephalogr Clin Neurophysiol*, 106(1): 52–63, 1998.

[16] Y.-C. Lai, M. G. Frei, I. Osorio, and L. Huang. Characterization of synchrony with applications to epileptic brain signals. *Phys Rev Lett*, 98(10): 108102, 2007.

[17] C.-C. Liu, P. M. Pardalos, W. A. Chaovalitwongse, D. S. Shiau, G. Ghacibeh V. A. Yatsenko, and J. C. Sackellares. Quantitative complexity analysis in multi-channel intracranial EEG recordings from epilepsy brains. *Journal of Combinatorial Optimization*, 15(3): 276-286, 2008.

[18] A. R McIntosh and F. Gonzalez-Lima. Structural equation modeling and its applica-

tion to network analysis in functional brain imaging. *Hum Brain Mapping*, 2:2–22, 1994.

[19] R. Moddemeijer. On estimation of entropy and mutual information of continuous distributions, signal processing. *Phys Rev Lett*, 16(3):223–248, 1989.

[20] F. Mormann, T. Kreuz, C. Rieke, R. G. Andrzejak, A. Kraskov, P. David, C. E. Elger, and K. Lehnertz. On the predictability of epileptic seizures. *Journal of Clinical Neurophysiology*, 116(3):569–587, 2005.

[21] T. I. Netoff and S. J. Schiff. Decreased neuronal synchronization during experimental seizures. *Journal of Neuroscience*, 22(16):7297–307, 2002.

[22] G. J. Ortega, L. Menendez de la Prida, R. G. Sola, and J. Pastor. Synchronization clusters of interictal activity in the lateral temporal cortex of epileptic patients: Intraoperative electrocorticographic analysis. *Epilepsia*, 49(2): 269-280, 2008.

[23] P. M. Pardalos and J. Xue. The maximum clique problem. *Journal of Global Optimization*, 4:301–328, 1992.

[24] M. Le Van Quyen, J. Foucher, J. P. Lachaux, E. Rodriguez, A. Lutz, J. Martinerie, and F. J. Varela. Comparison of hilbert transform and wavelet methods for the analysis of neuronal synchrony. *Joutnal of Neuroscience Methods*, 111(2):83–98, 2001.

[25] M. Le Van Quyen, J. Martinerie, M. Baulac, and F. Varela. Anticipating epileptic seizures in real time by non-linear analysis of similarity between eeg recordings. *NeuroReport*, 10:2149–2155, 1999.

[26] R. Quian Quiroga, A. Kraskov, T. Kreuz, and P. Grassberger. Performance of different synchronization measures in real data: a case study on electroencephalographic signals. *Phys Rev E Stat Nonlin Soft Matter Phys.*, 65 (4):041903, 2002.

[27] J. C. Sackellares, D.-S. Shiau, J. C. Principe, M. C. K. Yang, L. K. Dance, W. Suharitdamrong, W. A. Chaovalitwongse, P. M. Pardalos, and L. D. Iasemidis. Predictibility analysis for an automated seizure prediction algorithm. *Journal of Clinical Neurophysiology*, 23(6):509–520, 2006.

[28] C.E. Shannon. *A mathematical theory of communication,*. University of Illionis Press, 1963.

[29] N. K. Varma, R Kushwaha, A. Beydoun, W.J. Williams, and I. Drury. Mutual information analysis and detection of interictal morphological differences in interictal epileptiform discharges of patients with partial epilepsies. *Electroencephalogr Clin Neurophysiol*, 103(4):426–33, 1997.

[30] P. Velazquez, J. L. Dominguez, and R. L Wennberg. Complex phase synchronization in epileptic seizures: evidence for a devil's staircase. *Phys Rev E Stat Nonlin Soft Matter Phys*, 75(1):011922, 2007.

[31] C. G. Wasterlain. Neonatal seizures and brain growth. *Neuropaediatrie*, 9:213-228, 1978.

Chapter 15

Relating Subjective and Objective Pharmacovigilance Association Measures

Ronald K. Pearson

ProSanos Corporation
225 Market St., Suite 502, Harrisburg, PA, USA, 17101
ronald.pearson@prosanos.com

The field of pharmacovigilance is concerned with the detection and interpretation of associations between drugs and adverse medical events that may be related to the use of those drugs. These assocations can be measured in various ways, and this paper considers five: two are aggregate statistical measures derived from an entire adverse event database, two are case-specific objective measures, and one is a subjective measure related to the way adverse events are reported. Examination of the available data suggests that these measures are all interrelated, but in a complicated manner. This finding motivates the use of cluster analysis to explore these relationships, with the ultimate objective of constructing an *index of blame* that quantifies the tendency for some drugs to be subjectively blamed for adverse events even in the absence of objective evidence for an association with those events.

15.1. Introduction

Pharmacovigilance analysis attempts to understand and quantify the relationship between drugs and adverse events that they may have caused or exacerbated. In the U.S., the primary data source for pharmacovigilance analysis is the FDA's Adverse Event Reporting System (AERS) database [19], which is organized by Individual Safety Reports (ISR's). Each ISR describes a suspected adverse drug reaction experienced by a single patient, lising the drugs that patient was taking, the adverse reactions they reported, and a limited amount of additional data (e.g., patient age, gender, reporting source information, etc.). Each ISR typically lists several drugs and several adverse events, a fact that plays an important role in the results presented here.

Since the AERS database is built from spontaneous reports, there is an element of subjectivity inherent in the decision that a particular combination of drugs and medical conditions is (or is not) suspicious enough to report. Nevertheless, objective assessments of association are possible by comparing the relative frequencies with which different drug/adverse event combinations occur in the database. These *aggregate assessments* are based on the statistical measures discussed in Sec. 15.2. In addition, the AERS database also provides *subjective* association data, described in Sec. 15.3, and the numbers of drugs listed in each ISR provide the basis for defining two *case-specific* objective association measures, discussed in Sec. 15.4. It is demonstrated with two simple examples in Sec. 15.5 that the relationships between these five association measures is strongly drug-dependent.

The question of ultimate interest is how to define an *index of blame* that describes the tendency for a drug to be blamed for adverse reactions to a significantly greater or lesser extent than is warranted on the basis of the objective association measures. As a first step toward this goal, this paper presents the results of a cluster analysis of drugs, using correlations between different association measures over a fixed set of adverse events as their primary attributes. Detailed descriptions of the clustering method used, the drugs and adverse events considered, and the results obtained are presented in Sec. 15.6, and an interpretation of these results is given in Sec. 15.7. It is seen that the drugs considered here cluster naturally into three groups: a "high-blame" group, where subjective association measures between the drug and most adverse events are high, largely independent of the objective evidence for this association; a "low-blame" group, where subjective association measures are low, again largely independent of objective evidence; and an "appropriate blame" group, where subjective and objective association measures are in reasonable agreement. This result is discussed further in the summary, given in Sec. 15.8.

15.2. Aggregate Associations

Many different algorithms have been proposed for pharmacovigilance analysis. One of the earliest was the use of the Proportional Reporting Ratio (defined in Eq. (15.3) below) to detect unusually strong drug/adverse event associations advocated by Finney in 1974 [5]. Refinements of this approach were proposed by Evans *et al.* [4] to address some of the limitations described below, recognition of which also led to the development of Bayesian alternatives like the Gamma Poisson Shrinker (GPS) algorithm and its extensions by DuMouchel and co-workers [2, 3, 6] and the Bayesian Confidence Propagation Neural Network (BCPNN) method described by van Puijenbroek *et al.* [17]. In addition, nonstatistical alternatives have been developed, like the optimization-based method that Mammadov and co-workers have applied to the Australian Adverse Drug

Reaction Advisory Committee (ADRAC) database [13–15]. The basis for the results presented in this paper is another statistical approach based on an urn model of the adverse event database that is non-iterative like the PRR method and is therefore computationally simpler than the Bayesian and optimization-based approaches just described; further, this method has been found to be effective in predicting subsequent regulatory actions on the basis of the AERS database [10].

Given a specified drug (Drug A) and adverse event (Adverse Event B) listed in a spontaneous reporting database (e.g., the FDA's AERS database considered here), all of the statistically-based methods listed above are closely related to the two-way contingency table defined by the following four numbers of records:

1. N_a = the number listing Drug A,
2. N_b = the number listing Adverse Event B,
3. N_{ab} = the number listing both Drug A and Adverse Event B, and
4. N = the total number of records in the database.

These numbers yield the following simple estimates of the probabilities p_a of observing Drug A in a randomly selected record, p_b of observing Adverse Event B, and p_{ab} of observing both together:

$$\hat{p}_a = N_a/N, \quad \hat{p}_b = N_b/N, \quad \text{and} \quad \hat{p}_{ab} = N_{ab}/N. \tag{15.1}$$

In the absence of any association between drugs and adverse events, we expect the independence condition $p_{ab} = p_a p_b$ to hold. This observation motivates the *reporting ratio:*

$$R_{ab} = \frac{\hat{p}_{ab}}{\hat{p}_a \hat{p}_b} = \frac{N N_{ab}}{N_a N_b}, \tag{15.2}$$

which has the value 1 if the empirical probability estimates defined in Eq. (15.1) satisfy this independence condition. Unfortunately, terminology is not standard in the pharmacovigilance literature, so the quantity R_{ab} defined in Eq. (15.2) is called the *proportional reporting ratio* by Gould [8] and the *nonstratified relative report rate* by DuMouchel [2]. Confusing the situation further, most authors define the proportional reporting ratio as [4, 9]:

$$P_{ab} = \frac{N_{ab}(N - N_a)}{N_a(N_b - N_{ab})}. \tag{15.3}$$

The reporting ratio R_{ab} defined in Eq. (15.2) is considered here because it remains finite even in the limiting case where $N_{ab} = N_b$, in contrast to P_{ab} which is not well-defined for this case.

Nevertheless, it is well-known that even R_{ab} behaves badly in these limiting situations, motivating alternatives like DuMouchel's Bayesian shrinkage estimator

[2]. In particular, since $N_{ab} \leq \min\{N_a, N_b\}$, it follows that R_{ab} is bounded above by:

$$R_{ab} \leq \frac{N \cdot \min\{N_a, N_b\}}{\min\{N_a, N_b\} \cdot \max\{N_a, N_b\}} = \frac{N}{\max\{N_a, N_b\}}. \quad (15.4)$$

Further, note that this upper bound is achieved whenever N_{ab} is equal to its maximum possible value, $\min\{N_a, N_b\}$. Thus, in cases where both N_a and N_b are small, R_{ab} can become quite large. This situation commonly arises in practice when an unusual drug name (e.g., a misspelling) occurs only once in the database (implying $N_{ab} = N_a = 1$), in a record that also lists an unusual outcome (implying $N_b << N$). As a specific example, one "drug" listed in the portion of the AERS database considered here is "unspecified weed killer," which appears only once, in a record that lists the rare adverse event "murder." Since this adverse event appears only $N_b = 147$ times in $N = 462,936$ records, this combination has the huge reporting ratio value $R_{ab} = N/N_b \simeq 3149$.

Several approaches have been proposed to overcome this difficulty. One is the Bayesian shrinkage estimator of DuMouchel mentioned earlier [2]. Another is the use of the proportional reporting ratio P_{ab} with an associated χ^2 significance measure and minimum N_{ab} limits to down-weight small samples [4, 9]. A different approach is taken here [10], based on the reporting ratio R_{ab} and the *statistical unexpectedness* U_{ab}, defined as follows. Model the adverse event dataset as an urn of N balls, with the N_a balls corresponding to records that list Drug A colored black and the others colored white. In the absence of any association between Drug A and Adverse Event B, the records listing Adverse Event B may be viewed as a random sample of N_b balls drawn from the urn, of which N_{ab} are black. It is a standard result that this number should follow the hypergeometric probability distribution [11, Ch. 6]. If $R_{ab} > 1$, then N_{ab} is larger than the value expected under this independence assumption, and the significance of this difference can be assessed by computing the probability of observing a value as large as or larger than N_{ab} from the hypergeometric distribution defined by N, N_a and N_b. Conversely, if $R_{ab} < 1$, then N_{ab} is smaller than expected and the significance of this difference can be assessed by computing the probability of observing a value as small as or smaller than N_{ab}. The corresponding probability ρ_{ab} is easily computed via standard routines for the cumulative hypergeometric distribution, and the *statistical unexpectedness* is defined as the reciprocal of this probability, $u_{ab} = 1/\rho_{ab}$.

The reason for using u_{ab} instead of ρ_{ab} is that large u_{ab} values merit our attention, which is advantageous in the graphical display used here. Also, since the u_{ab} values span a wide numerical range, it is convenient to work instead with

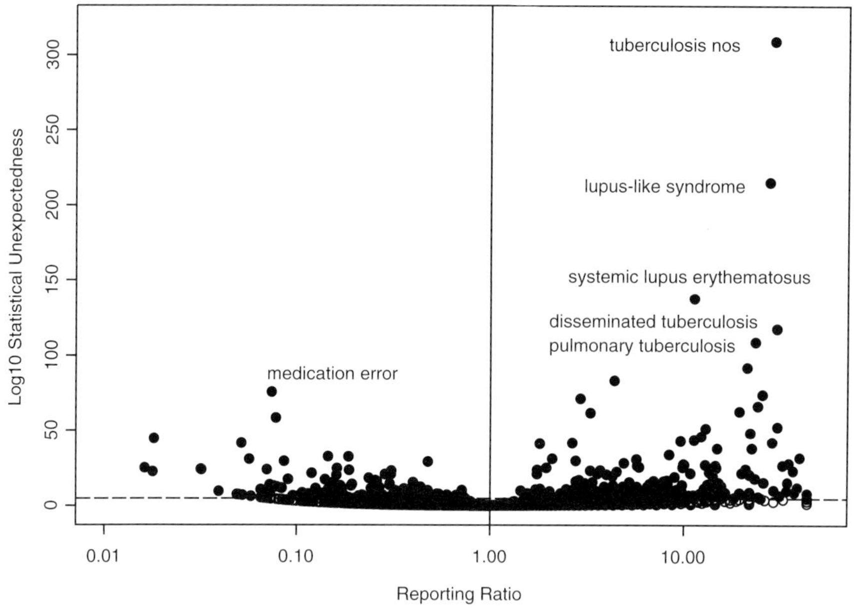

Fig. 15.1. Pharmacovigilance map for the drug infliximab, obtained by plotting U_{ab} vs. R_{ab} for each adverse event appearing in one or more records that list infliximab. The vertical line corresponds to $R_{ab} = 1$ and separates those adverse events showing a positive association with infliximab (points to the right of the line) from those showing a negative association (points to the left of the line). The dashed horizontal line corresponds to the Bonferroni significance limit discussed in the text.

the logarithm:

$$U_{ab} = \log_{10} u_{ab} = -\log_{10} \rho_{ab}. \tag{15.5}$$

Fig. 15.1 shows a plot of U_{ab} vs. R_{ab} for the drug infliximab and all adverse events that appear in AERS records listing this drug, plotted with a logarithmic scale on the horizontal axis. Each point represents a specific adverse event, and those lying in the upper right portion of the plot (large values of both U_{ab} and R_{ab}) exhibit a strong positive association with the drug (i.e., these adverse events appear significantly more often than expected by random chance alone). Conversely, points with $R_{ab} < 1$ correspond to adverse events that appear *less often* than expected by random chance alone, so points lying in the upper left portion of the plot exhibit strong negative associations with the drug infliximab.

Heeley *et al.* [9] display their results somewhat similarly, as a plot of the χ^2 statistic for each drug/adverse event combination against its associated P_{ab} value. An important extension included here is a simple correction for multiple comparison, based on the Bonferroni approximation [21]. That is, the uncorrected threshold for a single adverse event to be significant at the level α is $\rho_{ab} < \alpha$, corresponding to $U_{ab} > -\log_{10}\alpha$. In plots like Fig. 15.1, however, the question posed is which of M adverse events is significant, where M is typically much larger than 1 (in Fig. 15.1, $M = 3702$). The Bonferroni correction replaces α with α/M, increasing the significance threshold to $U_{ab} > log_{10}(M/\alpha)$. While it is known that the Bonferroni correction is conservative [21], experience has shown it to be extremely useful in focusing our attention on drug/adverse event combinations of practical importance [10].

15.3. Subjective Associations

The statistical association measures R_{ab} and U_{ab} are objective measures, based on the ensemble of drug/adverse event records available in spontaneous reporting repositories like the FDA's AERS database considered here. This particular database also includes subjective data on the association between each drug listed in an ISR and the adverse events listed with it. That is, every drug entry is classified into one of four categories: *primary suspect* (PS), *secondary suspect* (SS), *interacting* (I), or *concomitant* (C). Less than 0.1% of the drug records considered here are classified as interacting, with approximately 27% classified as primary suspect, approximately 14% as secondary suspect, and the remainder (58.3%) classified as concomitant. This classification is extremely inhomogoenous across the different drugs, however, suggesting the following numerical measure of subjective association between Drug A and Adverse Event B. For each ISR listing the drug and the adverse event, assign a *suspect classification* of 1 if the drug is classified as PS, SS, or I, and 0 if it is classified as C. The *suspect fraction* S_{ab} is then defined as the fraction of ISR's listing both Drug A and Adverse Event B that have suspect classifications of 1.

Fig. 15.2 presents a modified version of the pharmacovigilance map for the drug infliximab shown in Fig. 15.1, where the sizes of the points have been made proportional to S_{ab}. For comparison, the corresponding plot for the drug fluoxetine is shown in Fig. 15.3, where the point sizes are again proportional to S_{ab}. In both cases, points falling below the Bonferroni-corrected 5% significance threshold have been omitted to make the plots easier to interpret. Comparing these two plots, it is clear that S_{ab} is large for almost all adverse events associated with infliximab, even those adverse events exhibiting a negative objective association with the drug (i.e., adverse events with $R_{ab} < 1$). In contrast, the S_{ab} values for the drug fluoxetine exhibit a much wider range of variation across the adverse events than those for infliximab do. Also, note that the S_{ab} values for adverse events with $R_{ab} < 1$ are generally smaller than those for $R_{ab} > 1$. Both of these

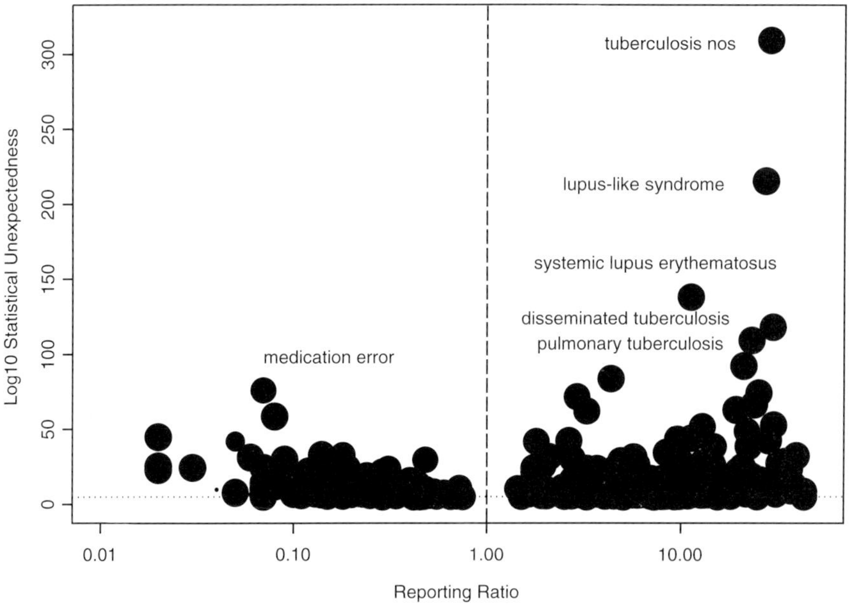

Fig. 15.2. Adverse event plot for the drug infliximab with the point sizes proportional to the suspect fraction S_{ab}.

observations suggest a greater agreement between the objective and subjective measures of association for fluoxetine than for infliximab.

15.4. Case-Specific Associations

The objective association measures R_{ab} and U_{ab} defined and discussed in Sec. 15.2 are aggregate measures, based on the entire dataset. In addition, it is possible to define two other *case-specific association measures* that are also objective measures of the relationship between Drug A and Adverse Event B, but which are based only on records lising the drug and the adverse event together. As noted in Sec. 15.1, a typical ISR lists several drugs and several adverse events, and in these ISR's, one drug is always specified as primary suspect. In ISR's that list only a single drug, a situation called a *pure play* in the pharmacovigilance community, that drug is necessarily the primary suspect. This observation motivates the following definition: the *pure play fraction* ψ_{ab} is the fraction of ISR's listing both Drug A and Adverse Event B that are pure plays (i.e., that list Drug A alone). It follows from this definition that $\psi_{ab} \leq S_{ab}$, establishing a relationship between this objective association measure and the subjective association measure discussed

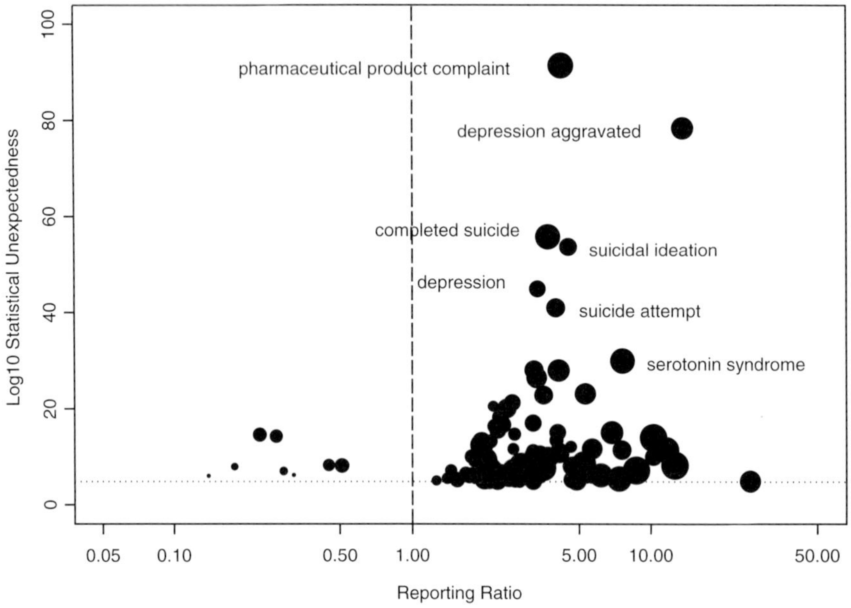

Fig. 15.3. Adverse event plot for the drug fluoxetine with the point sizes proportional to the suspect fraction S_{ab}.

in Sec. 15.3. The other case-specific association measure considered here is the *mean number of concomitant medications*, μ_{ab}, defined as the average number of other drugs appearing along with Drug A in all ISR's listing both Drug A and Adverse Event B. Note that if all of these ISR's are pure plays, it follows that $\mu_{ab} = 0$ and that $\psi_{ab} = S_{ab} = 1$; in fact, it is not difficult to show that $\mu_{ab} = 0$ if and only if $\psi_{ab} = S_{ab} = 1$.

15.5. Relations between Measures

The results just presented lead us to expect, for fixed Drug A, a positive correlation between S_{ab} and ψ_{ab} and negative correlations between μ_{ab} and both S_{ab} and ψ_{ab} as the Adverse Events B vary. Conversely, the pronounced differences in appearance between Figs. 15.2 and 15.3 suggest the possibility of poor agreement between the subjective association measure S_{ab} and the aggregate measures R_{ab} and U_{ab}, at least for some drugs (e.g., infliximab). Support for these suggestions is given in Figs. 15.4 and 15.5, which illustrate the relationships seen between two of these pairs of association measures for two different drugs.

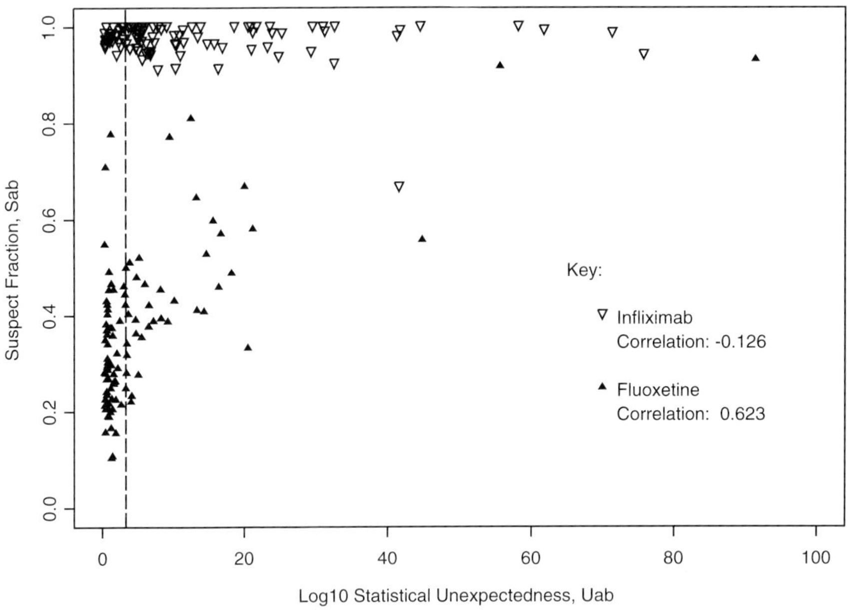

Fig. 15.4. Scatterplot of S_{ab} vs. U_{ab} values for two drugs: infliximab (open triangles) and fluoxetine (solid triangles).

Specifically, Fig. 15.4 shows the suspect fraction S_{ab} plotted against the statistical unexpectedness U_{ab} for two drugs and the 100 adverse events described in Sec. 15.6.1. The points represented as open triangles were obtained for the drug infliximab, while the points represented as solid triangles were obtained for the drug fluoxetine. The correlations between these two variables are listed in the lower right of this figure for each drug, and their difference is pronounced: the correlation for infliximab is -0.126, while that for fluoxetine is $+0.623$. The vertical dashed line at the left end of the plot represents the Bonferroni-corrected 5% significance limit, so only those points to the right of this line are deemed significant. If we compute the correlations solely from these significant adverse events, they become -0.107 for infliximab and $+0.690$ for fluoxetine. While these results differ slightly in detail, the are qualitatively the same: the degree of agreement between the subjective association measure S_{ab} and the objective association measure U_{ab} is much better for fluoxetine than it is for infliximab.

The results shown in Fig. 15.5 demonstrate that this conclusion also holds for the degree of agreement between the subjective association measure S_{ab} and the case-specific objective measure μ_{ab}. As noted at the beginning of this section, we expect a negative correlation between these variables, and this is observed for the

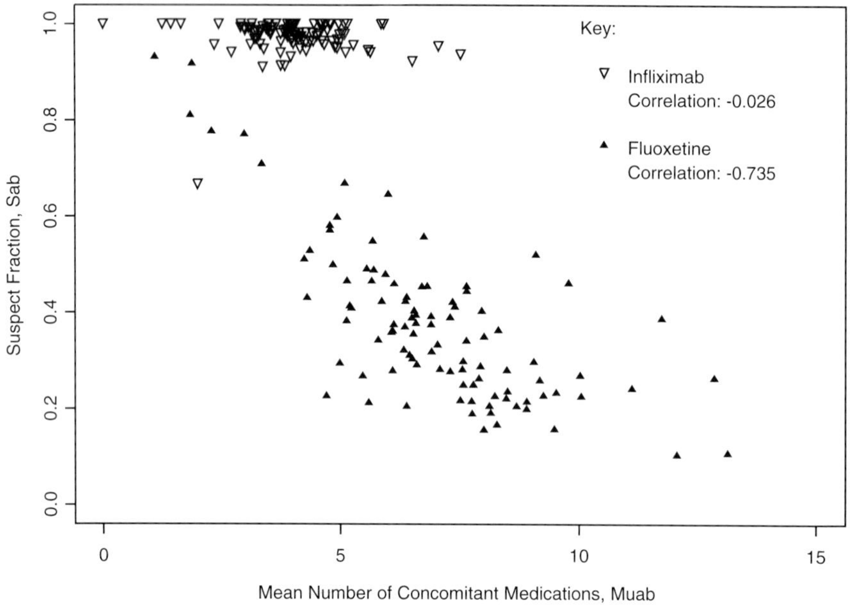

Fig. 15.5. Scatterplot of S_{ab} vs. μ_{ab} values for two drugs: infliximab (open triangles) and fluoxetine (solid triangles).

drug fluoxetine, where the scatter plot between S_{ab} and μ_{ab} is represented by solid triangles. The appearance of this scatter plot is consistent with the large negative correlation of -0.735 computed for this drug. The corresponding scatter plot for the drug infliximab is represented as open triangles and the apparent lack of dependence of S_{ab} on μ_{ab} seen in this plot is consistent with the very small negative correlation of -0.026 computed for this drug. These results suggest that, like the relationship between the subjective measure S_{ab} and the aggregate measures U_{ab} and R_{ab}, the relationship between S_{ab} and the case-specific objective measures ψ_{ab} and μ_{ab} are also strongly drug-dependent.

15.6. Clustering Drugs

The observations just presented motivate the following clustering case study. The goal is to cluster drugs on the basis of the relationships between the subjective and objective association measures defined and discussed in the preceeding sections of this paper. Sec. 15.6.1 describes how the drugs and adverse events were selected for this study, and Sec. 15.6.2 describes the clustering method and validation approach used to obtain clusters, assess their reasonableness, and determine the number of clusters present in the data. A brief summary of the results is given in Sec. 15.6.3, and an interpretation is presented in Sec. 15.7.

15.6.1. *The Case Study*

The AERS database is a large repository, updated quarterly, and the portion considered here corresponds to the 12 releases for the four quarters of the years 2001, 2002, and 2003, summarizing 462,936 adverse event reports. Altogether, these reports list 12,015 unique adverse reactions and 117,287 unique drug name entries, so it was not feasible to analyze the entire database. To obtain a representative subset, the following strategy was adopted. First, the adverse events were ranked in descending order of reporting frequency and the 100 most frequently reported adverse events were used as the basis for all of the drug/adverse event associations considered here. The top 25 of these adverse events are listed in Table 15.1, along with their frequencies (the term "nos" appearing in this table is an abbreviation for "not otherwise specified"). Altogether, these 100 terms account for 36.3% of the total reported adverse events in this portion of the AERS database, despite the fact that they constitute less than 1% of the distinct adverse events reported. In fact, the distribution of these events is extremely heavy-tailed, with over half (56.0%)

Table 15.1. The top 25 suspected adverse drug reactions and their frequencies of ocurrance in the AERS database.

No.	Adverse Reaction	Frequency
1	nausea	19428
2	drug ineffective	15817
3	pyrexia	14387
4	vomiting nos	12096
5	fatigue	10677
6	medication error	9970
7	condition aggravated	9941
8	diarrhoea nos	9385
9	dyspnoea nos	9313
10	pain nos	8757
11	asthenia	8368
12	abdominal pain nos	8332
13	chest pain	8323
14	rhabdomyolysis	7783
15	myalgia	7728
16	headache nos	7463
17	malaise	6907
18	myocardial infarction	6853
19	convulsions nos	6623
20	dizziness	6547
21	dyspnoea	6522
22	headache	6398
23	arthralgia	6358
24	constipation	6143
25	thrombocytopenia	6099

of the distinct terms appearing 10 times or less in the database, and almost three fourths (72.3%) appearing 30 times or less. These observations motivate the use of the top 100 adverse events in the case study presented here, all of which appear more than 3,000 times in the AERS database.

To obtain a collection of drugs that was representative of those appearing in the AERS database, a random sample of 50 text strings was drawn from the DRUG-NAME field of the AERS drug data files. Of these, 2 were exact duplicates of other drug names in the sample, and 12 others were omitted because they were either different names for the same drug (e.g., "quinapril" and "accupril"), dietary supplements (e.g., "multi-vitamins"), or otherwise un-interesting (e.g., "Eckerds brand lice shampoos"). The remaining 36 drugs were:

> accupril, aldioxa, aminocaproic acid, aspirin, atenolol, azasetron, bactrim, candesartan, carbamazepine, cisplatin, clonidine, clopidrogel, clozapine, coumadin, depamide, dipiperon, etanercept, fluoxetine, hydrochlorothiazide, ibuprofen, insulin, levodopa, lorazepam, luprolide, metoprolol, mianserin, morphine, mycophenolate, omeprazole, oxaprozin, oxycodone, paracetamol, pravastatin, prednisone, rofecoxib, and tamsulosin.

In addition to these 36 drugs, the following 15 were included because they were of independent interest:

> atorvastatin, bosentan, ciprofloxacin, cocaine, dutasteride, galantamine, gatifloxacin, infliximab, levofloxacin, lovastatin, moxifloxacin, simvastatin, tenofovir, valdecoxib, and zolpidem.

The association measures R_{ab}, U_{ab}, S_{ab}, ψ_{ab} and μ_{ab} were computed for each of these 51 drugs with each of the 100 most frequently ocurring adverse reactions in the AERS database.

Evaluating the five association measures considered here for each adverse event yields 500 numbers for each drug. Since the main question of interest is how these different association measures are related, the primary clustering variables used are the 10 correlations between each distinct pair of association measures as they vary over the 100 adverse events. In addition, since the subjective measure S_{ab} is of particular interest, the mean S_{ab} value over the adverse events is also included in the clustering variable set. The complete list of the eleven attribute variables used for clustering drugs is given in Table 15.2.

15.6.2. *The Clustering Approach*

The 51 drugs considered here were clustered using the eleven variables listed in Table 15.2 and the Partitioning Around Medoids (PAM) method described

Table 15.2. The eleven attribute variables used to cluster the 51 drugs considered in this study.

Variable	Definition
V_1	Mean S_{ab} value
V_2	Correlation between S_{ab} and R_{ab}
V_3	Correlation between S_{ab} and U_{ab}
V_4	Correlation between S_{ab} and ψ_{ab}
V_5	Correlation between S_{ab} and μ_{ab}
V_6	Correlation between R_{ab} and U_{ab}
V_7	Correlation between R_{ab} and ψ_{ab}
V_8	Correlation between R_{ab} and μ_{ab}
V_9	Correlation between U_{ab} and ψ_{ab}
V_{10}	Correlation between U_{ab} and μ_{ab}
V_{11}	Correlation between ψ_{ab} and μ_{ab}

by Kaufman and Rousseeuw [12], as implemented in the *S-plus* software package [18]. Given a matrix of M attribute values for each of N objects, this procedure partitions the objects into k clusters, where the number k is specified by the user. This procedure was chosen in large part because the results do not depend on the ordering of the data objects, in contrast to the more popular k-means procedure [12, p. 114]. The results presented here are based on the Manhattan dissimilarity measure; although a thorough comparison was not undertaken, comparable results were obtained with the more popular Euclidean dissimilarity measure for selected cases.

The number of clusters k present in the data was determined using the method described by Pearson *et al.* [16], based on two fundamental notions. The first is a measure of cluster quality that can be computed from any given data partitioning; many such measures have been described [1], but one that has been found to be particularly useful in this application [16] is the *silhouette coefficient* [12]. The second fundamental notion is the use of random permutations to destroy any structure that may be present in the data. This provides the basis for assessing the significance of a putative clustering: if the cluster structure is real, it should be largely destroyed by the random permutations, reducing the cluster quality substantially.

Combining these ideas leads to the following cluster validation procedure. First, the k-partition clusterings are obtained from the original dataset, for a range of k values. Next, independent random permutations are applied to each column of the attribute matrix and the same clustering procedure is applied to this modified dataset, to obtain k-partition clusterings over the same range of k values. This permutation procedure is repeated m times (in the results presented here, $m =$ 100) to obtain clusterings for m indpendently randomized datasets, for each value of k considered. Quantitative cluster quality measures are then computed for each clustering and the results obtained for the original dataset are compared, for each value of k, with the quality measures computed from the m randomizations. A good clustering is one whose quality measure lies significantly above the range of the quality measures computed from the randomized results, for the same value of k.

The silhouette coefficient used here as a measure of cluster quality is based on the idea that a good clustering should consist of *cohesive, well-separated* clusters. Given a partitioning $\mathcal{P}$ of N objects into k clusters, consider any fixed object i and let C_i denote the set of indices for all objects clustered together with object i. A useful measure of cohesion for this cluster is:

$$a(i) = \frac{1}{n_i} \sum_{j \in C_i} d_{ij}, \tag{15.6}$$

where n_i is the number of objects in cluster C_i and d_{ij} is the dissimilarity between objects i and j. Note that for a good clustering, $a(i)$ should be small for all i. To characterize the separation between clusters, let K_ℓ denote the ℓth neighboring cluster, distinct from C_i, for $\ell = 1, 2, \ldots, k-1$. Define $b(i)$ as the average dissimilarity between object i in cluster C_i and the objects in the closest neighboring cluster, given by:

$$b(i) = \min_{\ell} \left\{ \frac{1}{n_\ell} \sum_{j \in K_\ell} d_{ij} \right\}. \tag{15.7}$$

Here, for a good clustering, $b(i)$ should be large for all i. The silhouette coefficient $s(i)$ for object i is then defined as the following normalized difference between these two quantities:

$$s(i) = \frac{b(i) - a(i)}{\max\{a(i), b(i)\}}. \tag{15.8}$$

For $s(i)$ to be well-defined, the partitioning $\mathcal{P}$ must contain at least two clusters, and every cluster must contain at least two objects. Under these conditions, it is easily shown that $-1 \leq s(i) \leq 1$ for all i. From the previous observations, a good clustering should have $a(i) << b(i)$ for all i, implying $s(i) \simeq 1$ for all i. A useful measure of the overall quality of the partitioning $\mathcal{P}$ is therefore the average silhouette coefficient over all objects:

$$S = \frac{1}{N} \sum_{i=1}^{N} s(i). \tag{15.9}$$

Given this cluster quality measure, let $\mathcal{D}_0$ denote the original dataset and let $S_0(k)$ denote the value of S computed for the k-cluster partitioning of $\mathcal{D}_0$ obtained by a fixed clustering method $\mathcal{M}$. Let $\mathcal{D}_i$ denote the ith random permutation of $\mathcal{D}_0$, obtained as described above, and let $S_i(k)$ denote the value of S computed for the k-cluster partitioning of $\mathcal{D}_i$, obtained by the same clustering method $\mathcal{M}$. The working hypothesis here is that, if $\mathcal{D}_0$ exhibits k well-defined clusters, $S_0(k)$ should be significantly larger than $S_i(k)$ for all random permutations. One measure of significance is the (one-sided) empirical probability: if $S_0(k)$ is larger than all but $q-1$ of the m permutation values $S_i(k)$, the empirical probability of observing $S_0(k)$ by random chance is less than q/m. The other measure of significance used here is the z-score, computed from the mean $\bar{S}_k$ and standard deviation $\sigma_S(k)$ of the random permutation values $\{S_i\}$:

$$z_k = \frac{S_0(k) - \bar{S}_k}{\sigma_S(k)}. \tag{15.10}$$

Experience with simulation datasets having known cluster structures has shown that correct clusterings generally lead to significant results with respect to the empirical probability estimates (e.g., $S_0(k)$ exceeds all of the randomized values $S_i(k)$) and maximal or near maximal with respect to both z_k and $S_0(k)$ over the range of k considered [16]. Experience has also shown that in cases where no cluster structure exists (e.g., simulation datasets constructed from statistically independent random data vectors), none of the $S_0(k)$ values typically meet these significance criteria.

15.6.3. *Summary of the Results*

The clustering procedure described in Sec. 15.6.2 could be applied directly to the attribute matrix defined by the 11 variables listed in Table 15.2, but it has been shown using simulated datasets that the inclusion of extraneous variables can degrade clustering results badly [7, 16]. Thus, the approach taken here starts with the smallest interesting subset of these variables and proceeds in a manner analogous to stepwise regression, including each variable only if it improves the clustering. Since the original question motivating this work was the nature of the relationship between the subjective association measure S_{ab} and the objective measure U_{ab}, the smallest subset considered here includes variables V_1 (the mean S_{ab} value over the 100 adverse events considered) and V_3 (the correlation between S_{ab} and U_{ab}).

Fig. 15.6 shows the k-cluster partition results for k from 2 through 10, computed from this minimal variable set. The solid circles correspond to $S_0(k)$ and the boxplots describe the range of $m = 100$ random permutation results $\{S_i(k)\}$. Since none of these $S_0(k)$ values fall outside the ranges of the random permuta-

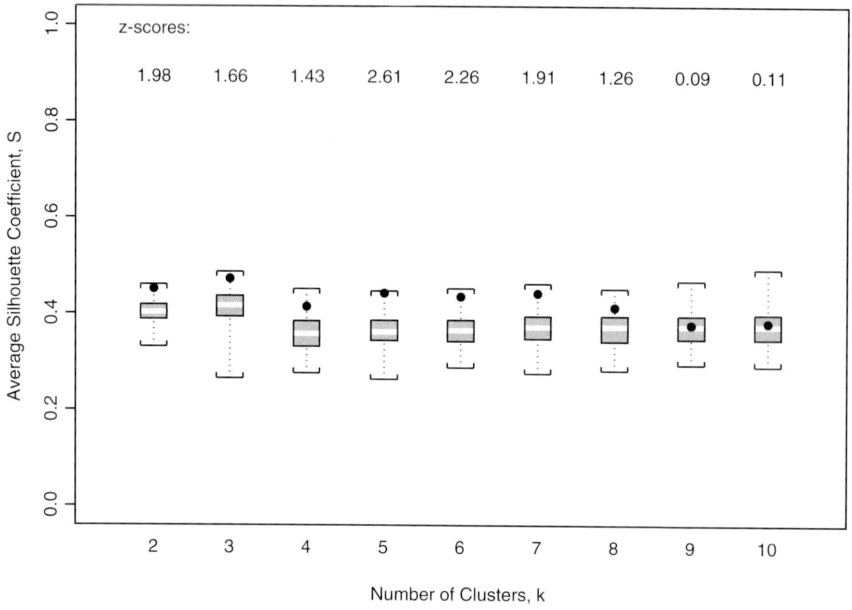

Fig. 15.6. Average silhouette coefficients vs. cluster size k computed from variables V_1 and V_3 (solid circles), together with boxplots of the ranges of average silhouette coefficients obtained for 100 independent random permutations of V_3.

tion values, it does not appear that the variables V_1 and V_3 alone provide the basis for a useful clustering of the data.

Including each of the nine remaining variables one by one, the best results are obtained when V_2 is added to the variable set. These results are shown in Fig. 15.7, which shows that both $S_0(3)$ and $S_0(4)$ exceed all of their associated random permutation values. Of these two results, the three-cluster partitioning exhibits both the larger silhouette coefficient and the larger z-score, so it is taken here as the *basic clustering*, which will serve as a reference case for all other clusterings of this dataset.

The results obtained by adding one variable at a time are summarized in Table 15.3, including the cases illustrated in Figs. 15.6 and 15.7. The optimum number of clusters is $k^* = 3$ in all cases, with the possible exception of the five-variable clustering on V_1, V_2, V_3, V_7, and V_9, where $k^* = 4$ achieves a slightly larger silhouette coefficient but a slightly smaller z-score. Detailed descriptions of these clusterings and their differences are given in Sec. 15.7, but two points are worth noting. First, the silhouette coefficient values S^* for $k = 3$ decrease monotonically as additional variables are included, while the associated z-scores

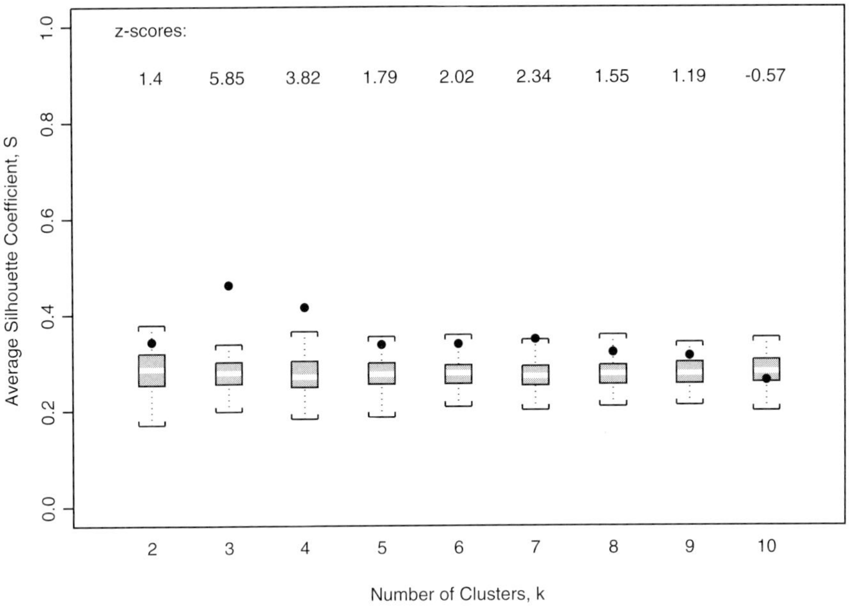

Fig. 15.7. Average silhouette coefficients vs. cluster size k computed from variables V_1, V_2 and V_3 (solid circles), together with boxplots of the ranges of average silhouette coefficients obtained for 100 independent random permutations of V_2 and V_3.

Table 15.3. Summary of the stepwise clustering results from Sec. 15.6.1, giving the variables included, the number of clusters k^*, the average silhouette coefficients S^*, and the z-scores, z^*.

Variables	k^*	S^*	z^*
V_1, V_3	–	–	–
V_1, V_2, V_3	3	0.461	5.85
V_1, V_2, V_3, V_9	3	0.438	7.69
V_1, V_2, V_3, V_7, V_9	3	0.389	8.81
	4	0.399	8.65
$V_1, V_2, V_3, V_7, V_9, V_{10}$	3	0.365	9.54

increase monotonically. In the example illustrating the influence of extraneous variables discussed by Pearson *et al.* [16], the silhouette coefficient values also decreased as additional — there extraneous — variables were added, but there the corresponding z-scores decreased there rather than increased, as here. The second point to note is the consistent optimality or near-optimality of $k^* = 3$ over all of the different variable subsets considered, suggesting stability but raising the question of how the actual clusterings compare, the point considered next.

15.7. Interpreting the Clusters

Examination of the drug classifications in the clustering results described in Sec. 15.6.3 shows that the partitions are all quite similar. Consequently, the following discussion first presents a detailed description of the basic clustering obtained for $k^* = 3$ from the three variables V_1, V_2, and V_3, and then briefly discusses how the results obtained from the other variable subsets differ from this reference case. The first cluster in the basic clustering contains the following 21 drugs:

> accupril, aminocaproic acid, aspirin, atenolol, azesetron, bactrim, candesartan, ciprofloxacin, clonadine, clopidrogel, dipiperon, hydrochlorothiazide, insulin, levodopa, lorazepam, metroprolol, mianserin, omeprazole, oxaprozin, prednisone, and zolpidem.

The general characteristics of this cluster are a low mean S_{ab} value (the median V_1 value is 0.161), and low correlations between S_{ab} and both R_{ab} (median correlation 0.048) and U_{ab} (median correlation 0.131). Informally, this cluster represents a collection of drugs that tend to have "low subjective suspicion," largely independent of objective evidence. The second partition in the basic clustering contains the following 13 drugs:

> aldioxa, atorvastatin, coumadin, depamide, fluoxetine, ibuprofen, lovastatin, morphine, oxycodone, paracetamol, pravastatin, simvastatin, and tamsulosin.

The defining characteristics of this cluster are moderate mean S_{ab} values (the median V_1 value is 0.301) and substantial correlations between S_{ab} and both R_{ab} (median correlation 0.426) and U_{ab} (median correlation 0.525). Informally, this cluster represents a collection of drugs whose subjective suspicion level seems to correlate reasonably well with objective evidence. Finally, the third partition in the basic clustering contains the 17 drugs:

> bosentan, carbamazepine, cisplatin, clozapine, cocaine, dutasteride, etanercept, galantamine, gatifloxacin, infliximab, levofloxacin, luprolide, moxifloxacin, mycophenolate, rofecoxib, tenofovir, and valdecoxib.

This cluster is characterized by a large mean S_{ab} value (the median V_1 value is 0.678), a moderate correlation between S_{ab} and R_{ab} (median correlation 0.353), and a low correlation between S_{ab} and U_{ab} (median correlation 0.079). Informally, the drugs in this cluster exhibit high subjective suspicion, largely independent of objective evidence.

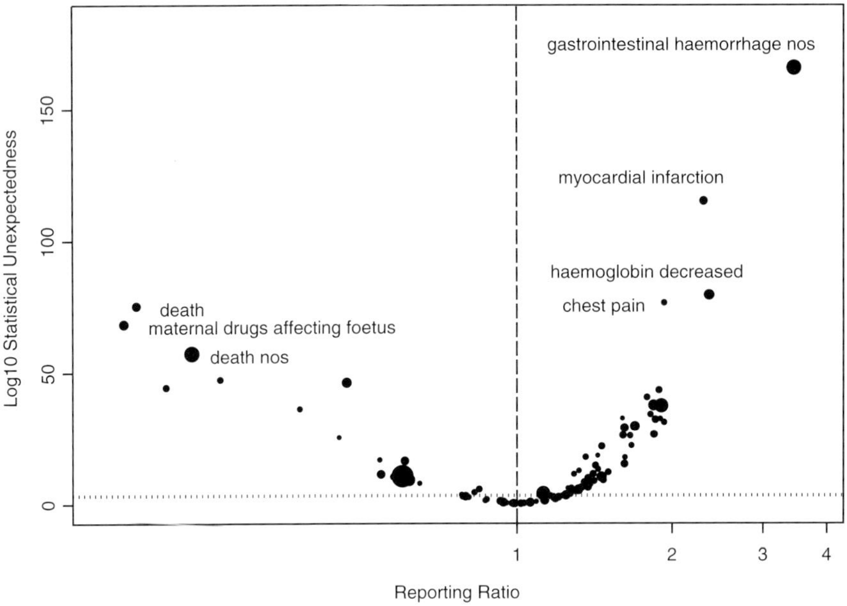

Fig. 15.8. Adverse event plot for the drug aspirin, based on the 100 most frequently occurring adverse events, with the point sizes proportional to the suspect fraction S_{ab}.

It is instructive to compare the appearance of the adverse event plots obtained for drugs from each of these groups with point sizes proportional to the subjective association measure S_{ab}. Fig. 15.8 presents this result for the drug aspirin from cluster no. 1 (the "low-blame group"). This plot should be compared with Figs. 15.2 and 15.3 discussed earlier, for the drugs infliximab (from cluster no. 3, the "high-blame group") and fluoxetine (from cluster no. 2, the "appropriate blame group"), respectively. It is clear from these plots that the points are uniformly large for infliximab, indicating a large S_{ab} value, mostly independent of U_{ab} and R_{ab}, and generally small for aspirin, indicating a small S_{ab} value, again mostly independent of U_{ab} and R_{ab}. In contrast, the point sizes vary significantly for fluoxetine, indicating that S_{ab} varies significantly with U_{ab} and R_{ab} for this drug.

The informal verbal descriptions given above for each of the three partitions in the basic clustering are represented graphically in Fig. 15.9, which gives side-by-side boxplot summaries by cluster for each of three variables on which this clustering is based. The left-most three boxplots show how the mean S_{ab} value varies for each cluster, demonstrating that this value is consistently larger for cluster 3 than for either of the other two clusters. Also, while the total range of S_{ab} values is essentially identical for clusters 1 and 2, the middle 50% of these values

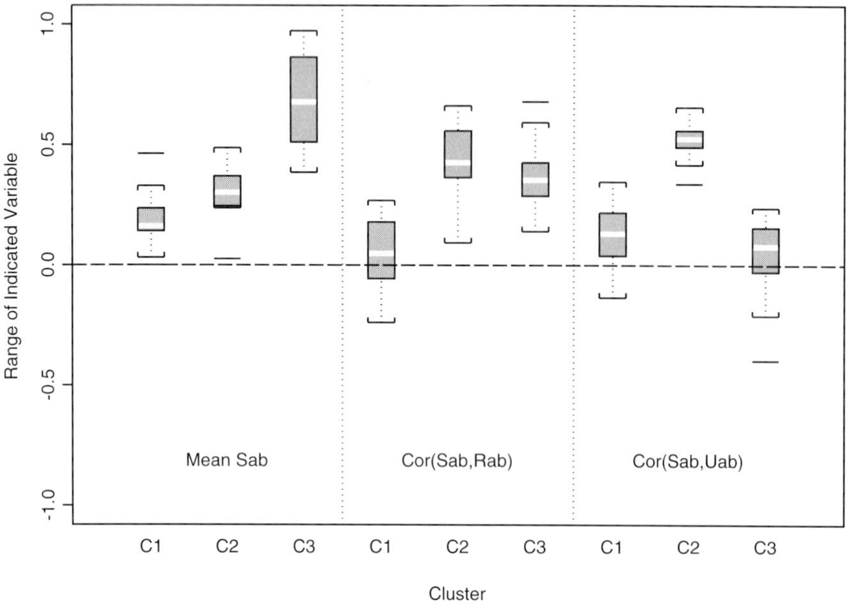

Fig. 15.9. Boxplot summaries of the variables V_1, V_2, and V_3 by cluster for partitioning with $k = 3$ obtained from these three variables.

is uniformly larger for cluster 2 than for cluster 1. Probably the main distingiuishing feature of cluster 2 is the range of its correlation values between S_{ab} and U_{ab}, which is consistently larger for cluster 2 than for either of the other clusters.

Table 15.4 summarizes the differences between the basic clustering just described and the other three-element partitions obtained using additional variables. It is clear that these differences are fairly minor, involving between three and five drugs. In all cases, the drugs aldioxa, mycophenolate, and zolpidem are placed in different clusters, which may be viewed as a persistent consequence of including variable V_9 (the correlation between U_{ab} and ψ_{ab}). In contrast, the changes in cluster assignment for dutasteride and tamsulosin caused by adding the variable V_7 (the correlation between R_{ab} and ψ_{ab}) may be regarded as transient, as they do not persist on the subsequent addition of variable V_{10} (the correlation between U_{ab} and μ_{ab}).

As noted in Sec. 15.6.3, in the results based on variables V_1, V_2, V_3, V_7, and V_9, the clustering for $k = 4$ exhibits a slightly larger mean silhouette coefficient than that for $k = 3$ ($S_4^* = 0.399$ vs. $S_3^* = 0.389$) and a slightly smaller z-score ($z_4^* = 8.65$ vs. $z_3^* = 8.81$). Thus, it is interesting to examine the differences between these two clustering assignments, which are summarized in Table 15.5,

Table 15.4. Changes in cluster assignment caused by the inclusion of additional variables, relative to the basic clustering that uses variables V_1, V_2, and V_3.

Variables Added	Drug	Basic Cluster	New Cluster
V_9	aldioxa	2	1
	mycophenolate	3	1
	zolpidem	1	2
V_7, V_9	aldioxa	2	1
	dutasteride	3	2
	mycophenolate	3	1
	tamsulosin	2	1
	zolpidem	1	2
V_7, V_9, V_{10}	aldioxa	2	1
	insulin	1	2
	mycophenolate	3	1
	zolpidem	1	2

Table 15.5. Cluster assignments for seven drugs in the basic clustering (the clustering for $k = 3$ based on variables V_1, V_2, and V_3), along with those for $k = 3$ and $k = 4$ based on variables V_1, V_2, V_3, V_7 and V_9.

Drug	Basic	$k = 3$	$k = 4$
aldioxa	2	1	2
dutasteride	3	2	3
fluoxetine	2	2	4
mycophenolate	3	1	1
oxycodone	2	2	4
tamsulosin	2	1	2
zolpidem	1	2	2

along with the corresponding assignments in the basic clustering. Five drugs exhibit different assignments between the five-variable clusterings with $k = 3$ and $k = 4$: aldioxa, dutasteride, and tamsulosin move back into their assigned partitions in the basic clustering, while fluoxetine and oxycodone form the new cluster for $k = 4$.

Fig. 15.10 gives a boxplot summary of the results for the five variable clustering with $k = 4$, in the same format as Fig. 15.9. It is clear from this plot that the key characteristic of the fourth partition in this clustering is the high correlation between the two aggregate measures R_{ab} and U_{ab} and both the subjective measure S_{ab} and the case-specific measure ψ_{ab}. In particular, the correlation between ψ_{ab} and the two aggregate measures is markedly greater than that seen for any of the other drugs in the case study, while the correlation between S_{ab} and these aggregate measures is among the highest seen for any of these drugs. Also, note that

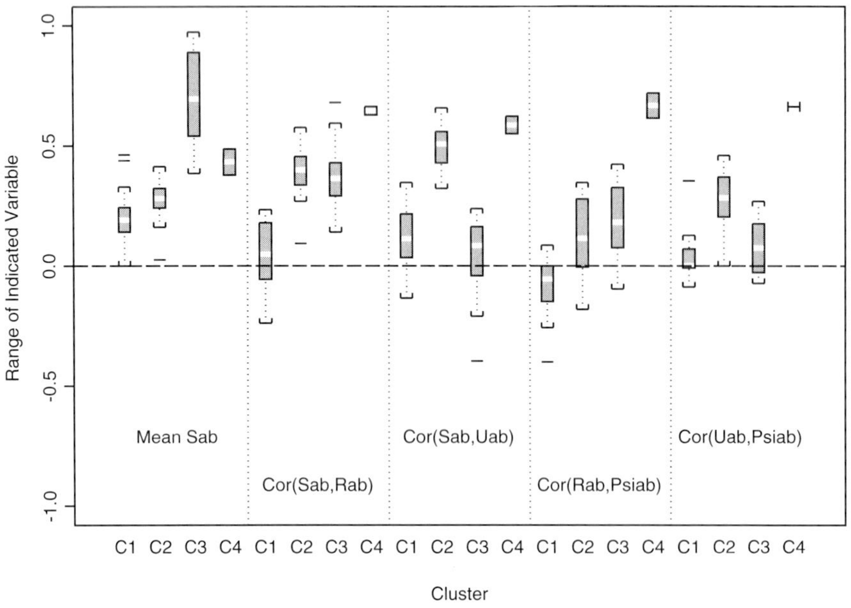

Fig. 15.10. Boxplot summaries of the variables V_1, V_2, V_3, V_7 and V_9 by cluster for the partitioning with $k = 4$ obtained from these five variables.

the two drugs forming this cluster are two of the three drugs with highest mean S_{ab} values from the second partition of the basic clustering. Thus, while the level of subjective blame associated with these two drugs is high, it is well supported by the objective evidence.

15.8. Summary

The fundamental problem of pharmacovigilance is to accurately assess the relationship between drugs and adverse medical reactions that may or may not be related to these drugs. This paper has presented a clustering study, aimed at elucidating the relationships among different objective and subjective association measures between drugs and adverse events in the FDA's Adverse Event Reporting System (AERS) database. As noted previously, the ultimate objective of this work is to define a quantitative *index of blame* that characterizes the tendency for drugs to be subjectively associated with adverse events to a greater or lesser extent than is warranted by the objective evidence. The results presented here consider a subjective association measure based on the classification of drugs as "suspect" or "concomitant" in the AERS Individual Safety Reports, along with four objective measures: two aggregate measures related to a simple urn model of the AERS database, and two case-specific measures based on the numbers of concomitant drugs listed in the database. These five association measures are described in Secs. 15.2 through 15.4, and their drug-dependent interrelationships are discussed briefly in Sec. 15.5.

The strength of this drug dependence motivated the clustering case study presented in Secs. 15.6 and 15.7, based on 36 drugs randomly selected from the AERS database and 15 others of independent interest. The five association measures noted in the previous paragraph were then computed for each drug and the 100 most frequently occurring adverse events in the AERS database. The attributes used for clustering were the correlations between the different association measures, along with the mean value of the subjective association measure. Using a permutation-based approach to determine the number of clusters [16] and a stepwise variable selection procedure, three clusters were identified: a "high-blame" cluster, characterized by a large average subjective association value but low correlations between this association measure and both of the objective aggregate association measures; a "low-blame" cluster, characterized by a small average subjective association value, again with low correlations between subjective and aggregate association measures; and an "appropriate blame" cluster, characterized by a moderate average subjective association measure but relatively high correlations between the subjective and aggregate measures.

The results presented here do not fully define the index of blame that motivated the work, but they do suggest that the approach described here goes in the right direction. For example, one reason for including the illegal drug cocaine in the study was the expectation that it would generally have a high subjective blame, representing a "positive control" for the classification; indeed, cocaine appears consistently in the high-blame group in all of the clusterings described here. Also, it is encouraging to note that all four of the statins considered (atorvastatin, lovastatin, pravastatin, and simvastatin) consistently appear together in the appropriate blame cluster. Similarly, it is not surprising that aspirin — a nonprescription drug in wide use for a very long time — belongs to the low-blame group in all clusterings. Indeed, it has recently been argued that, due to its long history of use, the adverse event profile for aspirin has often been overlooked by medical practioners, even though it is quite complicated and deserves careful clinical review [20]. More generally, it appears that the drugs in the low blame group identified here are mostly drugs like aspirin that have been widely used for a variety of conditions and have a long history of use.

At least four extensions of the results presented here suggest themselves. The first would be to expand the case study to additional association measures. In particular, one of the features of the AERS database that is not addressed in the present study is the fact that while each drug listed for an ISR has a unique subjective classification as Primary Suspect, Secondary Suspect, Interacting, or Concomitant, this designation does not indicate whether the classification refers to all of the adverse events listed for that ISR, or only a subset of these adverse events (and, if so, which subset). This point is important since, as noted, a typical ISR lists multiple drugs and multiple adverse events. Thus, one area to explore is the possibility of defining additional association measures based in part on the number of adverse events listed for an ISR, analogous to the mean number of concomitant medications μ_{ab} or the pure play fraction ψ_{ab} defined in Sec. 15.4 for drugs. The objective would be to explore the influence of the number of adverse events listed for an ISR on clustering results like those presented here. The second useful extension of the results presented in this paper would be to expand the case study to more drugs and possibly more adverse events. This expansion would begin to address the question of whether the clustering of drugs into "low blame," "appropriate blame," and "high blame" classes is adequate, or whether additional classifications become necessary as more drugs are added, possibly involving more variables. In particular, it would be interesting to see whether the four cluster solution discussed in Sec. 15.7 based on five variables increases in statistical significance on the addition of more drugs. A third possible extension would be to consider the drugs assigned to each of the three groups in the basic clustering developed here, comparing them on the basis of other possible explanatory variables like their frequency of occurrence in the AERS database, or the time since their introduction into the marketplace. The objective of this extension would be to determine whether it is possible to reliably assign drugs to different "blame clusters" on the basis of these other variables without performing the cluster analysis presented here. In particular, if it were possible to classify drugs *a priori* into these groups, this classification could be used as a basis for requiring additional evidence in classifying a "high blame drug" as suspect, or classifying a "low blame drug" as concomitant. Finally, a fourth extension of the work presented here would be to carefully examine anomalous cases like the drug ciprofloxacin: the three other drugs from the same class considered in this case study (gatifloxacin, levofloxacin, and moxifloxacin) are all assigned to the high-blame cluster, while ciprofloxacin is assigned to the low-blame cluster. One possible explanation for this difference is that ciprofloxacin is a much older drug than these other three, but this case needs to be examined further before any definitive conclusions can be drawn. Ultimately, it is hoped the work presented here and extensions like those just described will lead to an evidence-based approach to assigning drugs as suspect or concomitant, one that will yield assignments in better general agreement with objective association measures like those considered here.

References

[1] N. Bolshakova and F. Azuaje. Cluster validation techniques for genome expression data. *Signal Processing*, 83: 825–833, 2003.

[2] W. DuMouchel. Bayesian data mining in large frequency tables, with an application to the FDA spontaneous reporting system. *American Statistician*, 53: 177–190, 1999.

[3] W. DuMouchel and D. Pregibon. Empirical Bayes screening for multi item associations. *Proc. 7th ACM SIGKDD International Conf. Knowledge Discovery Data Mining*, pages 177–190, 2001.

[4] S. J. W. Evans, P. C. Waller, and S. Davis. Use of proportional reporting ratios (PRRs) for signal generation from spontaneous adverse drug reaction reports. *Pharmacoepidemiology Drug Safety*, 10: 483–486, 2001.

[5] D. J. Finney. Systemic signalling of adverse reactions to drugs. *Methods Inf. Med.*, 13: 1–10, 1974.

[6] D. M. Fram, J. S. Almenoff, and W. DuMouchel. Empirical Bayesian Data Mining for Discovering Patterns in Post-Marketing Safety Data. *Proc. SIGKDD'03*, pages 359–368, 2003.

[7] A. D. Gordon. *Classification, 2nd ed.*. Chapman and Hall, 1999.

[8] A. L. Gould. Practical pharmacovigilance analysis strategies. *Pharmacoepidemiology Drug Safety*, 12: 559–574, 2003.

[9] E. Heeley, L. V. Wilton, and S. A. W. Shakir. Automated Signal Generation in Prescription-Event Monitoring. *Drug Safety*, 25: 423–432, 2002.

[10] A. M. Hochberg, S. J. Reisinger, R. K. Pearson, D. J. O'Hara, and K. Hall. Using data mining to predict safety actions from FDA adverse event reporting system data. *Drug Information Journal*, 41: 633–643, 2007.

[11] N. L. Johnson, S. Kotz, and A. W. Kemp. *Univariate Discrete Distributions, 2nd ed.*, Wiley, 1993.

[12] L. Kaufman and P. J. Rousseeuw. *Finding Groups in Data*. Wiley, 1990.

[13] M. Mammadov and A. Banerjee. An optimization approach to identifying drugs responsible for adverse drug reactions. *Proc. 16th Australasian Workshop Combinatorial Algorithms*, pages 185–200, 2005.

[14] M. Mammadov and A. Banerjee. An optimization approach to the study of drug-drug interactions. *Proc. 16th Australasian Workshop Combinatorial Algorithms*, pages 201–216, 2005.

[15] M. Mammadov, A. Rubinov, and J. Yearwood. The study of drug-reaction relationships using global optimization techniques. *Optimization Methods and Software*, 22: 99–126, 2007.

[16] R. K. Pearson, T. Zylkin, J. S. Schwaber, and G. E. Gonye. Quantitative evaluation of clustering results using computational negative controls. *Proc. 4th SIAM International Conf. Data Mining*, Lake Buena Vista, FL, pages 188–199, April, 2004.

[17] E. P. van Puijenbroek, A. Bate, H. G. M. Leufkens, M. Lindquist, R. Orre, and A. C. G. Egberts. A comparison of measures of disproportionality for signal detection

in spontaneous reporting systems for adverse drug reactions. *Pharmacoepidemiology Drug Safety*, 11: 3–10, 2002.

[18] *S-Plus 6: Guide to Statistics*, vol. 2. Insightful Corp., Seattle, WA, 2001.

[19] A. Trontell. How the US Food and Drug Administration defines and detects adverse drug events. *Current Therapeutic Research*, 62: 641–649, 2001.

[20] E. Van De Graaff and S. R. Steinhubl. Complications of oral antiplatelet medications. *Current Cardiol. Rep.*, 3: 371–379, 2001.

[21] P. Westfall and S. Young. *Resampling-Based Multiple Testing*. Wiley, 1993.

Chapter 16

A Novel Clustering Approach: Global Optimum Search with Enhanced Positioning

Meng Piao Tan

Department of Chemical Engineering
Princeton University
Princeton, NJ 08544-5263, USA
E-mail address: mtan@princeton.edu

Christodoulos A. Floudas*

Department of Chemical Engineering
Princeton University
Princeton, NJ 08544-5263, USA
Tel: (609) 258-4595
E-mail address: floudas@titan.princeton.edu

Cluster analysis of genome-wide expression data from DNA microarray hybridization studies is a useful tool for identifying biologically relevant gene groupings. It is hence important to apply a rigorous yet intuitive clustering algorithm to uncover these genomic relationships. In this study, we describe a novel clustering algorithm framework based on a variant of the Generalized Benders Decomposition, denoted as the Global Optimum Search [2, 19, 21, 23, 51] which includes a procedure to determine the optimal number of clusters to be used. The approach involves a pre-clustering of data points to define an initial number of clusters and the iterative solution of a Linear Programming problem (the primal problem) and a Mixed-Integer Linear Programming problem (the master problem), that are derived from a Mixed Integer Nonlinear Programming problem formulation. Badly-placed data points are removed to form new clusters, thus ensuring tight groupings amongst the data points and incrementing the number of clusters until the optimum number is reached. We apply the proposed clustering algorithm to experimental DNA microarray data centered on the Ras signaling pathway in the yeast Saccharomyces Cerevisiae and compare the results to that obtained with some commonly-used clustering algorithms. Our algorithm comes up favorably against these algorithms in the aspects of intra-cluster similarity and

*Author to whom all correspondence should be addressed

inter-cluster dissimilarity, often considered two key tenets of clustering. Furthermore, our algorithm can predict the optimal number of clusters, and the biological coherence of the predicted clusters is analyzed through gene ontology.

16.1. Introduction

The aim of cluster analysis is to establish a set of clusters such that the data points in a cluster are more similar to one another than they are to those in other clusters. The clustering problem is old, can be traced back to Aristotle, and has already been studied quite extensively by 18th century naturalists such as Buffon, Cuvier, and Linne [28]. Since then, clustering has been used in many disciplines, such as market research, social network analysis, and geology, thus reflecting its broad appeal and utility as a key step in exploratory data analysis [37]. In market research for instance, cluster analysis is widely used when working with multivariate data from surveys and test panels. Market researchers use cluster analysis methods to segment and determine target markets, and position new products. Cluster analysis is also used in the service of market approaches to the establishment of business enterprise value. [38] addresses the potential role and utility of cluster analysis in transfer pricing practices. Given the importance of clustering, a substantial number of books, such as [17, 29, 36], as well as review papers, such as [71] have been published on this subject.

In biology, clustering provides insights into transcriptional networks, physiological responses, gene identification, genome organization, and protein structure. Genome-wide measurements of mRNA expression levels have provided an efficient and comprehensive means of gathering information on genetic functions and transcriptional networks. However, extracting useful information from the resulting large data sets first involves organizing genes by their pattern and/or intensity of expression in order to define those genes that are co-regulated. Such information provides a basis for extracting regulatory motifs for transcription factors driving the diverse expression patterns, allowing assembly of predictive transcriptional networks [3]. This information also provides insights into the functions of unknown genes, since functionally related genes are often co-regulated [68]. Furthermore, clustered array data provides identification of distinct categories of otherwise indistinguishable cell types, which can have profound implications in processes such as disease progression [63]. In sequence analysis, clustering is used to group homologous sequences into gene families. Examining characteristic DNA fragments helps in the identification of gene structures and reading frames. In protein structure prediction, clustering the ensemble of low energy conformers is used to identify the best possible protein structures.

Two popular similarity metrics are correlation and Euclidean distance. The latter is often popular, since it is intuitive, can be described by a familiar distance function, and satisfies the triangular inequality. Clustering methods that employ asymmetric distance measures [45, 53] are probably more difficult to intuitively comprehend even though they may be highly suited to their intended applications. The earliest work on clustering emphasized visual interpretations for the ease of study, resulting in methods that utilize dendograms and color maps [11]. Other examples of clustering algorithms include: (a) Single-Link and Complete-Link Hierarchical Clustering [36, 62], (b) K-Means Algorithm and its family of variants, such as the K-Medians [30, 46, 49, 73, 74], (c) Reformulation Linearization-based Clustering [1, 60], (d) Fuzzy Clustering [6, 15, 55, 59] , (e) Quality Cluster Algorithm (QTClust) [32] , (f) Graph-Theoretic Clustering [26, 70, 72], (g) Mixture-Resolving Clustering Method [13, 37], (h) Mode Seeking Algorithms [37], (i) Artificial Neural Networks for Clustering [9, 42] such as the Self-Organizing Map (SOM) [43] and a variant that combines the SOM with hierarchical clustering, the Self-Organizing Tree Algorithm (SOTA) [31], (j) Information-Based Clustering [56, 61, 67] , (k) Stochastic Approaches [40, 48, 50].

In some instances such as gene expression data, clustering results could be affected when there exists a number of experimental conditions in which gene activity, even amongst those known to be closely associated, are uncorrelated. For this reason, there has also been a number of biclustering algorithms that allows for simultaneous clustering of the rows and columns of a matrix. Though first described in [10, 29] were the first to apply it to gene expression data by defining a score for each candidate bicluster and developing heuristics to solve the resultant constrained optimization problem. Since then, much of the research on biclustering has been directed towards biological applications. Additional examples of biclustering algorithms include: (a) Coupled two-way clustering [24], (b) Iterative Signature Algorithm [5, 34], (c) Statistical-Algorithmic Method for Bicluster Analysis (SAMBA) [66], (d) Plaid Model [44], (e) Spectral biclustering approaches such as the study assuming a hidden checkerboard-like structure within the expression matrix reported in [41], (f) Probabilistic models such as that reported in [58], (g) the method of [4] in defining a bicluster as an order preserving submatrix. More recently, [8] present a fractional 0-1 programming approach to identify selected features for consistent biclustering, and [14] develop a rigorous biclustering approach OREO, which is based on the optimal reordering of rows and columns in a data matrix using a network flow model.

In this paper, we present a novel Mixed-Integer Nonlinear Programming (MINLP)-based clustering algorithm, the Global Optimal Search with Enhanced Positioning (EP_GOS_Clust), which is robust yet intuitive [64, 65]. This algorithm is significant in that it is able to progressively identify and weed out outlier data points. Also, our algorithm contains a convenient method to predict the optimal number of clusters. We compare our algorithm with several approaches commonly used in clustering biological microarray data, namely K-methods, QT-Clust., SOM, and SOTA. We use two assessment criteria to assess the results: the intra-cluster and inter-cluster error sums. We also examine the difference between the two error sums. In an optimal cluster configuration, the intra-cluster error sum is to be minimized and the inter-cluster error sum to be maximized. In this respect, we show that our proposed algorithm compares favorably. We also incorporate a methodology to predict the optimal number of clusters. In addition, in view of the context of the particular test dataset used, we compare the strength of biological coherence uncovered by the various approaches using Gene Ontology resources, and also the level of correlation between data points with the same cluster. We base our comparative study on actual DNA microarray data, though our algorithm can be readily utilized for data from other applications.

16.2. Methods

16.2.1. *Experimental Data*

For the clustering studies described in this report, we used experimental microarray data derived from a study in the role of the Ras/protein kinase A pathway (PKA) on glucose signaling in yeast [69]. These experiments analyzed mRNA levels in samples extracted from cells at various times following stimulation by glucose or following activation of either Ras2 or Gpa2, which are small GTPases involved in the metabolic and transcriptional response of yeast cells to glucose [57]. These experiments were performed in wild type cells and cells defective in PKA activity. Clustering these microarray data has proven to be a critical step in using the data to develop a predictive model of a topological map of the signaling network surrounding the Ras/PKA pathway [47].

Levels of RNA for each of the 6237 yeast genes in each of the RNA samples from the above experiments were measured using Affymetrix microarray chips and analyzed by the Affymetrix software. Each of the eight test and control experiments consisted of four time points over a hour period, yielding 32 data points for each of the 6237 genes. We used the Affymetrix MicroArray Suite 5.0, which analyzes the consensus of intensities of hybridization of an RNA to the collection of perfect match probes for a gene on the array, relative to the intensities of hybridization to single mismatch probes, to further determine whether a signal for a specific RNA in a sample was reliable (P or present), unreliably low (A or absent), or ambiguous (M). Before clustering the array data, we filtered the data to remove unreliable data. In particular, we retained all genes for which all the time points were present (4105 genes), all the genes for which greater than 50% of the time points were present, and all the genes for which the present/absent calls exhibited a biologically relevant pattern (e.g. PAAA for the four time points in the experiment, suggested repression of gene expression over the course of the experiment). In all, we retained 5652 genes. The expression patterns for these genes are then z-normalized over each gene.

16.2.2. *Theoretical and Computational Framework*

16.2.2.1. *Notation*

We denote the measure of distance for a gene i, for i = 1,.,n having k features (or dimensions), for k = 1,.., s as a_{ik}. Each gene 32-time point expression pattern is transformed into a 24-dimensional vector, for which each vector element indicates the change in normalized expression level between time points for each gene, a_{ik}. Each gene is to be assigned to only one (hard clustering) of c possible clusters, each with center z_{jk}, for j = 1,.,c. The binary variables w_{ij} indicates whether gene i falls within cluster j (w_{ij} = 1, if yes; w_{ij} = 0, if no). We then pre-cluster the data to expedite the computational resources required to solve the hard clustering problem by (i) identifying genes with similar experimental responses, and (ii) removing outliers deemed not to be significant to the clustering process. To provide just adequate discriminatory characteristics so that the genes can be pre-clustered properly, we reduce the expression vectors into a set of representative variables [+, o, -]. The (+) variable represents an increase in expression level compared to the previous time point, the (-) variable represents a decrease in expression level from the previous time point, and the (o) variable represents an expression level that does not vary significantly ($\pm$ 10% of change across the time points). We could have used other comparative metrics such as distance or correlation to pre-cluster the genes, though at this first pass stage, using the representative variables [+, o, -] lends more ease and produces pre-clusters of similar quality. Obviously the pre-clustering process of choice can differ across datasets to be clustered, and we choose the approach most expeditious to our data of interest.

16.2.2.2. *Hard Clustering by Global Optimization*

The global optimization approach seeks to minimize the Euclidean distances between the data points and the centers of their assigned clusters as:

$$\min \quad \sum_{i=1}^{n}\sum_{j=1}^{c}\sum_{k=1}^{s} w_{ij}(a_{ik}-z_{jk})^2 \qquad \text{(Problem 1)}$$

$$s.t. \qquad \sum_{j=1}^{c} w_{ij} = 1, \forall i = 1, ..., n$$

w_{ij} are binary variables, z_{jk} are continuous variables

There are two sets of variables in the problem, w_{ij} and z_{jk}. While the bounds of w_{ij} are clearly 0 and 1, that of z_{jk} is obtained by observing the range of a_{ik} values.

$$z_{jk}^{L} = min[a_{ik}], \forall k = 1, ..., s$$

$$z_{jk}^{U} = max[a_{ik}], \forall k = 1, ..., s$$

The pre-clustering work suggests that some of the genes need only be restricted to some number of known clusters, since it can be determined (for instance by distance and correlation metrics) that certain genes are exceedingly dissimilar from some of the pre-clusters and thus have virtually zero probability of being clustered there. This restriction can be described by introducing an additional binary parameter $suit_{ij}$. A data point deemed to belong uniquely to just one cluster will only have $suit_{ij} = 1$ for only one value of j and zero for the others, whereas a data point restricted to a few clusters will have $suit_{ij} = 1$ for only those clusters. This reduces the computational demands of the problem. The introduction of the $suit_{ij}$ parameters also obviates the need for constraints that prevent the redundant re-indexing of clusters.

Together with the first-order optimality condition (FOC) (i.e. the vector distance sum of all genes within a cluster to the cluster center should be intuitively zero), the formulation becomes:

$$\min \quad \sum_{i=1}^{n}\sum_{k=1}^{s} a_{ik}^2 - \sum_{i=1}^{n}\sum_{j=1}^{c}\sum_{k=1}^{s}(suit_{ij}(a_{ik}w_{ij}z_{jk}) \qquad \text{(Problem 2)}$$

$$s.t. \qquad (suit_{ij})(z_{jk}\sum_{i=1}^{n} w_{ij} - \sum_{i=1}^{n} a_{ik}w_{ij}) = 0, \forall j \forall k$$

$$\sum_{j=1}^{c}(suit_{ij})w_{ij} = 1, \forall i$$

$$1 \leq \sum_{i=1}^{n}(suit_{ij})w_{ij} \leq n-c+1$$

$$w_{ij} = 0-1, \forall i, \forall j$$

$$z_{jk}^{L} \leq z_{jk} \leq z_{jk}^{U}, \forall j, \forall k$$

The first set of constraints are the FOC, the second demand that each gene can belong to only one cluster, and the third state that there is at least one and no more than (n-c+1) data points in a cluster. Note also that the $\sum_{i=1}^{n}\sum_{k=1}^{s} a_{ik}^2$ term in the objective function of Problem 2 is a constant and can be dropped, though for the sake of completeness we will retain the term throughout the subsequent formulations in the paper. Problems 1 and 2 are Mixed Integer Nonlinear Programming (MINLP) problems with bilinear terms in the objective function and the first set of constraints. To handle the nonlinearities formed by the product of variables w_{ij} and z_{jk}, new variables y_{ijk} along with additional constraints [19] are defined as follows:

$$y_{ijk} = w_{ij} z_{jk} \tag{16.1}$$

$$z_{jk} - z_{jk}^U (1 - w_{ij}) \leq y_{ijk} \leq z_{jk} - z_{jk}^L (1 - w_{ij}) \tag{16.2}$$

$$z_{jk}^L w_{ij} \leq y_{ijk} \leq z_{jk}^U w_{ij}, \forall i, \forall j, \forall k \tag{16.3}$$

The introduction of y_{ijk} and the additional constraints reduces the formulation to an equivalent Mixed-Integer Linear Programming (MILP) problem, but results in an inordinately large number of variables. Thus, there is a need for new approaches to address large datasets.

16.2.2.3. *The GOS Algorithm for Clustering*

The introduction of the bilinear variable y_{ijk} results in a large number of variables to be considered. In a problem with over 2000 data points, each having 24 features, to be placed into over 380 clusters, the number of variables to be considered numbers over 18 million. Without introducing the y_{ijk} variables will leave the problem in a nonlinear form. Mixed-integer nonlinear programming (MINLP) problems are considered extremely difficult. Theoretical advances and prominent algorithms for solving MINLP problems are addressed in [19, 20, 22].

The general form of a MINLP problem is:

$$\begin{aligned} \min \quad & C(x, y) && \text{(Problem 3)} \\ s.t. \quad & h(x, y) = 0 \\ & g(x, y) \leq 0 \\ & y \in (0, 1)^m, x \in \Re^n \end{aligned}$$

Here, x represents the continuous variables in real space and y, the integer variables. For simplicity here, y is assumed to be binary. In addition, C(x,y) is the objective function, h(x,y) represents the set of equality constraints, and g(x,y) is the set of inequality constraints. We propose here a variant of the Generalized

Benders Decomposition (GBD) algorithm [21], denoted as the Global Optimum Search (GOS). For brevity, only a outline of the GOS algorithm is presented here, while a more detailed description can be found in [2, 19, 21, 23, 51].

In brief, the GBD method decomposes the problem into a primal problem and the master problem. The former optimizes the continuous variables while fixing the integer variables and provides an upper bound solution, while the latter optimizes the integer variables while fixing the continuous variables and provides a lower bound solution. The two sequences of upper and lower bounds are iteratively updated until they converge in a finite number of iterations. In addition, the GOS algorithm assumes that (i) the optimal solution of the primal problem together with the relevant Lagrange multipliers can be used to determine the support functions, (ii) f(x,y) and g(x,y) are convex functions in y for every fixed x, and (iii) h(x,y) are linear functions in y for every x. An outline of the GOS algorithm is as follows:

Step 1 - Solving the primal problem

The primal problem results from fixing the binary variables to a particular 0-1 combination. Here, w_{ij} is fixed and z_{jk} is solved from the resultant linear programming (LP) problem. In addition, the solution also includes the relevant Lagrange multipliers. The objective function obtained is the upper bound solution. The general form of the feasible problem:

$$\begin{aligned} \min_x \quad & C(x, y^k) && \text{(Problem 4)} \\ s.t. \quad & h(x, y^k) = 0 \\ & g(x, y^k) \leq 0 \\ & x \in \Re^n \end{aligned}$$

If the primal problem is found to be infeasible, the inactive (i.e., inequality) constraints are relaxed by introducing slack variables α and then solving for α, as well as z_{jk} and the Lagrange multipliers. In this event, no new upper bound solution is found. The infeasible problem to be solved has the form:

$$\begin{aligned} \min \quad & \sum \alpha && \text{(Problem 5)} \\ s.t. \quad & h(x, y^k) = 0 \\ & g(x, y^k) \leq \alpha, \text{ one } \alpha \text{ value for each inactive constraint} \\ & \alpha \geq 0 \end{aligned}$$

Step 2 - Solving the relaxed master problem

The master problem is essentially the problem projected onto the y-space (i.e., that of the binary variables). To expedite the solution of this projection, the dual representation of the master is used. This dual representation is in terms of the supporting Lagrange functions of the projected problem. It is assumed that the optimal solution of the primal problem as well as its Lagrange multipliers can be used for the determination of the support function. Also, the support functions are gradually built up over each successive iteration. The relaxed master problem to be solved is hence:

$$\begin{aligned} \min_{y,\mu_B} \quad & \mu_B \qquad \text{(Problem 6)} \\ s.t. \quad & \mu_B \geq L(x^k, y, \lambda^k, \mu^k), k = 1, K \\ & 0 \geq \overline{L}(x^l, y, \overline{\lambda}^l, \overline{\mu}^l), l = 1, L \\ L(x^k, y, \lambda^k, \mu^k) &= f(x^k, y) + \lambda^k h(x^k, y) + \mu^k g(x^k, y) \\ \overline{L}(x^l, y, \overline{\lambda}^l, \overline{\mu}^l) &= \overline{\lambda}^l h(x^l, y) + \overline{\mu}^l g(x^l, y) \end{aligned}$$

In accordance to the general format of the problem, f(x, y) is the objective function, h(x, y) are the active constraints, and g(x, y) are the inactive constraints. 'x' represents the solutions of the continuous variables (i.e., z_{jk}) from the primal problem and 'y' represents the binary variables (i.e., w_{ij}) to be determined in the relaxed master. 'λ' represents the Lagrange multipliers for the active constraints and 'μ' represents the Lagrange multipliers for the inactive constraints. The superscript 'k' represents values from the feasible primal problems and the superscript 'l' (and the over-bar) represents values from the infeasible primal problems. The master problem is to be solved as a MIP (mixed integer programming) problem.

The solution loop then returns to step 1 and the process is repeated. For each excursion into step 1, the primal could be either feasible or infeasible. With each successive iteration, a new support function is added to the list of constraints for the master problem. Thus in a sense, the support functions for the master problem build up with each iteration, forming a progressively tighter envelope and gradually pushing up the lower bound solution until it converges with the upper bound solution.

Since fixing x to the solution of the corresponding primal problem may not necessarily produce valid support functions, the master solution obtained at each iteration is checked against the current lower bound solution so that the latter is updated only if the master solution is higher than the current lower bound solution.

For our clustering problem, with fixed starting values for w_{ij}, the primal problem becomes:

$$\begin{aligned} &\min_{z_{jk}} \textstyle\sum_{i=1}^{n} \sum_{k=1}^{s} a_{ik}^2 - \sum_{i=1}^{n} \sum_{j=1}^{c} \sum_{k=1}^{s} a_{ik} w_{ij}^* z_{jk} \quad \text{(Problem 7.1)} \\ &s.t. \quad z_{jk} \textstyle\sum_{i=1}^{n} w_{ij}^* - \sum_{i=1}^{n} a_{ik} w_{ij}^* = 0, \forall j \forall k \\ &\quad z_{jk}^L \le z_{jk} \le z_{jk}^U, \forall j, \forall k \end{aligned}$$

The primal problem is a Linear Programming (LP) problem. All the other constraints drop out since they do not involve z_{jk}, which are the variables to be solved in the primal problem. Besides z_{jk}, the Lagrange multipliers λ_{jk}^m for each of the constraints above is obtained. The objective function is the upper bound solution. These are substituted into the master problem, which becomes:

$$\begin{aligned} &\min_{y,\mu_B} \mu_B \quad \text{(Problem 7.2)} \\ s.t. \quad &\mu_B \ge \textstyle\sum_{i=1}^{n} \sum_{k=1}^{s} a_{ik}^2 - \sum_{i=1}^{n} \sum_{j=1}^{c} \sum_{k=1}^{s} a_{ik} w_{ij} z_{jk}^* + \ldots \\ &\textstyle\sum_{j=1}^{c} \sum_{k=1}^{s} \lambda_{jk}^{m^*} (z_{jk}^* \sum_{i=1}^{n} w_{ij} - \sum_{i=1}^{n} a_{ik} w_{ij}), m = 1, M \\ &\textstyle\sum_{j=1}^{c} w_{ij} = 1, \forall i \\ &1 \le \textstyle\sum_{i=1}^{n} w_{ij} \le n - c + 1, \forall j \\ &w_{ij} = 0 - 1, \forall i, \forall j \end{aligned}$$

The master problem solves for w_{ij} and μ_B, and results in a lower bound solution (i.e., the objective function). The master problem is a Mixed Integer Linear Programming (MILP) problem. The w_{ij} solutions are cycled back into the primal problem and the process is repeated until the solution converges. Thus, there is no longer a need for the variables y_{ijk}, which substantially reduces the number of variables to be solved. Also, after every solution of the master problem, where a solution set for w_{ij} is generated, an integer cut is added for subsequent iterations to prevent redundantly considering that particular solution set again. The cut is expressed as:

$$\sum_{i \in [n | w_{ij} = 1]}^{n} w_{ij} - \sum_{i \in [n | w_{ij} = 0]}^{n} w_{ij} \le n - 1 \tag{16.4}$$

Note that the initial condition w_{ij} for the primal problem can be generated either by solving the above problem as a relaxed MINLP problem or by randomly generating starting w_{ij} values. For the former, the w_{ij} solution is then rounded up and used as the initial condition for the GOS algorithm. It is found that well over 95% of the w_{ij} solution from the MINLP problem adopt [0,1] solutions anyway.

In addition, if it is not certain that the initial w_{ij} values form an optimal solution, such as the case of randomly generated w_{ij} values, it is then not included in subsequent integer cuts. It is important to note that while the GOS algorithm tends to give good optimal solutions, it does not have a theoretical guarantee of returning a globally optimum solution. Hence the issue of providing the algorithm with a quality initialization point is important and will be addressed later in the paper. Note also that in a typical GBD algorithm, there may be an infeasible primal problem, for which the problem statement would have to be reformulated accordingly. In this case, since there is only one set of continuous variable (i.e., z_{jk}) to be solved in the primal problem, and there is always a feasible assignment of points to clusters, leading to the calculation of the cluster centers, all primal problems are feasible.

16.2.2.4. *Determining the Optimal Number of Clusters*

Most clustering algorithms do not contain screening functions to determine the optimal number of clusters. Yet this is important to evaluate the results of cluster analysis in a quantitative and objective fashion. On the other hand, while it is relatively easy to propose indices of cluster validity, it is difficult to incorporate these measures into clustering algorithms and appoint thresholds on which to define key decision values [27, 36]. Some of the indices used to compute cluster validity include the Dunn's validity index [16], the Davis-Bouldin validity index [12], the Silhouette validation technique, the C index [33], the Goodman-Kruskal index [25], the Isolation index [52], the Jaccard index [35], and the Rand index [54]. We note that the optimal number of clusters occurs when the inter-cluster distance is maximized and the intra-cluster distance is minimized. We adapt the concept of a clustering balance [39], where it has been shown to have a minimum value when intra-cluster similarity is maximized and inter-cluster similarity is minimized. This provides a measure of how optimal is a certain number of clusters used for a particular clustering algorithm. We introduce the following:

$$\text{Global Center, } z_k^o = \frac{1}{n}\sum_{i=1}^{n} a_{ik}, \forall k \tag{16.5}$$

$$\text{Intra-Cluster Error Sum, } \Lambda = \sum_{i=1}^{n}\sum_{j=1}^{c}\sum_{k=1}^{s} w_{ij}\|a_{ik} - z_{jk}\|_2^2 \tag{16.6}$$

$$\text{Inter-Cluster Error Sum, } \Gamma = \sum_{j=1}^{c}\sum_{k=1}^{s} \|z_{jk} - z_{jk}^o\|_2^2 \tag{16.7}$$

From here, [39] proposed a clustering balance parameter, which is the α-weighted sum of the two error sums.

$$\text{Clustering Balance, } \varepsilon = \alpha\Lambda + (1-\alpha)\Gamma \tag{16.8}$$

We note here that the rightful α-ratio is 0.5. We note that the factor α should balance the contributive weights of the two error sums to the clustering balance. At extreme cluster numbers, that is, the largest and smallest number possible, the sum of the intra-cluster and inter-cluster error sums at both cluster numbers should be balanced. In the minimal case, all the data points can be placed into a single cluster, in the case of which the inter-cluster error sum is zero and the intra-cluster error sum can be calculated with ease. In the maximal case, each data point forms its own cluster, in the case of which the intra-cluster error sum is zero and the inter-cluster error sum can be easily found. Obviously the intra-cluster error sum in the minimal case and inter-cluster error sum in the maximal case are equal, suggesting that the most appropriate weighting factor to use is in fact 0.5.

This suggest that for any clustering algorithm including that using the GOS algorithm, one can deduce the optimal number of clusters by performing multiple repetitions of the clustering process over a suitably large range of cluster numbers and watching for the clustering gain or clustering balance turning points.

16.2.3. *Proposed Algorithm*

The discussion thus far points to the GOS formulation as a suitable clustering algorithm. But for it to be effective, the formulation must be provided with a good initialization point. Also, we want to expeditiously incorporate the approach to predict the optimal number of clusters into a clustering algorithm. With these considerations in mind, we propose the following GOS clustering algorithm with enhanced data point positioning (EP_GOS_Clust).

Gene Pre-Clustering: We choose to pre-cluster genes based on the feature pattern representation of their expression vectors. This conforms well to the intuitive notion that two co-expressed genes similarly-shaped expression patterns, rather than comparing the magnitudes of the two series of measurements [18]. In our 24-dimensional expression vectors, only genes with two or less different expression vector points from one another are pre-clustered together. Many of these genes end up belonging to more than one pre-cluster. Since their specific membership is in question, we take the clusters formed only by uniquely-clustered genes. As a result, we find 388 genes uniquely placed into 157 clusters.

We note here in particular when pre-clustering by finding complete cliques, which means that the pre-clustered genes that belong uniquely to only one cluster, or in other words, there is a link between every gene within the same cluster, we could have iterated the process using various levels of pre-clustering criteria. When the criterion is overly lenient, a large number of pre-clusters are formed, but most of the genes will belong to multiple pre-clusters, and the number of maximal cliques formed is small. On the other hand, an unnecessarily strict cut-off results in a small number of pre-clusters, thus not accurately reflecting the extent of relatedness between the data. In pre-clustering over a range of cut-off values, we would then be able to select the optimum criterion as the point where the maximum number of complete cliques is formed [64].

Iterative Clustering: We let the initial cluster set be defined by the unique genes pre-clustered in the previous step and compute the cluster centers. We next compute the distance between each of the remaining genes and these initial centers and as a good initialization point placed these genes into the nearest cluster based on:

$$\min[\sum_{k=1}^{s}(a_{ik} - z_{jk}^{initial})^2, \forall j], \forall i \notin unique$$

We then create a rank-order list for each of the remaining genes for its distance to each of the initial clusters, and for each gene allow its suitability in 4 nearest clusters via its $suit_{ij}$ parameters. For this particular dataset, a separate study (results not shown here) has indicated that the clustering results cease to change significantly once the number of $suit_{ij}$ values for each gene exceed 4. The initialization point and the $suit_{ij}$ parameter assignments are then utilized in the primal problem of the GOS algorithm as described in Problem 7.1 to solve for z_{jk}. These, together with the Lagrange multipliers, are inputted into the master problem (Problem 7.2) to solve for w_{ij}. The primal problem gives an upper bound solution and the master problem provides a lower bound. The optimal solution is obtained when the lower and upper bounds converge. Then, the worst placed gene based on:

$$\max[\sum_{j=1}^{c}\sum_{k=1}^{s}(a_{ik} - z_{jk}^{updated})^2, \forall i \notin unique]$$

is removed and used as a seed for a new cluster with center z^{new}. This gene has already been subjected to the initial search for membership so there is no reason for it to belong to any one of the older clusters. Based on z^{new} and $z^{updated}$ (updated without the worst-placed gene), the iterative steps are repeated, by selecting a new initialization point, assignment $suit_{ij}$ parameters, and running the GOS algorithm again. With these iterations, the number of clusters builds up from the initial number defined by the pre-clustering work, until the optimal number of clusters is attained. Our proposed clustering methodology can be summarized by the schematic in Fig. 16.1.

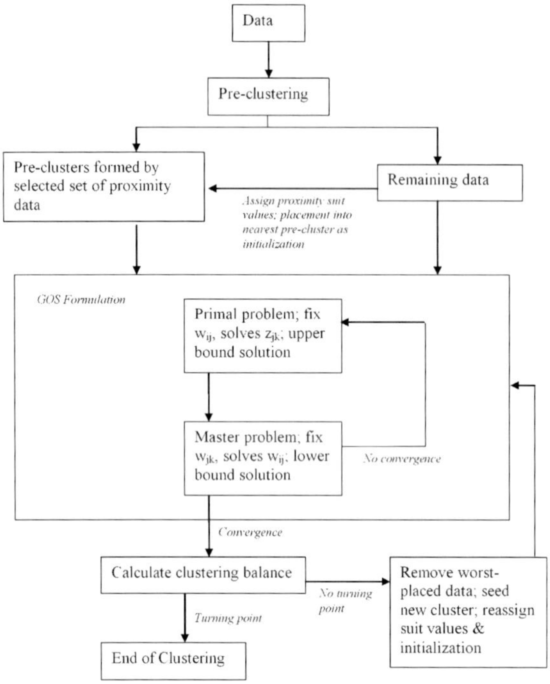

Fig. 16.1. Schematic flowchart of the EP_GOS_Clust algorithm. Although the formulation in the paper has been given for DNA microarray data, the algorithm framework can be adapted for clustering any numeric data.

16.3. Results and Discussion

16.3.1. *Description of Comparative Study*

We will work with the 5652 genes obtained previously. The clustering algorithms to be compared are (a) K-Means, (b) K-Medians, (c) K-Corr, where the Pearson correlation coefficient is the distance metric, (d) K-CityBlock, where the distance metric is the city block distance, or the 'Manhattan' metric, which is akin to the north-south or east-west walking distance in a place like New York's Manhattan district, (e) K-AvePair, where the cluster metric is the average pair-wise distance between members in each cluster, (f) QTClustering, (g) SOM, (h) SOTA, (i) GOS I, where genes with up to 7 different feature points are pre-clustered, initial clusters are defined by uniquely-placed genes, and each gene is placed into its nearest cluster as the initialization point, and (j) EP_GOS_Clust, for which genes are pre-clustered if they have 2 or less different feature points and can be uniquely clus-

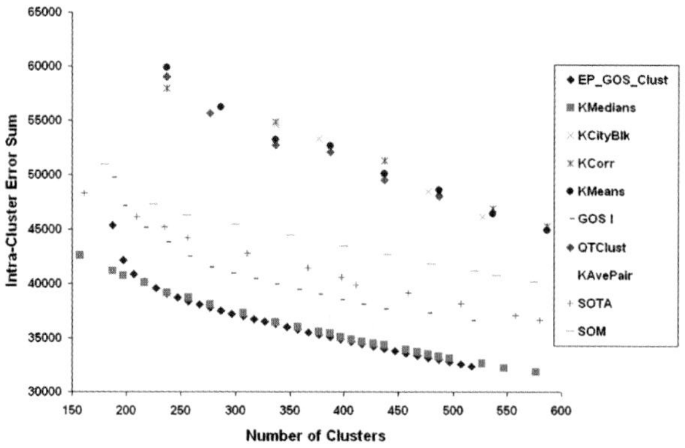

Fig. 16.2. Comparison of intra-cluster error sum from the clustering of 5652 yeast genes based on DNA expression levels in glucose pathway experiments, using different clustering algorithms. Each gene contains 36 time points, or a 24-dimensional feature vector. The intra-cluster error sum measures the extent of dissimilarity between objects within the same cluster, and should be minimized.

tered. For convenience, the comparison involving SOM and SOTA will only be carried out at the optimal cluster number predicted for the EP_GOS_Clust. Since the K-family of clustering approaches are sensitive to the initialization point, we run each 25 times and use only the best result.

16.3.2. *Intra-cluster Error Sum*

Data points in the same cluster should be as similar as possible; hence the intra-cluster error sum should be minimized. From Fig. 16.2, it can be seen that the best performing clustering algorithms are the K-Medians and the EP_GOS_Clust. In fact, other than within regions of low cluster number, the EP_GOS_Clust outperforms all the other algorithms. One reason for the efficacy of the K-Medians at low cluster numbers is due to it using the data median to compute cluster centers. This circumvents the distorting effects of outlier data points, which particularly affects algorithms that use random initialization points, such as K-Means. It is also notable that the GOS I performs admirably even though the pre-clustering allows genes with up to 30% difference in feature points to be grouped together. This reflects the rigor of the subsequent steps in assigning $suit_{ij}$ parameters, the GOS clustering, and the process of incrementing the cluster number. The clustering results also show up the inadequacy of QTClust. It groups genes together till the cluster reaches a pre-determined tolerance. The algorithm then determines the number of clusters to use. A different tolerance criterion needs to be specified in order to obtain a different cluster number. This implies that the process

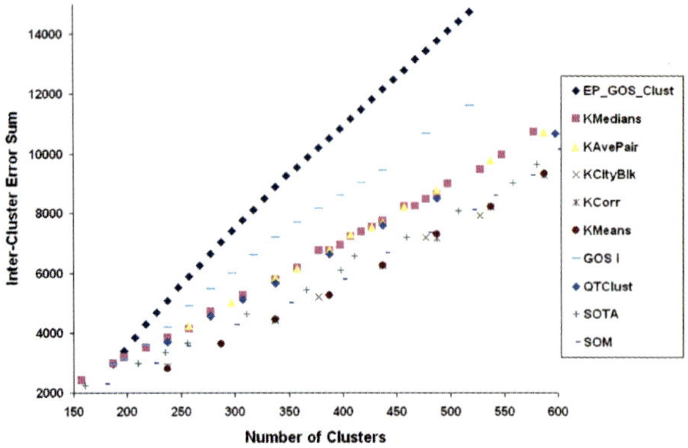

Fig. 16.3. Comparison of inter-cluster error sum from the clustering of 5652 yeast genes based on DNA expression levels in glucose pathway experiments, using different clustering algorithms. Each gene contains 36 time points, or a 24-dimensional feature vector. The inter-cluster error sum measures the extent of dissimilarity between clusters, and should be maximized.

of probing for the optimal number of clusters using QTClust uses clusters of inconsistent qualities. We further look in detail at the clustering results obtained by QTClust and note that genes with up to 14 different feature points (60% of all feature points) are in fact clustered together.

16.3.3. *Inter-cluster Error Sum*

This error sum indicates how different clusters are from one another and is given by:

$$\sum_{j=1}^{c}\sum_{k=1}^{s}(z_{jk} - z^{o}_{jk})^2 \, .$$

This is another measure of cluster quality, and it is desirable for the error sum to be maximized. The inter-cluster error sum for the clustering of 5652 genes is shown in Fig. 16.3. Here, the EP_GOS_Clust outperforms all the other cluster algorithms. In using the intra-cluster error sum as the objective function and demanding that the worst-fitting gene be extracted to seed new clusters, the EP_GOS_Clust explicitly seeks a minimal intra-cluster error sum and implicitly searches for a configuration that maximizes the inter-cluster error sum. Note that while K-Medians does well in obtaining a minimal intra-cluster error sum, it performs averagely in discerning dissimilar clusters. This is due to the 'localized' nature of the K-family of clustering methods, where there is a tendency to become 'stuck' within a limited vicinity of the initialization point for most data structures.

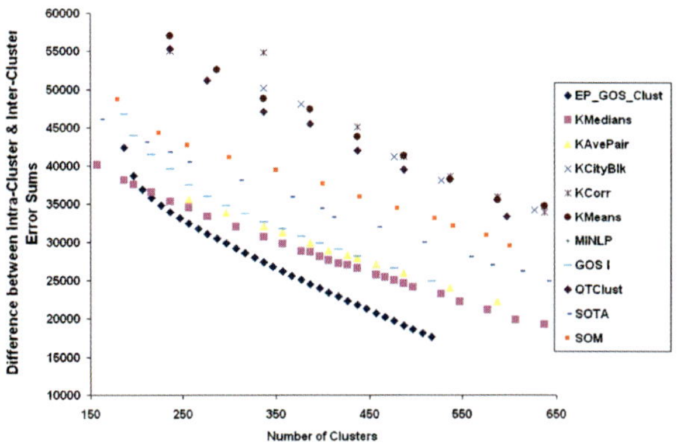

Fig. 16.4. Comparison of the difference between error sums from the clustering of 5652 yeast genes based on DNA expression levels in glucose pathway experiments, using different clustering algorithms. Each gene contains 36 time points, or a 24-dimensional feature vector. This comparison allows an overview of the extent of overall 'error-ness' for the clusters formed and should be minimized.

16.3.4. *Difference between Intra-cluster and Inter-cluster Error Sums*

We look also at an overall measure of clustering quality the difference between the intra-cluster and inter-cluster error sums. Since it is desirable for the former to be minimized and the latter maximized, an effective and rigorous clustering algorithm will have a low value for this difference. The results are shown in Fig. 16.4. Again, the EP_GOS_Clust is the best performer except for certain regions of low cluster number, where the K-Medians dominate with its capability to handle outlier data points. At higher cluster numbers however, the EP_GOS_Clust identifies and isolates these outlier data points into new clusters and subsequently its clustering performance overtakes that of the K-Medians.

16.3.5. *Optimal Number of Clusters*

We compute the optimal number of clusters by applying a suitable weighting factor to the two error sums and then finding the clustering balance. The EP_GOS_Clust predicts the lowest number of optimal clusters. From Fig. 16.5, it can be seen that EP_GOS_Clust predicts 237 clusters. On the other hand, K-Means and KCorr, for instance, predict the optimal number of clusters to be around 700, while K-Medians puts the number at around 450. Together with the quality of the EP_GOS_Clust from the previous comparisons, we infer the superior 'economy' of the EP_GOS_Clust in producing tighter data groupings by utilizing a lower number of clusters, as it is actually possible to achieve tight groupings by using a large number of clusters, even with an inferior clustering algorithm.

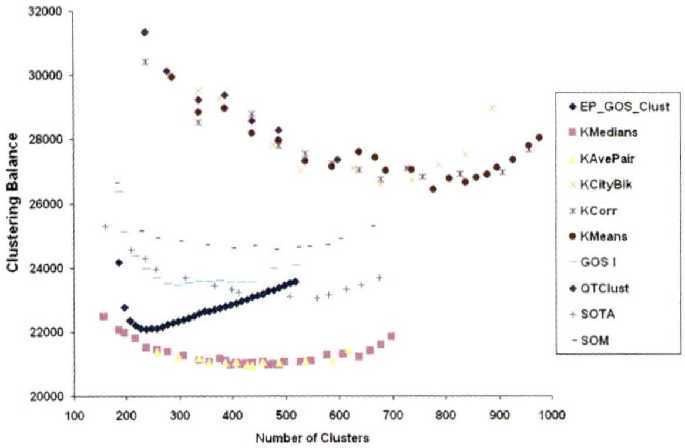

Fig. 16.5. Prediction of the optimal number of clusters, as shown by the turning point in the cluster balance, by different clustering algorithms from the clustering of 5652 yeast genes based on DNA expression levels in glucose pathway experiments. Each gene contains 36 time points, or a 24-dimensional feature vector.

16.3.6. *Coherence and Biological Relevance*

Often the most intuitive and convenient manner of evaluating the robustness of a clustering approach is to visually inspect the cluster tightness. For brevity, Fig. 16.6 depicts the expression time course for 4 sample clusters. We use the

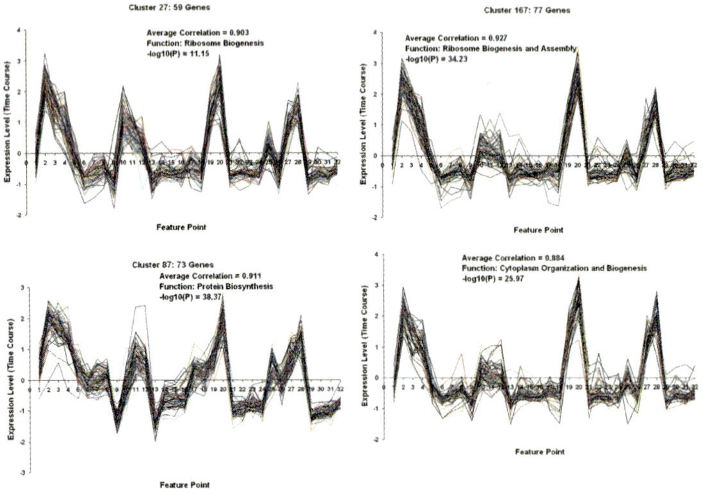

Fig. 16.6. Gene expression time course plots for 4 sample clusters, found using the EP_GOS_Clust algorithm.

Table 16.1. Comparison of cluster correlation from the clustering of 5652 genes based on DNA expression levels in glucose pathway experiments. The comparison shows tha average correlation coefficients across all clusters for each clustering algorithm, the maximum and minimum coefficient, as well as the standard deviation of the coefficients to give a sense of the spread of correlation displayed by the clusters. The table also shows the optimal number of clusters predicted by each clustering approach. The shaded row contains the results for EP_GOS_Clust and the top three performers for each correlation performance indicator is marked with an asterisk.

(Clustering Method)	Optimal Cluster Number	Correlation Coefficient			
		Average	Maximum	Minimum	Standard Deviation
EP_GOS_Clust	237	0.617*	0.938*	0.264*	0.128*
KMedians	445	0.615	0.937	0.197	0.134
KCityBlk	665	0.398	0.760	-0.159	0.149
KCorr	665	0.630*	0.931	0.239*	0.119*
KMeans	775	0.614	0.959*	0.072	0.131
GOS I	295	0.590	0.933	0.202	0.148
KAvePair	452	0.567	0.909	0.156	0.141
SOTA	540	0.604	0.925	0.378*	0.122*
SOM	485	0.623*	0.968*	0.202	0.156

largest clusters formed, as well as clusters formed in the later stages of the procedure and smaller-sized clusters to show good consistency and lack of size-bias. The plots clearly show the tightness of the clustering throughout. To more conveniently demonstrate the overall tightness of the clusters uncovered by the EP_GOS_Clust as compared to other methods, we also find the Pearson correlation coefficients of all the clusters uncovered.

Table 16.1 provides a summary of the cluster correlations, where in particularly the average correlation coefficient and the standard coefficient reflect the overall tightness for all clusters. It can be seen that the EP_GOS_Clust compares very well with other clustering methods in producing highly correlated clusters, even against methods such as K-Corr that already explicitly uses correlation as a metric for clustering and the correlation hunting SOM (see Table 16.1). The data in the table for each clustering algorithm is obtained at the respective optimal number of clusters predicted by the clustering balance.

We also evaluate our clusters by performing a functional search using the Gene Ontology (GO) term finder on the SGD website (http://www.yeastgenome.org). The biological coherence of each cluster is scored according to the percentage of its genes covered by annotations significantly enriched in the cluster. From our

Table 16.2. Gene Ontology comparison between clusters found by different clustering approaches. The table compares the $-\log_{10}(P)$ values of the clusters, which reflect the level of annotative richness, as well as the proportion of yeast genes that fall into biologically significant clusters. The latter is important in 'presenting' the maximal amount of relevant genetic information for follow-up work in areas such as motif recognition and regulatory network inference. The shaded row contains the results for EP_GOS_Clust and the top three performers for each performance indicator is marked with an asterisk.

		$-\log_{10}$(P) Comparison		% Genes (Total 5652)	
		Average	Standard Deviation	In Clusters with $-\log_{10}$(P) values >= 4	In Clusters with $-\log_{10}$(P) values >= 3
(Clustering Method)	EP_GOS_Clust	4.40*	0.37	32.82*	64.92*
	KMedians	4.27*	0.34*	30.83*	62.23*
	KCityBlk	3.69	0.49	27.53	56.68
	KCorr	4.15*	0.39	32.59*	60.08*
	KMeans	3.45	0.41	25.11	55.20
	GOS I	3.84	0.42	28.19	57.75
	KAvePair	3.77	0.48	25.18	54.43
	SOTA	3.67	0.31*	30.20	58.86
	SOM	3.94	0.35*	30.47	59.24

results, 91% of the genes group into clusters with p-values under 0.01 and 87% of the genes fall into clusters with p-values under 0.005, which is a significant indication of clustering quality. Table 16.2 shows that the EP_GOS_Clust performs well against other clustering algorithm in obtaining clusters with good overall p-values (expressed as $-log_{10}(p)$ values in this table) and the proportion of genes that are placed into significantly coherent clusters, which we consider to be two broad tenets in assessing the strength of biological coherence. We would like to point out that the EP_GOS_Clust procedure isolates errant data points as the clustering progresses. Thus, in further analysis of the clusters we have good justification to consider these data points as being irrelevant.

16.3.7. *Additional Constraints for Large Datasets*

It is interesting to note that a close examination of the clustering results within each GOS iteration reveals that the cluster size distribution does not change significantly over successive iterations This suggests that we can analyze the intermediate results from a particular run and introduce additional constraints on the number of clusters allowable in each size class without significantly compromising on the optimality of the final solution. Using $1 \leq \sum_{j=1}^{n} w_{ij} \leq n-c+1$ as the constraint for cluster size can unnecessarily increase the problem size. Hence, we can further tighten the constraint for cluster size as a plausible strategy for further expediting the clustering of even larger data sets.

Indexing the new cluster size ranges by l, for l = 1,.,d, we introduce a new binary variable w'_{jl}, which equals one if cluster j belongs to size class l, and zero if otherwise. The additional constraints are then formulated as follows:

$$\sum_{l=1}^{d} w'_{jl} = 1, \forall j \tag{16.9}$$

$$n_c - \epsilon \leq \sum_{j=1}^{c} w'_{jl} \leq n_c + \epsilon, \forall l \tag{16.10}$$

$$\sum_{l=1}^{d} d_{l,min} \leq \sum_{i=1}^{n} w_{ij} \leq \sum_{l=1}^{d} d_{l,max}, \forall j \tag{16.11}$$

The first set of constraints allows each cluster into only one size class. The second constraint restricts the number of clusters allowable in each class. The parameter ε is judiciously picked to allow a reasonable range over the number of clusters in each class, for instance 10%. Finally, the third constraint bounds the size of the clusters allowed in each class. These additional constraints also involve the variable w_{ij} but not z_{jk}; hence they are all included into the master problem.

16.4. Conclusion

In our study, we propose a novel clustering algorithm (EP_GOS_Clust) based on a Mixed-Integer Nonlinear Programming (MINLP) formulation. We test our proposed algorithm on a substantially large dataset of gene expression patterns from the yeast Saccharomyces Cerevisiae, and show that our method compares favorably (if not outperforms) with other clustering methods in identifying data points that are the most similar to one another as well as identifying clusters that are the most dissimilar to one another. We also show that the EP_GOS_Clust is capable of uncovering tightly-correlated clusters. Given the nature of the test datasets, we too show that the EP_GOS_Clust does well in uncovering clusters with good biological coherence. In addition, we demonstrate the utility of the pre-clustering procedure and a methodology that works in concert with the algorithm itself to predict the optimal number of clusters. For consistency, we repeated our study on other DNA microarray datasets based on the various glucose signaling pathways in the yeast Saccharomyces Cerevisiae (other results not reported here) and obtained similar result trends.

16.5. Computational Resources

All optimization formulations are written in GAMS (General Algebraic Modeling System) [7] and solved using the commercial solver CPLEX 8.0. GAMS is a high level modeling system specifically designed for mathematical optimization. It consists of a language compiler and an integrated high performance solver such as CPLEX, DICOPT, or XPRESS.

Acknowledgements

The authors gratefully acknowledge support from the National Science Foundation and the National Institutes of Health.

References

[1] W. P. Adams and H.D. Sherali. Linearization strategies for a class of zero-one mixed integer programming problems. *Operations Research*, 38:217–226, 1990.

[2] A. Aggarwal and C. A. Floudas. Synthesis of general distillation sequences - nonshare separations. *Computers & Chemical Engineering*, 14(6):631–653, 1990.

[3] M. Beer and S. Tavazoie. Predicting gene expression from sequence. *Cell*, 117:185–198, 2004.

[4] A. Ben-Dor, B. Chor, R. Karp, and Z. Yakhini. Discovering local structure in gene expression data. In *Proc. RECOMB'02*, pages 49–57. ACM Press, 2002.

[5] S. Bergmann, J. Ihmels, and N. Barkai. Iterative signature algorithm for the analysis of large-scale gene expression data. *Phys. Rev. E Stat. Nonlin. Soft Matter Phys.*, 67:03190201–03190218, 2003.

[6] J. C. Bezdek. *Pattern Recognition with Fuzzy Objective Function Algorithms*. Plenum Press, New York, 1981.

[7] A. Brooke, D. Kendrick, and A Meeraus. *GAMS: A User's Guide*. The Scientific Press, San Francisco, CA, 1988.

[8] S. Busygin, O. A. Prokopyev, and P. M. Pardalos. Feature selection for consistent biclustering via fractional 0-1 programming. *Journal of Combinatoial Optimization*, 10:7–21, 2005.

[9] G. Carpenter and S. Grossberg. Art3: Hierarchical search using chemical transmitters in self-organizing patterns recognition architectures. *Neural Networks*, 3:129–152, 1990.

[10] Y. Cheng and G. Church. Biclustering of expression data. In *Proc. ISMB'00*, pages 93–103. AAAI Press, 2000.

[11] J. Claverie. Computational methods for the identification of differential and coordinated gene expression. *Human Molecular Genetics*, 8:1821–1832, 1999.

[12] D. L. Davis and D. W. Bouldin. A cluster separation measure. *IEEE Trans. Pattern Anal. Machine Intell.*, 1(4):224–227, 1979.

[13] A. P. Dempster, N. M. Laird, and D. B. Rudin. Maximum likelihood from incomplete data via the em algorithm. *J. Royal Stat. Soc. B*, 39(1):1–38, 1977.

[14] P. A. DiMaggio, S. R. McAllister, C. A. Floudas, X. J. Feng, J. D. Rabinowitz, and H. A. Rabitz. A network flow model for biclustering via optimal reordering of data matrics. *Journal of Global Optimization*, In Press.

[15] J. C. Dunn. A fuzzy relative of the isodata process and its use in detecting compact well-separated clusters. *Journal of Cybernetics*, 3:32–57, 1973.

[16] J. C. Dunn. Well separated clusters and optimal fuzzy partitions. *Journal of Cybernetics*, 4:95–104, 1974.

[17] M. A. Duran and P. L. Odell. *Cluster Analysis: A Survey*. Springer Verlag, 1974.

[18] M. B. Eisen, P. T. Spellman, P. O. Brown, and D. Botstein. Cluster analysis and display of genome-wide expression patterns. *Proc. Nat. Acad. Sci. U.S.A.*, 95(25):14863–14868, 1998.

[19] C. A. Floudas. *Nonlinear and Mixed-Integer Optimization: Fundamentals and Applications*. Oxford University Press, 1995.

[20] C. A. Floudas. *Deterministic Global Optimization: Theory, Algorithms, and Applications*. Kluwer Academic Publishers, 2000.

[21] C. A. Floudas, A. Aggarwal, and A. R. Ciric. Global optimum search for non convex nlp and minlp problems. *Comp. & Chem. Eng.*, 13(10):1117–1132, 1989.

[22] C. A. Floudas, I. G. Akrotirianakis, S. Caratzoulas, C. A. Meyer, and J. Kallrath. Global optimization in the 21st century: Advances and challenges. *Computers and Chemical Engineering*, 29:1185–2002, 2005.

[23] C. A. Floudas and A. R. Ciric. Strategies for overcoming uncertainty in heat exchanger network synthesis. *Computers & Chemical Engineering*, 13(10):1133–1152, 1989.

[24] G. Getz, E. Levine, and E. Domany. Coupled two-way clustering analysis of gene microarray data. *Proc. Nat. Acad. Sci. U.S.A.*, 97:12079–12084, 2000.

[25] L. Goodman and W. Kruskal. Measures of associations for cross-validations. *J. Am. Stat. Assoc.*, 49:732–764, 1954.

[26] J. C. Gower and G. J. S. Ross. Minimum spanning trees and single-linkage cluster analysis. *Appl. Stat.*, 18:54–64, 1969.

[27] M. Halkidi, Y. Batistakis, and M. Vazirgiannis. Cluster validity methods: Part 1. *SIGMOD Record*, 31(2):40–45, 2002.

[28] P. Hansen and B. Jaumard. Cluster analysis and mathematical programming. *Mathematical Programming*, 79:191–215, 1997.

[29] J. A. Hartigan. *Clustering Algorithms*. John Wiley & Sons, 1975.

[30] J. A. Hartigan and M. A. Wong. Algorithm AS 136: A k-means clustering algorithm. *Appl. Stat.-J. Roy. St. C.*, 28:100–108, 1979.

[31] J. Herrero, A. Valencia, and J. Dopazo. A hierarchical unsupervised growing neural network for clustering gene expression patterns. *Bioinformatics*, 17(2):126–136, 2001.

[32] L. J. Heyer, S. Kruglyak, and S. Yooseph. Exploring expression data: Identification and analysis of co-expressed genes. *Genome Res.*, 9:1106–1115, 1999.

[33] L. Huber and J. Schultz. Quadratic assignment as a general data-analysis strategy. *British Journal of Mathematical and Statistical Psychologie*, 29:190–241, 1976.

[34] J. Ihmels, G. Friedlander, S. Bergmann, O. Sarig, Y. Ziv, and N. Barkai. Revealing modular organization in the yeast transcriptional network. *Nature Genetics*, 31(4):370–377, 2002.

[35] P. Jaccard. The distribution of flora in the alpine zone. *New Phytologist*, 11:37–50, 1912.

[36] A. K. Jain and R. C. Dubes. *Algorithms for Clustering Data*. Prentice-Hall Advanced Reference Series. Prentice-Hall, Inc., 1988.

[37] A. K. Jain, M. N. Murty, and P. J. Flynn. Data clustering: A review. *ACM Computing Surveys*, 31(3):264–323, 1999.

[38] R. E. Johnson. The role of cluster analysis in assessing comparability under the us transfer pricing regulations. *Business Economics*, April 2001.

[39] Y. Jung, H. Park, D. Du, and B. L. Drake. A decision criterion for the optimal number of clusters in hierarchical clustering. *Journal of Global Optimization*, 25:91–111, 2003.

[40] S. Kirkpatrick, C. D. Gelatt, and M. P. Vecchi. Optimization by simulated annealing. *Science*, 220(4598):671–680, 1983.

[41] Y. Kluger, R. Barsi, J. T. Cheng, and M. Gerstein. Spectral biclustering of microarray data: Coclustering genes and conditions. *Genome Res.*, 13(4):703–716, 2003.

[42] T. Kohonen. *Self Organization and Associative Memory*. Springer Information Science Series. Springer Verlag, 1989.

[43] T. Kohonen. *Self-Organizing Maps*. Springer Verlag, Berlin, 1997.

[44] L. Lazzeroni and A. Owen. Plaid models for gene expression data. Technical report, Stanford University, 2000.

[45] F. Leisch, A. Weingessel, and E. Dimitriadou. Competitive learning for binary valued data. In L. Niklasson, M. Bod'en, and T. Ziemke, editors, *Proceedings of the 8th International Conference on Artificial Neural Networks (ICANN 98)*, volume 2, pages 779–784, Skovde, Sweden, 1998. Springer.

[46] A. Likas, N. Vlassis, and J. L. Vebeek. The global k-means clustering algorithm. *Pattern Recognition*, 36:451–461, 2003.

[47] X. Lin, C. Floudas, Y. Wang, and J. R. Broach. Theoretical and computational studies of the glucose signaling pathways in yeast using global gene expression data. *Biotechnology and Bioengineering*, 84(7):864–886, 2003.

[48] A. V. Lukashin and R. Fuchs. Analysis of temporal gene expression profiles: Clustering by simulated annealing and determining the optimal number of clusters. *Bioinformatics*, 17(5):405–414, 2001.

[49] J. McQueen. Some methods for classification and analysis of multivariate observations. In *Proceedings of the Fifth Berkeley Symposium on Mathematical Statistics and Probability*, pages 281–297, 1967.

[50] N. Metropolis, A. Rosenbluth, M. Rosenbluth, A. Teller, and E. J. Teller. Equation of state calculations by fast computing machines. *J. Chem. Phys.*, 21:1087–1092, 1953.

[51] G. E. Paules and C. A. Floudas. Apros-algorithmic developmeng for discrete-continuous optimization problems. *Operations Research*, 37(6):902–915, 1989.

[52] E. J. Pauwels and G. Fregerix. Finding salient regions in images: Non-parametric clustering for image segmentation and grouping. *Computer Vision and Image Understanding*, 75:73–85, 1999.

[53] P. Pipenbacher, A. Schliep, S. Schneckener, A. Schonhuth, D. Schomburg, and R. Schrader. Proclust: Improved clustering of protein sequences with an extended graph-based approach. *Bioinformatics*, 18(Supplement 2):S182–191, 2002.

[54] W. M. Rand. Objective criteria for the evaluation of clustering methods. *Journal of American Statistical Association*, pages 846–850, 1971.

[55] E. H. Ruspini. A new approach to clustering. *Inf. Control*, 15:22–32, 1969.

[56] Dhillon I. S. and Y. Guan. Information theoretic clustering of sparse co-occurrence data. In *Proceedings of the Third IEEE International Conference on Data Mining (ICDM)*, 2003.

[57] L. Schneper, K. Dvel, and J. R. Broach. Sense and sensibility: Nutritional response and signal integration in yeast. *Current Opinion in Microbiology*, 7(6):624–630, 2004.

[58] E. Segal, B. Taskar, A. Gasch, N. Friedman, and D. Koller. Rich probabilistic models for gene expression. *Bioinformatics*, 17:S243–S252, 2001.

[59] H. D. Sherali and J. Desai. A global optimization rlt-based approach for solving the fuzzy clustering approach. *Journal of Global Optimization*, 33(4):597–615, 2005.

[60] H. D. Sherali and J. Desai. A global optimization rlt-based approach for solving the hard clustering problem. *Journal of Global Optimization*, 32(2):281–306, 2005.

[61] N. Slonim, G. S. Atwal, G. Tkacik, and W. Bialek. Information based clustering. *Proc. Nat. Acad. Sci. U.S.A.*, 102(51):18297–18302, 2005.

[62] R. R. Sokal and C. D. Michener. A statistical method for evaluating systematic relationships. *Univ. Kans. Sci. Bull.*, 38:1409–1438, 1958.

[63] T. Sorlie, R. Tibshirani, J. Parker, T. Hastie, J. S. Marron, A. Nobel, S. Deng, H. Johnsen, R. Pesich, S. Geisler, J. Demeter, C. M. Perou, P. E. Lonning, P. O. Brown, A. L. Borresen-Dala, and D. Botstein. Repeated observations of breast tumor subtypes in independent gene expression data sets. *Proc. Nat. Acad. Sci. U.S.A.*, 100:8418–8423, 2003.

[64] M.P. Tan, J. R. Broach, and C. A. Floudas. Evaluation of normalization and pre-clustering issues in a novel clustering approach: Global optimum search with enhanced positioning. *J. Bioinform. Comput. Biol.*, 5:895–913, 2007.

[65] M.P. Tan, J. R. Broach, and C. A. Floudas. A novel clustering approach and prediction of optimal number of clusters: Global optimum search with enhanced positioning. *Journal of Global Optimization*, 39:323–346, 2007.

[66] A. Tanay, R. Sharan, M. Kupeic, and R. Shamir. Revealing modularity and organiztion in the yeast molecular network by integrated analysis of highly heterogeneous genomewide data. *Proc. Nat. Acad. Sci. U.S.A.*, 101(9):2981–2986, 2004.

[67] N. Tishby, F. Pereira, and W. Bialek. The information bottleneck method. In *Proceedings of the 37th Annual Allerton Conference on Communication. Control, and Computing*, pages 368–377, 1999.

[68] O. G. Troyanskaya, K. Dolinski, A. B. Owen, R. B. Altman, and D. Botstein. A bayesian framework for combining heterogeneous data sources for gene function prediction (in saccharomyces cerevisiae). *Proc. Nat. Acad. Sci. U.S.A.*, 100:8348–8353, 2003.

[69] Y. Wang, M. Pierce, L. Schneper, C. G. Guldal, X. Zhang, S. Tavazoie, and J. R. Broach. Ras and Gpa2 mediate one branch of a redundant glucose signaling pathway in yeast. *Plos Biology*, 2(5):610–622, 2004.

[70] Z. Wu and R. Leahy. An optimal graph theoretic approach to data clustering: Theory and its application to image segmentation. *IEEE Transactions on Pattern Recognition and Machine Intelligence*, 15(11):1101–1113, 1993.

[71] R. Xu and D. Wunsch II. Survey of clustering algorithms. *IEEE Transactions on Neural Networks*, 16(3):645–678, 2005.

[72] C. T. Zahn. Graph theoretical methods for detecting and describing gestalt systems. *IEEE Trans. Comput.*, C-20:68–86, April 1971.

[73] B. Zhang. Generalized k-harmonic means: Boosting in unsupervised learning. Technical report, Hewlett-Packard Research Laboratory, October 2000.

[74] B. Zhang, M. Hsu, and U. Dayal. K-harmonic means - a data clustering algorithm. Technical report, Hewlett-Packard Research Laboratory, June 1999.

Index